WITHDRAWN

LIVERPOOL JMU LIBRARY

Advances in
COMPUTERS
VOLUME 81

Advances in
COMPUTERS

EDITED BY

MARVIN V. ZELKOWITZ

Department of Computer Science
University of Maryland
College Park, Maryland
USA

VOLUME 81

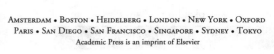

Amsterdam • Boston • Heidelberg • London • New York • Oxford
Paris • San Diego • San Francisco • Singapore • Sydney • Tokyo
Academic Press is an imprint of Elsevier

Academic Press is an imprint of Elsevier
32 Jamestown Road, London, NW1 7BY, UK
Radarweg 29, PO Box 211, 1000 AE Amsterdam, The Netherlands
30 Corporate Drive, Suite 400, Burlington, MA 01803, USA
525 B Street, Suite 1900, San Diego, CA 92101-4495, USA

First edition 2011

Copyright © 2011 Elsevier Inc. All rights reserved

No part of this publication may be reproduced, stored in a retrieval system or transmitted in any form or by any means electronic, mechanical, photocopying, recording or otherwise without the prior written permission of the publisher Permissions may be sought directly from Elsevier's Science & Technology Rights Department in Oxford, UK: phone (+44) (0) 1865 843830; fax (+44) (0) 1865 853333; email: permissions@elsevier.com. Alternatively you can submit your request online by visiting the Elsevier web site at http://elsevier.com/locate/permissions, and selecting *Obtaining permission to use Elsevier material*

Notice
No responsibility is assumed by the publisher for any injury and/or damage to persons or property as a matter of products liability, negligence or otherwise, or from any use or operation of any methods, products, instructions or ideas contained in the material herein

Library of Congress Cataloging-in-Publication Data
A catalog record for this book is available from the Library of Congress

British Library Cataloguing-in-Publication Data
A catalogue record for this book is available from the British Library

ISBN: 978-0-12-385514-5

ISSN: 0065-2458

For information on all Academic Press publications
visit our web site at elsevierdirect.com

Printed and bound in USA

11 12 13 10 9 8 7 6 5 4 3 2 1

Working together to grow libraries in developing countries

www.elsevier.com | www.bookaid.org | www.sabre.org

ELSEVIER   BOOK AID International   Sabre Foundation

Contents

CONTRIBUTORS ix
PREFACE xiii

VoIP Security: Vulnerabilities, Exploits, and Defenses

Xinyuan Wang and Ruishan Zhang

1. Introduction 3
2. VoIP Basics 5
3. Security Requirements of VoIP 11
4. VoIP Built-in Security Mechanisms 13
5. Vulnerabilities of Existing VoIP and Security Threats 16
6. Attacks on Currently Deployed VoIP 20
7. Related Works in VoIP Security Research 42
8. Technical Challenges in Securing VoIP 43
9. Conclusions 46
   Acknowledgments 47
   References 47

Phone-to-Phone Configuration for Internet Telephony

Yiu-Wing Leung

1. Introduction 52
2. Sparse Telephone Gateway for Internet Telephony 55
3. Packet Loss Recovery 81

4. Summary 100
   References 100

SLAM for Pedestrians and Ultrasonic Landmarks in Emergency Response Scenarios

Carl Fischer, Kavitha Muthukrishnan, and Mike Hazas

1. Localization for Emergency Responders: State of the Art 104
2. Implementation: Multimodal Sensing and Algorithms 116
3. System Evaluation 128
4. Conclusion 157
   References 158

Feeling Bluetooth: From a Security Perspective

Paul Braeckel

1. Bluetooth Details 162
2. BT Exploitation 189
3. BT Auditing Tools 218
4. Security Recommendations 228
5. Conclusion 235

Digital Feudalism: Enclosures and Erasures from Digital Rights Management to the Digital Divide

Sascha D. Meinrath, James W. Losey, and Victor W. Pickard

1. Introduction 238
2. Enclosures Along OSI Dimensions 245

CONTENTS

3. The Need for Open Technology 266
4. Policy Recommendations for the New Critical Juncture
   in Telecommunications 270
5. The Need for a New Paradigm 280
   References 283

Online Advertising

Avi Goldfarb and Catherine Tucker

1. Introduction 290
2. What Is Different About Online Advertising? 292
3. Different Types of Online Advertising 298
4. Consequences of Measurability and Targetability on
   the Online Advertising Industry 303
5. Regulation 306
6. Conclusions and Avenues for Future Research 310
   References 311

AUTHOR INDEX 317
SUBJECT INDEX 325
CONTENTS OF VOLUMES IN THIS SERIES 337

Contributors

Paul Braeckel possesses over 11 years of software and product development experience, specifically within Computer Networking and Security, and using a wide variety of programming languages, platforms, and technologies. Currently, he leads the development of Ephemeral Credentialing Products as Project Manager at the Identity Theft and Financial Fraud Research and Operations Center in Las Vegas, Nevada, directed by Dr. Hal Berghel. He specializes in object-oriented development of cyber-security applications using .NET languages and his software development experience is broad, ranging from desktop business applications to complex web applications. Paul holds an M.S. in Computer Science from the UNLV and a B.S. in Mechanical Engineering from Washington University in St. Louis and is working on his Ph.D. in Informatics at the University of Nevada, Las Vegas. He may be contacted through the following e-mail address: paulb314@gmail.com.

Carl Fischer is a researcher and Ph.D. candidate in the School of Computing and Communications at Lancaster University, UK. He holds an engineering degree (2006) from Supélec and a Masters (2006) from the University of Rennes, both in France. He is interested in infrastructure-less tracking and navigation systems. Carl can be contacted at fischer@comp.lancs.ac.uk.

Avi Goldfarb is associate professor of marketing in the Rotman School of Management at the University of Toronto. He received his Ph.D. in economics from Northwestern University. His research has explored brand value, behavioral modeling in industrial organization, and the impact of information technology on marketing, on universities, and on the economy. Professor Goldfarb has published more than 30 articles in a variety of outlets. He is coeditor at *Journal of Economics and Management Strategy* and an associate editor of *Quantitative Marketing and Economics, Information Economics and Policy, International Journal of Industrial Organization,* and *Management Science.*

Mike Hazas is an academic fellow and lecturer in the School of Computing and Communications at Lancaster University. Mike received his Ph.D. degree from the

University of Cambridge (2003) for his work on broadband ultrasonic location systems. Since then, Mike has worked on position and context sensing hardware and signal processing, with a focus on evaluation using real deployments. Mike has served on the programme committees for a number of leading conferences on pervasive computing, including the International Conference on Ubiquitous Computing (UbiComp) and the European Conference on Wireless Sensor Networks (EWSN). Mike can be contacted at hazas@comp.lancs.ac.uk.

Yiu-Wing Leung received his B.Sc. and Ph.D. degrees from the Chinese University of Hong Kong in 1989 and 1992, respectively. His Ph.D. advisor was Prof. Peter T. S. Yum. He has been working in the Department of Computer Science of the Hong Kong Baptist University, and now he is a full professor. His research interests include two major areas: (1) networking and multimedia which include the design and optimization of wireless networks, optical networks, and multimedia systems, and (2) cybernetics and systems engineering which include evolutionary computing and multiobjective programming. He has published over 70 journal papers in these areas. Email address: ywleung@comp.hkbu.edu.hk.

James W. Losey is a program associate with the Open Technology Initiative at the New America Foundation. He focuses on connectivity and technology at the community, national, and international level. Most recently he has published in *Slate* and *IEEE Spectrum*, as well as resources on federal broadband stimulus opportunities and analyses of the National Broadband Plan. He can be reached at lesey@newamerica.net.

Sascha D. Meinrath is the director of the New America Foundation's Open Technology Initiative and is a well-known expert on community wireless networks, municipal broadband, and telecommunications policy. Sascha is a cofounder of Measurement Lab, a distributed server platform for the deployment of Internet measurement tools, and coordinates the Open Source Wireless Coalition, a global partnership of open source wireless integrators, researchers, implementors, and companies. Sascha has worked with Free Press, the Cooperative Association for Internet Data Analysis (CAIDA), the Acorn Active Media Foundation, the Ethos Group, and the CUWiN Foundation. He can be reached at meinrath@newamerica.net.

Kavitha Muthukrishnan is currently a postdoctoral researcher at the Embedded Software Group, Delft University of Technology, The Netherlands. She obtained her Ph.D. degree in Computer Science in 2009 from the University of Twente. Her Ph.D. thesis was entitled "Multimodal localisation: Analysis, Algorithms and Experimental Evaluation." She was associated with the Embedded Interactive Systems group at

Lancaster University as a visiting researcher in 2008. Prior to that, she obtained her Master's degree in Electrical Engineering from the University of Twente, the Netherlands in 2004 and her Bachelor's degree in Engineering (specializing in Electronics and Communication) from the University of Madras, India in 2000. Her research interests include location sensing systems, localization algorithms, ubiquitous computing, sensor networks, SLAM methods, sensor fusion, and evaluation of location sensing technologies and algorithms. Kavitha can be contacted at k.muthukrishnan@tudelft.nl.

Victor W. Pickard is an assistant professor in the Department of Media, Culture, and Communication at the Steinhardt School of New York University. He received his doctorate from the Institute of Communications Research at the University of Illinois. His research explores the intersections of U.S. and global media activism and politics, media history, democratic theory, and communications policy and has been published in over two dozen scholarly journal articles, essays, and book chapters. He is currently finishing a book on the history and future of news media. His e-mail address is vwp201@nyu.edu.

Catherine Tucker is the Douglas Drane Career Development Professor in IT and management and assistant professor of marketing at MIT Sloan School of Management. She is interested in understanding how networks, privacy concerns, and regulation affect marketing outcomes. She received an undergraduate degree in politics, philosophy, and economics from Oxford University and a Ph.D. in economics from Stanford University.

Xinyuan Wang is an associate professor in the Computer Science Department at George Mason University. He received his Ph.D. in Computer Science from North Carolina State University in 2004. His main research interests are around computer network and system security including malware analysis and defense, attack attribution, anonymity and privacy, VoIP security, digital forensics. He has developed the first interpacket timing based packet flow watermarking scheme that is provably robust against timing perturbation. He has first demonstrated that it is feasible to track encrypted, anonymous peer-to-peer VoIP calls on the Internet. In his later work, he has demonstrated the fundamental limitations of existing low-latency anonymous communication systems in the presence of timing attack and developed the first practical attack that has "penetrated" the Total Net Shield—the "ultimate solution in online identity protection" of www.anonymizer.com with less than 11 min worth of Internet traffic. He is a recipient of the 2009 NSF Faculty Early Career Development (CAREER) Award. He may be contacted through the following e-mail address: xwangc@gmu.edu.

Ruishan Zhang received his Ph.D. in Computer Science and Engineering from Shanghai Jiaotong University for his dissertation on "Security in Mobile *Ad Hoc* Networks" in 2006. He spent 1½ years as a postdoctoral researcher at George Mason University, investigating the security of the SIP protocol and deployed SIP-based VoIP systems. Then he worked at Helsinki Institute for Information Technology until December 2009, focusing on VoIP spam prevention and P2P SIP security. Currently, his research interests include VoIP security, fuzz testing, reverse engineering, 3G network security, and peer-to-peer network security. He may be contacted through the following e-mail address: zhangruishan@gmail.com.

Preface

Welcome to Volume 81 of the *Advances in Computers*. This series, continuously published since 1960, is entering its sixth decade of publication, and it is the oldest series covering the development of the computer industry. Today there is no doubt that the dominant force in computing is the Internet; therefore, the theme of this volume is "the Internet and mobile technology." Whereas the design goal for the original ARPANET in the 1960s was to be able to reliably link together computers at various locations, this concept has evolved to where the computer-to-computer connection is taken for granted, and the current goals are to free the user from being tied down to a specific location. Therefore, mobility is a current research topic that has led to an explosion of mobile computing devices. We no longer have cellular telephones, but instead have small mobile computers that are able to communicate via telephony. This leads to numerous security and related issues to provide the reliability and integrity needed in today's world. In this volume, we present six chapters that address various aspects of these issues.

In the first chapter, "VoIP Security: Vulnerabilities, Exploits, and Defenses" by XinyuanWang and Ruishan Zhang, the authors discuss telephony via Voice over IP (VOIP). Rather than having a fixed wire connecting a telephone to the central switching office, VOIP works by using the Internet to send voice packets on the Internet along with assorted other Internet traffic such as e-mail, video, or Web pages. But if voice is carried along as digital packets, what level of security must there be to avoid the issues that plague other Internet traffic, such as spamming, phishing, hijacking, eavesdropping, etc.? In this chapter, the authors explain the general workings of VOIP transmission and then discuss the various strategies for dealing with security problems with this technology.

Yiu-Wing Leung in chapter, "Phone-to-Phone Configuration for Internet Telephony," addresses Internet telephony, the topic of the first chapter, with a different perspective. In the first chapter, the focus is on a communication line from one computer through the Internet to a receiving computer. But people are very mobile. How does one use a mobile telephone to provide VOIP services? One approach is for a service provider (e.g., local telephone company) to provide a local telephone number (a telephone gateway) a mobile phone user can call. At this local telephone number, the service provider connects to the Internet to send the call to a distant location, where it again is sent to a local mobile phone at the distant location. This allows users to use both computer-to-computer communications and mobile phone-to-mobile phone communications (e.g., Skype). The issues of optimizing traffic and minimal costs are the focus of this chapter.

In the third chapter, "SLAM for Pedestrians and Ultrasonic Landmarks in Emergency Response Scenarios," Carl Fischer, Kavitha Muthukrishnan, and Mike Hazas look at the issues in determining location from a mobile device. In particular, they are looking at the needs in emergency situations of determining location where visual clues are missing (such as inside a burning building). While most cellular telephones now contain GPS receivers, "darkness, smoke, fire, power cuts, water, and noise can all prevent a location system from working, and heavy protective clothing, gloves, and facemasks make using a standard mobile computer impossible." In this chapter, they discuss several existing systems for solving this problem, focusing on their simultaneous localization and mapping or SLAM method.

By now everyone is familiar with Bluetooth, that ubiquitous technology that allows one to connect one device to another device over short distances (e.g., microphone and earpiece to cellular phone without need to hold telephone). But what is Bluetooth, how does it work, and more importantly, what security exploits does it permit? In "Feeling Bluetooth—From a Security Perspective" by Paul Braeckel in the fourth chapter, the author discusses Bluetooth and provides insights into the kinds of security risks one has in using this technology.

The fifth chapter is titled "Digital Feudalism: Enclosures and Erasures from Digital Rights Management to the Digital Divide" and is written by Sascha D. Meinrath, James W. Losey, and Victor W. Pickard. One of the side effects of the Internet is that an increasing number of aspects of our daily lives are becoming digital and communicated over the Internet. From telephony (e.g., first two chapters in this volume), GPS (third chapter), wi-fi, radio, picture and video transmission to numerous other technologies, all are vying for space on the network bandwidth. This leads to a congestion problem—who has access to this bandwidth? Do all share equally (e.g., "net neutrality"), or do some applications take precedence over other technologies (e.g., real-time video over e-mail)? Can or should one pay more for better access? These are the questions that this chapter addresses.

In the last chapter, "Online Advertising" by Avi Goldfarb and Catherine Tucker, the authors discuss an important feature of the Internet, one without which the Internet would not have existed, and that is advertising. Running the Internet, supporting ISPs (Internet Service Providers), and paying for the various Web sites and search engines that exist all take money. While some organizations use membership fees to support their activities, advertising has become the dominant method for paying for Internet access. But how does Internet advertising work? Who pays what and why? The general model of how Internet advertising works is the focus of this chapter.

I hope that you find these chapters of use to you. I also want to say that I have enjoyed producing these volumes. I have been series editor of the *Advances in Computers* since 1993, and Volume 81 is the 41st volume I have worked on in 19 years. The 2011 volumes will be my last; however, the series will continue under new competent leadership. I hope that you will continue to find this series of use to you in your work.

Marvin Zelkowitz
College Park, Maryland

VoIP Security: Vulnerabilities, Exploits, and Defenses

XINYUAN WANG

Department of Computer Science, George Mason University, Fairfax, Virginia, USA

RUISHAN ZHANG

Department of Computer Science, George Mason University, Fairfax, Virginia, USA

Abstract

Telephone network is an important part of the critical information infrastructure. Traditional Public Switched Telephony Network (PSTN) has been shown to be reliable and hard to be tampered with by normal people. The general public has put a lot of trust on landline telephone, and they are relying on voice communication for many critical and sensitive information (e.g., emergency 911 calls, calls to financial institutions) exchange.

Voice over IP (VoIP) is an emerging technology that allows voice calls to be carried over the public Internet instead of traditional PSTN. While more and more voice calls are shifting from PSTN to VoIP, most people are not aware of the security vulnerabilities introduced by VoIP and they keep trusting VoIP the same as traditional PSTN.

In this chapter, we systematically study the security issues of VoIP and present the state of the art of VoIP security. Specifically, we discuss the security requirements of VoIP, people's expectations of VoIP, and existing VoIP security mechanisms. We present the identified vulnerabilities of existing VoIP, known and potential exploits of those VoIP vulnerabilities. We discuss not only the impact on the VoIP infrastructure itself but also the implications to the VoIP users. We discuss the inherent technical challenges and open problems in securing VoIP.

1. Introduction 3
2. VoIP Basics 5
   2.1. VoIP Signaling 5
   2.2. SIP Overview 5
   2.3. VoIP Transport 9
3. Security Requirements of VoIP 11
   3.1. Authenticity 11
   3.2. Confidentiality, Privacy, and Anonymity 12
   3.3. Integrity 12
   3.4. Temper Resistance and Availability 12
4. VoIP Built-in Security Mechanisms 13
   4.1. VoIP Signaling Security Mechanisms 13
   4.2. VoIP Transport Security Mechanisms 15
5. Vulnerabilities of Existing VoIP and Security Threats 16
   5.1. Vulnerabilities of Existing VoIP 16
   5.2. Security Threats to VoIP 17
6. Attacks on Currently Deployed VoIP 20
   6.1. Registration Hijacking 20
   6.2. Transparent Detour of Selected VoIP Calls on the Internet . . . 22
   6.3. Transparent Redirection of Selected VoIP Calls 26
   6.4. Manipulating and Hijacking Call Forwarding Setup 28
   6.5. Voice Pharming 32
   6.6. Remote MITM 34
7. Related Works in VoIP Security Research 42
8. Technical Challenges in Securing VoIP 43
   8.1. Open Architecture of VoIP 43
   8.2. Real-time Constraints of VoIP 43
   8.3. Multiple Protocols 44
   8.4. E-911 44
   8.5. Firewall Traversal 45
   8.6. NAT Traversal 45
9. Conclusions 46
   Acknowledgments 47
   References 47

1. Introduction

Voice over IP (VoIP) is a technology that allows people to make voice phone calls across the public Internet instead of traditional Public Switched Telephony Network (PSTN). VoIP not only makes voice communication cheaper but also enables many functionalities (e.g., free choice of area code, e-mail notification of voice mail) that were not possible in traditional PSTN. In the past 10 years, VoIP has experienced phenomenal growth and more and more voice calls are carried at least partially over the Internet using VoIP technologies. A study by ABI [1] predicted that the number of residential VoIP subscribers worldwide will increase from 38 million in 2006 to more than 267 million by 2012.

One of the most basic and fundamental requirements of any VoIP services is that they must be reliable and trustworthy. When people subscribe or use any VoIP service, they have actually put a lot of implicit trust on it. For example, when people make phone calls, they intuitively trust that their calls will reach the intended callee once they dial the correct phone number and no one but the intended callee will receive their calls. When people talk over the established phone session, they trust that their conversation and any PIN number pressed will reach the intended receiver unaltered. In addition, people would expect that their calls will not be wiretapped without proper legal authorization. Based on this trust, voice communication has been used for exchanging much critical and sensitive information (e.g., emergency 911 calls, calls to customer service of financial institutions). The general public are used to giving out their SSN, credit card number, and PIN when they interact with the interactive voice response (IVR) system before they are connected to a service representative of their financial institution. Furthermore, people are comfortable to give out their credentials (e.g., SSN, account number, authentication code) to the service representative of their financial institution over the phone even if they do not personally know the service representative.

Now suppose a VoIP user Alice is planning to buy a house, and she wants to cash out some of her Google stock options for that. Because of the large amount of money involved, Alice prefers talking to a broker over the phone than using the Web. So she dials the 1-800 number shown in her TD AMERITRADE statement, and left a message asking for a call back since no one is available at the time. A few minutes later, a call with TD AMERITRADE callerID comes in, and a representative named Bob says he is returning the call to Alice. Alice is quite technical savvy, and she insists on getting Bob's extension number and calling him back. After hanging up the incoming call, Alice calls the official number of TD AMRITRADE with Bob's extension. Once Alice reaches Bob again, she feels at ease and requests to exercise 5000 shares of her Google stock options and wire the expected $500,000 profit to her

Citibank checking account. Of course, she gives Bob all the credentials (e.g., broker account number, PIN, checking account number) needed for the transaction.

After calling the broker, Alice calls Hilton San Diego Bayfront to make a room reservation with her credit card for her vacation there. Before she left for San Diego, she sets up her home phone to forward any incoming call to her cell phone.

A few days later, Alice gets back from her vacation at San Diego, and she is ready for the closing of her house purchase. Alice checks her Citibank checking account just to make sure it has the expected balance of $500,000 dollars. Surprisingly, Alice finds her Citibank checking account has zero balance. Alice calls TD AMRITRADE again to see if it has transferred the money to her Citibank checking account. This time, Alice talks to Chris, and Chris confirms the completion of the exercise of her 5000 shares of Google stock options. Alice is shock to hear that over $500,000 profit has been transferred to some foreign bank accounts, all of which have been authorized by Alice. Soon later, Alice receives a call from her credit card company reporting suspicious activity of buying expensive jewelry at France in the last 24 h. Later in the month, Alice receives the bill from her VoIP service provider, and she was shock again to see the charge of calls to Inmarsat East Atlantic—Aero at the rate of $12.62 per minutes since she has never called any Inmarsat number.

The hypothetical incidents described above might sound too far-fetched. As we show in this chapter, attackers could indeed make these happen to the targeted VoIP users. Specifically, the attacker can transparently divert Alice's call to TD AMRITRADE to the bogus representative who claims to be Bob. Since Alice dials the authentic phone number of TD AMRITRADE, she trusts she reaches a true TD AMRITRADE representative and gives all the necessary credential to complete the stock option exercise transaction and money wiring. With such credentials, the attacker can call TD AMRITRADE to exercise the stock option but wire the profit to their accounts at some foreign banks. In addition, the attacker could become the man-in-the-middle (MITM) even if he is thousands of miles away from where Alice lives, and capture Alice's credit card number when Alice calls Hilton Hotel. When Alice sets up the call forwarding of her home VoIP phone, the attacker can transparently set the forwarding number to some Inmarsat number in Europe.

The objective of this chapter is to systematically overview the security issues of VoIP and present the state of the art of VoIP security. Specifically, we discuss the security requirements of VoIP, people's expectations of VoIP, and existing VoIP security mechanisms. We present the identified vulnerabilities of existing VoIP, known and potential exploits of those VoIP vulnerabilities. We discuss not only the impact on the VoIP infrastructure itself but also the implications to the VoIP users. We discuss the inherent technical challenges and open problems in securing VoIP.

The rest of the chapter is organized as follows. In Section 2, we overview the basics of VoIP. In Section 3, we present the security requirements of VoIP.

In Section 4, we overview the built-in security mechanism of current VoIP protocols. In Section 5, we elaborate the vulnerabilities of existing VoIP protocols and the security threats to VoIP. In Section 6, we present some recently identified attacks on currently deployed VoIP. In Section 7, we discuss related works in VoIP security research. In Section 8, we elaborate the technical challenges in securing VoIP. We conclude in Section 9.

2. VoIP Basics

The core of VoIP consists of two functionalities: the *signaling* and the *transport*. The signaling is responsible for setting up, managing, and tearing down VoIP calls, and the transport is responsible for delivering the encoded voice over the Internet. In the following sections, we overview the VoIP signaling and transport, respectively.

2.1 VoIP Signaling

Existing VoIP signaling protocols include Session Initiation Protocol (SIP) [2], H.323, and Media Gateway Control Protocol (MGCP) [3]. SIP is an RFC standard from the Internet Engineering Task Force (IETF), and it is a generic signaling protocol for establishing sessions in an IP network. H.323, however, is an ITU standard that was originally designed to provide multimedia communication over LANs, and it is suited for interworking between IP and ISDN. MGCP [3] is a signaling protocol for controlling telephony gateways from external call control elements called media gateway controllers or call agents. Currently, SIP is the dominant signaling protocol for VoIP.

2.2 SIP Overview

SIP [2] is a general purpose, application layer signaling protocol used for creating, modifying, and terminating multimedia sessions (e.g., VoIP calls) among Internet endpoints. Specifically, SIP provides the following functionalities for establishing and terminating communication sessions:

- *Learn*, determine the location, availability of remote communicating parties. This involves registration, call routing, and call redirection.
- *Establish*, manage sessions between endpoints. This involves call setup, transfer, and termination.
- *Negotiate* the media capability among the endpoints. This involves using the Session Description Protocol (SDP) [4].

While SIP can be used to establish a number of different multimedia communications such as video conference, instant messaging, we focus on how SIP establishes VoIP communications only.

SIP defines the signaling interaction between: the *user agent* (UA) and the *SIP servers*. An UA represents an endpoint of the communication (i.e., a SIP phone) which is usually owned or used by a VoIP user. The UA that initiates the VoIP call is the *user agent client* (UAC) and the UA that receives the VoIP call is the *user agent server* (UAS). VoIP servers are maintained by the VoIP service provider to manage the VoIP calls. Based on its functionality, a VoIP server can be *proxy server*, *redirect server*, *registrar server*, and *location server*. The proxy server is the intermediate server that relays the signaling messages between the caller and the callee. The registrar server accepts registration from subscribers about their current locations. The location server maintains the current location (i.e., IP address) of the registered UAs. The redirect server provides the UA client with an alternative set of contact addresses on behalf of the UA server.

In SIP network, each user is identified by a SIP *Uniform Resource Identifiers* (URIs), which is similar to an e-mail address. Suppose there are two UAs *UA-A* and *UA-B* belong to domain *Alpha.com* and *Beta.com*, respectively, both of which have their own proxy servers. Figure 1 shows the SIP message flow of a typical and successful call setup and tear down without authentication.

2.2.1 SIP Messages

Based on client-server model, SIP uses *request* and *response* messages to establish sessions between two or more endpoints. To establish, manage or tear down a VoIP session, UAC will send to SIP server or UAS a SIP *request message* identified by one of the 13 SIP method names. The most commonly used SIP methods are the following:

- `REGISTER` is used by the UAC to inform its current location to the SIP server.
- `INVITE` is used by the UAC to initiate a call session.
- `ACK` acknowledges the successful receipt of some SIP messages.
- `BYE` terminates an established session.
- `CANCEL` quits from the ongoing setup of a session.
- `OPTIONS` queries the capability of a server.

Upon receiving SIP request message, the SIP server or UAS will, when appropriate, reply with one or more SIP *response messages* identified by the following status codes:

- `1xx Provisional` indicates that the request has been successfully received, and it is in the process of processing the request. For example, `100 Trying`, `180 Ringing`.

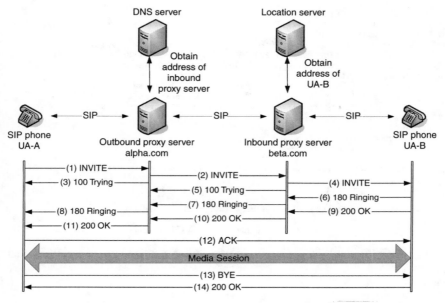

FIG. 1. SIP message flow without digest authentication.

- 2xx Success indicates the successful completion of the action requested. For example, 200 Ok.
- 3xx Redirection indicates that the request needs to be redirected to someone else. For example, 301 Moved Permanently, 302 Moved Temporarily.
- 4xx Client Error indicates that the request is invalid. For example, 404 Not Found, 410 Gone, 403 Forbidden.
- 5xx Server Error indicates that the server cannot fulfill certain valid request. For example, 503 Service Unavailable.
- 6xx Global Failure indicates that the request cannot be fulfilled at any server. For example, 600 Busy Everywhere.

Figures 2 and 3 show the INVITE and 200 OK messages of a call from a PSTN phone (703-xxx-9398) to our AT&T VoIP phone (703-xxx-0461). The sample SIP messages include important fields such as To, From, CSeq, Call-ID, Max-Forward, Via, Request-URI, and Contact as well as part of the message body. Specifically, the message body part of the INVITE and 200 OK messages includes the IP addresses and UDP port numbers that the caller and callee choose to

```
Session Initiation Protocol
    Request line:   INVITE sip:703***0461@192.168.1.188:5060;user=phone SIP/2.0
    Method:   INVITE
    Message Header
       Via: SIP/2.0/UDP 12.194.224.134:5060;branch=z9hG4bKmv39713030o13hglj4s1.1
       Call-ID: SDopf0b01-9f4f824d3dc750fde8c2db1462d304ec-cgg1e32
       CSeq: 1 INVITE
       From:"full name" sip:703***9398@12.194.224.137>; tag=123456
       To: <sip:703***0461@12.194.224.134:5060;user=phone;transport=udp>
       Contact: <sip:703***9398@12.194.224.134:5060; transport=udp>
       Max-Forwards: 69
       Content-Length: 293
       Content-Type: application/sdp
Session Description Protocol
    Session Description Protocol Version (v): 0
    Owner/Creator, Session Id (o): - 197366797 197366797 IN IP4 12.194.224.134
    ...
    Connection Information (c): IN IP4 12.194.224.134
    ...
    Media Description, name and address (m): audio 22148 RTP/AVP 2 18 0 100
```

FIG. 2. An example of SIP INVITE message.

```
Session Initiation Protocol
    Status line:   SIP/2.0 200 OK
    Status-Code:   200
    Message Header
       Via: SIP/2.0/UDP 12.194.224.134:5060;branch=z9hG4bKmv39713030o13hglj4s1.1
       Call-ID: SDopf0b01-9f4f824d3dc750fde8c2db1462d304ec-cgg1e32
       CSeq: 1 INVITE
       From: "full name"<sip:703***9398@12.194.224.137>;tag=123456
       To: <sip:703***0461@12.194.224.134:5060;user=phone>;tag=468c0
       Contact:703***0461<sip:703***0461@192.168.1.188:5060;user=phone>
       Allow: REFER,UPDATE,INFO,MESSAGE,OPTIONS
       Content-Type: application/sdp
       Content-Length: 169
Session Description Protocol
    Session Description Protocol Version (v): 0
    Owner/Creator, Session Id (o): 703***0461 0 0 IN IP4 192.168.1.188
    ...
    Connection Information (c): IN IP4 192.168.1.188
    ...
    Media Description, name and address (m): audio 10000 RTP/AVP 2 100
```

FIG. 3. An example of SIP 200 OK message.

use for the upcoming real-time transport protocol (RTP) voice stream. Figure 2 shows that the caller is expecting to receive RTP voice stream from the callee at IP address 12.194.224.134 on port 22148. Figure 3 shows that the callee is expecting to receive RTP voice stream from the caller at IP address 192.168.1.188 on port 10000.

2.2.2 SIP Message Flow

SIP is based on an HTTP-like request/response model. Figure 1 shows a typical SIP message flow of a call setup and tear down without authentication. When the caller (UA-A) initiates a call to the callee (UA-B), it first sends an INVITE message to its outbound proxy server at domain Alpha.com. After receiving the INVITE message, the outbound server locates the inbound server at domain Beta.com via Domain Name Service (DNS), and forwards the INVITE message to the inbound server. Meanwhile, the outbound proxy server sends back a 100 TRYING message to UA-A to indicate that the outbound proxy has received the request and it is forwarding the INVITE message to the destination. Upon receiving the INVITE message, the inbound proxy server obtains the current location (i.e., IP address) of UA-B by querying the location service, then it delivers the INVITE message to UA-B. After receiving the INVITE message, UA-B begins to ring and replies with a 180 Ringing message to UA-A so that the caller can hear the ringback tone. When the callee picks up the phone, UA-B replies a 200 OK message to indicate that the call has been answered. Upon receiving the 200 OK message, UA-A stops the ringback tone and replies an ACK message to UA-B. After UA-B receives the ACK message, the three-way handshake is completed and the VoIP call is established. Note the message bodies of the INVITE message and 200 OK message contain the negotiated media session parameters (e.g., codec, IP address, and port number of the RTP stream) that are specified in the SDP [4]. Then UA-A and UA-B send RTP voice streams to each other based on the negotiated media session parameters. At the end of the call, UA-A (UA-B) hangs up and sends a BYE message to its peer. After receiving the BYE message, UA-B (UA-A) replies with a 200 OK message and stops sending its RTP stream. Upon receiving the 200 OK message, UA-A (UA-B) stops sending its RTP stream. Then the VoIP call is terminated.

2.3 VoIP Transport

2.3.1 RTP Overview

The RTP [5,6] is an application layer protocol for end-to-end delivery of real-time data such as interactive audio or video over multicast or unicast network services. While RTP normally runs on top of UDP, it may be used with other underlying transport protocols such as TCP. RTP supports data transfer to multiple destinations if the underlying network supports multicast.

RTP contains two complementing functionalities: *real-time data transport* and *data transport monitoring and control*, which are provided by the RTP and the RTP control protocol (RTCP), respectively. RTP itself, however, does not provide any guarantee on timely delivery or other quality-of-service. It relies on the underlying network services to do so. Specifically, RTP does not guarantee packet delivery or prevent out-of-order delivery. The RTP sequence number allows the receiver to detect out-of-order delivery and to reconstruct the original packet sequence.

2.3.2 RTP Packet Format

Table I shows the RTP packet structure. The first 12 bytes are mandatory in every RTP packet, and the following optional part (e.g., the list of contributing source (CSRC) identifiers) is present only when inserted by a mixer. Some important fields are as follows:

1. *Version (V): 2 bits*. This field identifies the version of RTP. The version defined by RFC 3550 is two (2).
2. *Payload type (PT): 7 bits*. The PT field identifies the format of the RTP payload.
3. *Sequence number: 16 bits*. The sequence number increases by one for each RTP data packet sent. The sequence number allows the receiver to detect packet loss and restore the original packet sequence even if the RTP packets are received out-of-order.
4. *Timestamp: 32 bits*. The timestamp is the sampling instant of the first byte in the RTP data packet.
5. *SSRC: 32 bits*. The SSRC field identifies the synchronization source, which is chosen randomly such that no two synchronization sources within the same RTP session will have the same SSRC identifier.

Figure 4 shows an example of RTP packet.

TABLE I
RTP HEADER FORMAT

+Bits	0–1	2	3	4–7	8	9–15	16–31
0	V=2	P	X	CC	M	PT	Sequence number
32	Timestamp							
64	Synchronization source (SSRC) identifier							
96	Contributing source (CSRC) identifiers (optional)							
...	Extension header (optional)							
...	Data							

```
Real-Time Transport Protocol
    10.. ....    = Version: RFC 1889 Version (2)
    ..0. ....    = Padding: False
    ...0 ....    = Extension: False
    .... 0000    = Contributing source identifiers count: 0
    1... ....    =  Marker: True
    Payload type: ITU-T G.721 (2)
    Sequence number: 15467
    Timestamp: 408338420
    Synchronization Source identifier: 3740388056
    Payload: 7B58B1B3A34F3D3CB1C1D24E1C2DBFD2F34F2D1CADD1E33F...
```

FIG. 4. An example of RTP packet.

3. Security Requirements of VoIP

Voice communication is fundamental to the normal operation of our society and the daily lives of billions of people. The general public have put a lot of implicit trust in voice communication, and they have been relying on it for many critical (e.g., emergency 911 calls) and sensitive (customer service calls of financial institutions) information exchange. The general public are used to giving out their SSN, credit card number, and PIN when they interact with the IVR system before they are connected to a service representative of their financial institution. Furthermore, people are comfortable to give out their credentials (e.g., SSN, account number, authentication code) to the service representative of their financial institution over the phone even if they do not personally know the service representative.

When using VoIP, the general public would expect similar level of security and trustworthiness of landline voice communication. In fact, security is consistently among the top concern of VoIP [7]. In this section, we clarify the security requirements of VoIP.

3.1 Authenticity

During a phone call, both the caller and the callee are uniquely identified by their phone numbers. Therefore, when people make phone calls, they intuitively trust that their calls will reach the intended callee once they dial the correct phone number and no one but the intended callee will receive their calls. For example, when people dial 911, they simply trust that their 911 calls, once connected, will reach the appropriate PSAP (Public Safety Answering Point). When people receive phone calls, they would expect the callerID to be authentic so that they can decide whether to accept the phone call accordingly. In addition, when people retrieve voice mails, they would expect those recorded voice mails are really from the number and the caller shown.

3.2 Confidentiality, Privacy, and Anonymity

Given that phone calls are widely used for exchanging sensitive information, people would expect that no one other than the intended caller and callee could listen the content of their voice communication. For example, when people call their financial institutions, they often need to give out their account number and PIN in order to authenticate themselves. Such information conveyed over the phone call is confidential and should not be divulged to anyone else.

In addition to confidentiality, people sometimes want to hide their identity when making or receiving phone calls. For example, a survey or vote conducted over the voice phone call should protect the identity of the participants. In this case, the existence of the phone call is known to the third party, but the caller and/or callee remains anonymous.

Furthermore, there are situations that people want to hide the very existence of the phone call itself. For example, Alice wants to call Bob, but she does not want anyone else to know their call has ever happened.

3.3 Integrity

When people talk over the established phone session, they want to make sure that their conversation and any number pressed will reach the intended receiver unaltered. For example, when someone interacts with the IVR system of a financial institution, he wants to be ensured that the IVR will receive the exact PIN number he send over the phone, and the balance information returned from the IVR cannot be modified by any malicious third party without being detected. In addition, we would expect the call history to be accurate in that it shows all and only the calls made (or received) by the subscriber for the exact duration of the calls. For those VoIP calls that are charged on a per minute basis (e.g., certain international call, 900 calls), the VoIP service providers want to prevent any service fraud such that they will charge all the billable VoIP services for the duration they have provided the services. However, VoIP subscribers want to make sure that they are only charged for the calls they have made for the actual call duration and there is no overcharge.

3.4 Temper Resistance and Availability

As any other service built upon the Internet, VoIP is subject to attacks from the Internet. On one hand, the VoIP service should be as open as the Internet in that it should allow anyone from anywhere to call anyone at anywhere even if the caller and the callee do not know each other at all. On the other hand, the VoIP service should be resilient to malicious attacks in that it allows legitimate and authenticated subscribers to make and receive calls in the presence of attack. Here, temper resistance applies to the VoIP components owned/operated by the VoIP service provider (e.g., VoIP proxy servers) and the VoIP subscribers (e.g., VoIP phone), as well as the underlying VoIP protocols. Ideally, the VoIP services should maintain all the authenticity, confidentiality, anonymity, privacy, integrity, and availability even when they are under attack.

4. VoIP Built-in Security Mechanisms

Existing VoIP has limited built-in security mechanisms to protect both VoIP signaling and VoIP transport. In this section, we overview current VoIP built-in security mechanisms.

4.1 VoIP Signaling Security Mechanisms

VoIP signaling security mechanisms aim to protect the authenticity, integrity, and confidentiality of the signaling messages. SIP does not define its own security mechanism, and it reuses existing security mechanisms for HTTP, SMTP whenever possible. SIP security mechanisms have the following two building blocks:

- *Authentication* which ensures the authenticity and integrity of the protected parts of the protect SIP messages.
- *Encryption* which provides the confidentiality of the protected SIP messages.

Unlike most Internet applications (e.g., Web) whose traffic can be protected from end-to-end, SIP message cannot be protected from end-to-end as a whole. This is because certain fields of certain SIP messages need to be examined, modified, or inserted by the intermediate SIP proxies. For example, SIP specification requires each intermediate SIP proxy to update the `Request-URI, Max-Forward` fields, add `Via` field when forwarding SIP request messages (e.g., `INVITE`). Therefore, the whole SIP message protection can only be applied on a hop-to-hop basis. However, end-to-end protection can be applied to those SIP messages fields that do not need to be changed during the SIP message routing.

4.1.1 SIP Digest Authentication

SIP digest authentication aims to provide stateless authentication and replay protection of selected SIP messages based on challenge–response paradigm. Assuming the two parties involved in the authentication share a secret password, SIP digest authentication reuses the HTTP digest authentication [8] with very minor customization.

Specifically, when a SIP proxy or UAC receives a SIP request message (e.g., REGISTER), it may ask the message sender to authenticate itself by sending it a challenge within the authentication request. The SIP request message sender calculates a response to the challenge and resends the SIP request message with the response as the authentication credential. To prevent the authenticated SIP request message from being replayed by the adversary who intercepts the authenticated SIP message, the challenge always includes a random nonce number. The response is essentially the result of some one-way function of the following fields:

$$\text{Response} = F(\text{nonce}, \text{username}, \text{password}, \text{realm}, \text{SIP} - \text{method}, \text{request} - \text{URI}). \quad (1)$$

Assuming the password is a shared secret between the SIP message sender and receiver, it is infeasible for the adversary to come up with the correct response. The one-way property of the hash function prevents the adversary from figuring out the secret password based on the captured response. Because the response is specific to random nonce, which is generated for each SIP message to be authenticated, the response can only be used once.

4.1.2 Transport Layer Security (TLS)

TLS [9] is a network security protocol that provides reliable, end-to-end transport between Internet applications. It is built on top of some reliable transport protocol such as TCP, and it consists of two sublayers: the *TLS Record Protocol* and the *TLS Handshake Protocol*.

The TLS Record Protocol establishes a secure, reliable channel for higher level protocols such as the TLS Record protocol. Specifically, it provides data confidentiality with symmetric key encryption, and data integrity with keyed MAC (e.g., SHA-1). The TLS Handshake Protocol provides the following functionalities:

- Negotiates the cipher suite, which includes the encryption and hash algorithms, authentication and key establishment methods, to be used.
- Authenticate the participating parties (e.g., client and server) and the exchanged data. In most e-commerce applications, servers are nearly always authenticated, and clients are rarely authenticated by TLS.
- Establish fresh, shared secret used by various encryption and authentication algorithms.

TLS is suited for securing SIP on a hop-by-hop basis. The SIP RFC [2] mandates that SIP proxy servers, redirect servers, and registrars must support TLS. However, UAs are strongly recommended to be able to initiate TLS.

4.1.3 IPsec

IPsec [10] is a suite of protocols that provides the following point-to-point security services at the IP layer: (1) *authentication of data origin* (2) *data integrity* (3) *data confidentiality* (4) *partial flow confidentiality* and (5) *resistance against replay-attack*. IPsec's security services are provided via two extended headers: *Authentication Header* (*AH*) and *Encapsulation Security Payload* (*ESP*). IPsec supports two modes of operation: *tunnel mode* and *transport mode*. The transport mode encapsulates the payload of original IP packet, and the tunnel mode encapsulates the whole IP packet.

IPsec is "attached" upon IP layer but below any layer upon IP. It is transparent to any layer upon IP in that anything running upon IP can use IPsec without any changes. However, SIP specification does not require any SIP component to run upon IPsec.

4.1.4 S/MIME

S/MIME [11] provides confidentiality, integrity, authentication, and nonrepudiation of origin protection on MIME data. Since certain fields of certain SIP messages need to be examined, modified, and added by the intermediate SIP proxy, S/MIME can only be applied to those volatile fields of SIP messages directly. However, S/MIME can be applied to MIME bodies within SIP message body.

It is possible to provide some confidentiality and integrity protection on the SIP message header by SIP message tunneling—encapsulating a whole SIP message inside an S/MIME and embed it in the outer layer SIP message body.

4.2 VoIP Transport Security Mechanisms

The VoIP transport can be protected by general transport security mechanisms such as IPsec and TLS. Alternatively, the VoIP transport can be protected by VoIP specific transport security protocols such as SRTP.

4.2.1 Secure Real-Time Transport Protocol

The secure real-time transport protocol (SRTP) [12] is an extension to RTP which provide confidentiality, message integrity, and replay protection to the RTP and RTCP traffic. Except the SRTCP integrity protection, which is mandatory, these three security protections are optional and independent from each other.

TABLE II
SRTP HEADER FORMAT

+Bits	0–1	2	3	4–7	8	9–15	16–31
0	V=2	P	X	CC	M	PT	Sequence number
32	Timestamp						
64	Synchronization source (SSRC) identifier						
96	Contributing source (CSRC) identifiers (optional)						
...	RTP extension header (optional)						
...	Payload...						
...	Payload...				RTP padding		RTP pad count
...	SRTP MKI (optional)						
...	authentication tag (recommended)						

Table II shows the format of SRTP packet. Since SRTP is an extension of RTP, the first 12 bytes in the SRTP packet is the same as that in the RTP packet. Therefore, an SRTP packet is an RTP packet with SRTP specific information stored in the extended fields.

5. Vulnerabilities of Existing VoIP and Security Threats

Since VoIP is an application upon IP, it inherits the inherent vulnerabilities of current Internet. For example, VoIP UAs and servers are susceptible to denial-of-service (DoS) attacks and the VoIP traffic may be mislead or corrupted by DNS hijacking. In this section, we leave aside those vulnerabilities that are general to the Internet protocols and instead focus on the vulnerabilities that are specific and inherent to VoIP protocols.

5.1 Vulnerabilities of Existing VoIP

Existing SIP authentication is based on HTTP digest. However, SIP authentication does not provide full integrity protection of SIP messages. Specifically,

- SIP authentication only applies to three SIP request messages `INVITE`, `BYE`, and `REGISTER`, and leaves other SIP messages unprotected. This allows the

adversary to freely modify and forge those unprotected SIP messages (e.g., `ACK`, `CANCEL`, `OPTION`) without being detected.
- SIP authentication only protects three fields of those protected SIP messages: `request-URI`, `username`, and `realm`. This leaves all other fields such `From`, `To` unprotected.
- Most currently deployed SIP-based VoIP services (e.g., Vonage, AT&T) only authenticate selected SIP messages from the UAC (i.e., SIP phone) to the SIP server, and leave all the SIP messages from the SIP server to the UAC unprotected.

Although SIP specification suggests using TLS or IPsec to protect the SIP messages and the VoIP streams between the VoIP phones and the VoIP servers, hardly any currently deployed VoIP services (e.g., Vonage, AT&T) uses TLS or IPsec. These enable the adversary to modify, spoof many SIP messages and their fields without being detected.

In addition, hardly any currently deployed VoIP services uses SRTP to protect the voice streams between the VoIP phones and the VoIP servers. This leaves the voice streams open for eavesdropping, modification, and spoofing.

5.2 Security Threats to VoIP

Due to lack of full integrity and confidentiality protection, existing VoIP are susceptible to many security threats.

- *Service stealing*. By exploiting the vulnerabilities, it is possible for someone to use the VoIP services without paying the necessary fees or call at other's expenses. Apparently, service stealing is a big concern for the VoIP service providers as well as VoIP subscribers.
- *Service disruption*. By attacking the VoIP infrastructure such as the VoIP servers, the adversary could potentially make the VoIP services unavailable to large number of VoIP subscribers. Taking out the VoIP servers has very obvious impact to the overall VoIP service and is easy to detect. A more stealthy form of service disruption is against individual VoIP subscriber. For example, the attacker could prevent the targeted VoIP user from making or receiving calls. Such an individualized service disruption is much harder for the VoIP service provider to detect as individual service disruption has neglectable impact on the overall VoIP service and it could indeed happen naturally.
- *Call eavesdropping*. Since the VoIP traffic is carried over the public Internet, it is much easier for the adversary to do wiretap or traffic analysis on VoIP than PSTN. For example, the adversary can set up a rogue wireless access point and become an MITM between any wireless users who use his rogue wireless access

point and any Internet destinations they communicate with over the wireless link. This allows the attacker to "listen" all the unencrypted VoIP conversations.

- *Call hijacking.* By exploiting the VoIP vulnerabilities, attackers could potentially hijack VoIP calls and become the caller or callee of hijacked calls. For example, if the attacker has somehow registered himself as the (victim) VoIP subscriber Alice, he will receive all the VoIP calls to Alice and leave Alice unaware of the incoming VoIP calls. Alternatively, the attacker could potentially redirect the VoIP calls for Alice to someone else. In these cases, the VoIP call actually reaches to a third party while the caller thinks he reaches Alice. At the same time, Alice does not even know the caller has called her. If the attacker has registered himself as the VoIP subscriber Alice, he could make calls pretending Alice as the VoIP service provider thinks the attacker is Alice.
- *Call interception and modification.* Instead of hijacking targeted VoIP calls, attackers can also intercept and modify the unencrypted VoIP voice streams in real time. Such call modification can be implemented by a malicious MITM between the targeted VoIP user and the VOIP server. For example, suppose a VoIP user wants to set up the call forwarding number to his cell phone number X, the MITM can modify the call forwarding number from X to some other number Y by modifying the VoIP traffic from the VoIP user to the VoIP server. In this case, the number received by the VoIP server is Y even if the number keyed by the VoIP user is X. To trick the VoIP user, the MITM can further modify the number confirmation voice stream from the VoIP server such that it says the number received by the VoIP server is X. Therefore, call interception and modification can trick the targeted VoIP users and cause hard-to-resolve discrepancy between the VoIP users and the VoIP service providers.
- *VoIP fraud.* VoIP fraud essentially exploits people's trust on voice communication. Instead of asking people to visit some bogus Web site and input their credentials such as account number and password, voice phishing (i.e., vishing) asks people to call a number (e.g., a claimed number of their financial institution). When the victim calls the bogus phone number, he/she may be prompted to input his/her credit card number, account number, PIN. Since we are used to giving out these credentials when interact with the IVR of our financial institution over the phone, voice phishing is more likely to harvest victim's sensitive information. In fact, voice phishing attacks against PayPal and Santa Barbara Bank & Trust customers were reported [13,14] back in 2006. The voice phishing attacks urged people to call a bogus (VoIP) phone number (805-xxx-xxxx) and input their 16-digit credit card number to the IVR system "in order to prevent any fraudulent activity from occurring." VoIP makes it very easy to spoof the IVR system such that the prompt will be exactly the same as authentic.

- *VoIP annoyance.* Unlike traditional PSTN calls which are carried on a relatively closed PSTN network, VoIP calls are carried on the public Internet. This enables anyone with Internet access to generate potentially bogus VoIP calls. Spam over Internet Technology (SPIT) is an emerging threat to all VoIP (as well as traditional landline phone) users. Now it is technically easy and economically cheap to generate large volume of VoIP spam calls to large number of phone numbers from a single computer connected to the Internet. Because people are used to pay immediate attention to voice phone call, SPIT could become much more annoying than e-mail spam. Because SPIT is real-time voice stream and real-time voice recognition is a well-known hard problem, it is much harder to filter or block SPIT than e-mail spam.

In addition, attackers can ring any targeted VoIP phone by simply sending one forged INVITE packet. All the attacker needs to know is the phone number and the IP address of the targeted VoIP phone. Because the attacker can forge any callerID, the victim cannot rely on callerID to block such annoyance VoIP call. It is easy for an attacker to write a program to periodically send a forged INVITE packet, with different callerID and forge source ID address, to some targeted VoIP phone. Just image how annoying if your VoIP phone rings with different callerID every 10 min day and night.

- *Involuntary involvement in attacking others.* To address the SPIT problem, some VoIP vendors adapted voice Turing test to tell if the incoming call is from a human or machine. Specially, when the VoIP phone receives an incoming call, it will first respond with some simple question in somewhat distorted voice that is easy for human, but hard for machine, to recognize and answer. The incoming call will reach the callee only if the caller has answered the question correctly. While such a voice Turing test is very effective in distinguishing human from machine, it does introduce a new vulnerability that can be exploited by the attacker to launch DoS attacker against other Internet host.

To launch DoS attack against some Internet host X, the attacker can simply generate large number of spoofed INVITE packets using X's IP address as the (spoofed) source IP address, and send them to large number of VoIP phones with voice Turing test protection. According to the voice Turing test protocol, large number of VoIP phones think Internet host X is trying to call them, and each of them will send a voice stream to X. If too many VoIP phones send traffic to X at the same time, X will be overwhelmed. In this DoS attack, the large number of VoIP phones are innocent in that they are simply acting according to the voice Turing test protocol but somehow they are exploited by the attacker. From X's point of view, it is very difficult to find out the real attacker behind large number of VoIP phones.

- *VoIP based botnet.* A botnet is a network of compromised computers (i.e., bot) controlled by an attacker (i.e., botmaster). The botmaster needs some command and control (C&C) channel to control potentially large number of bots. Traditional botnets use IRC text channel as the C&C channel, which is easy to filter and block. Because VoIP traffic is very popular and it is hard to recognize and filter in real time, it is an attractive alternative for the C&C channel of botnet. For example, Skype [15] is the most popular softphone and it makes most international calls. However, Skype traffic is encrypted using AES with 256-bit key and it could penetrate virtually any firewall. In fact, recent research has shown that attacker could indeed use Skype protocol to build stealthy and resilient botnet [16].

Now more and more people are using VoIP softphone in their laptop and handheld. While VoIP softphone bring convenience, it does introduce new security vulnerabilities to the host running it. There has been a lot of suspicion about why Skype software uses so many malware-like obfuscation techniques (e.g., self-modifying code, antidebugging logic, cover traffic) and what exactly it does in addition to normal Skype call.

6. Attacks on Currently Deployed VoIP

Despite recent development in VoIP defense, currently deployed VoIP systems have serious security flaws and they are vulnerable to many attacks that have far reaching impact on VoIP users.

In this section, we examine the currently deployed VoIP systems and seek to find out what the active adversary could do to transparently divert selected VoIP calls. We choose to use the VoIP services of Vonage, AT&T, and Gizmo [17] in our empirical investigation. According to Telephia's recent survey [18], Vonage and AT&T are the No. 1 (53.9%) and the No. 2 (5.5%), respectively, in U.S. VoIP market share. Gizmo, on the other hand, "is the best-known open-standards softphone project" [19]. Note all the VoIP exploits in our investigation were against our own phones rather than the VoIP infrastructure. At no time did we send any traffic to affect any other VoIP subscribers or violate any service agreement.

6.1 Registration Hijacking

Registration hijacking refers to the action of an attacker to register himself as the targeted VoIP user. If successful, all the incoming calls to the victim VoIP user will be routed to the VoIP phone chosen by the attacker rather than the victim's

VoIP phone. In other words, the attacker rather than the victim will receive all the incoming calls to the victim. In this section, we describe how attacker could hijack the VoIP registration and discuss why currently deployed systems are vulnerable.

Figure 5 shows two basic scenarios of registration hijacking depending whether the attacker can intercept the traffic of the targeted VoIP phone. Figure 5A shows the first scenario, where the attacker has an MITM that can intercept all the VoIP traffic between the targeted VoIP phone and the VoIP server. Assume the VoIP user is at SIP phone A, which registers itself by periodically (e.g., once every 60 s) sending REGISTER message to the SIP registrar server.

The MITM can intercept the REGISTER message from SIP phone A, and forward it to attacker's SIP phone B. Phone B can simply change the IP address in fields Via

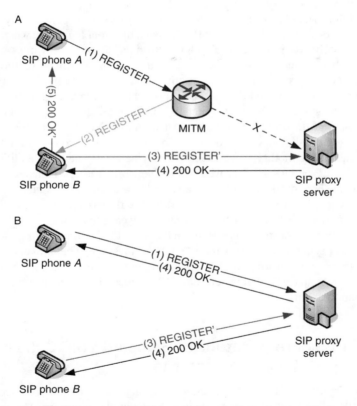

FIG. 5. Registration hijacking attacks against SIP phones. (A) Registration hijacking via MITM. (B) Registration hijacking without MITM.

and `Contact` of the intercepted `REGISTER` message to phone B's IP address and send the slightly changed `REGISTER` message to the SIP server. After the SIP server receives and authenticates the `REGISTER` message, it will bind the IP address in the `Contact` field to the URI of the VoIP user. From now on, all the calls toward the VoIP user will be routed to phone B. The SIP server will acknowledge the successful registration by replying with a `200 OK` message to phone B. Phone B can slightly modify the `200 OK` message and send it to phone B with spoofed IP address of the SIP server. This will trick phone A into believing that its registration is successful.

Figure 5B shows the second scenario, where the attacker has no MITM to intercept the VoIP traffic of the targeted VoIP traffic. Assuming the attacker knows the telephone number of the targeted VoIP user, he can forge a plausible `REGISTER` message and periodically send it to the SIP server from phone B. Since phone A also periodically sends the `REGISTER` message to the SIP server, there is a race condition here. From the SIP server's point of view, it periodically receives `REGISTER` messages for the same VoIP user from different locations, which is quite normal for VoIP users. If attacker's phone B sends its `REGISTER` message more frequently than phone A does, it will hijack the registration most of time. If the attacker sends the spoofed `REGISTER` message too frequently, the SIP server may ignore some of his and the victim's `REGISTER` messages. Although frequent `REGISTER` messages with interleaving IP addresses in the `Contact` field looks anomalous, the SIP server has no idea about where the VoIP user really is.

Since the SIP authentication does not protect fields `Contact` and `Via`, the attacker can freely modify these fields without being detected. In addition, most residential VoIP phones are deployed behind Network Address Translation (NAT). Therefore, the IP address in fields `Contact` and `Via` are likely unroutable private IP addresses (e.g., 192.168.x.x) since it is hard for the IP phone that is behind NAT to find out its public IP address. In order to route VoIP calls to the registered VoIP user, what the VoIP registrar really needs to remember is the public source IP address of the received packet that contains the `REGISTER` message. This prevents the VoIP registrar from authenticating the IP address in the `Contact` and `Via` fields of the `REGISTER` message. Therefore, current SIP protocol is inherently vulnerable to registration hijacking.

6.2 Transparent Detour of Selected VoIP Calls on the Internet

In this section, we show how an active adversary could detour any selected SIP-based VoIP call through any remote device chosen by the adversary. The goal of the remote transparent detour is to divert the RTP voice stream of the selected call through an arbitrary node (the remote device) on the Internet before it reaches its final destination.

During the SIP call setup process, the caller and callee can choose where (i.e., at what IP, on what port) they want to receive the upcoming RTP voice stream and they inform the other party about their choices via the INVITE and 200 OK messages, respectively. Since the RTP endpoint information (i.e., IP address, port number) is specified in the SDP part of INVITE and 200 OK messages which is not protected by the SIP digest authentication at all, the active adversary is free to manipulate the RTP endpoint information. Due to performance consideration, some VoIP service providers (e.g., Vonage) may choose to use different servers for the SIP signaling and the RTP voice stream. Consequently, SIP phone will initiate its RTP stream to any IP address and port number specified in the SDP part of the INVITE or 200 OK messages. However, the SIP server may remember the IP address of any registered SIP phone. However, the SIP server cannot insist on sending its RTP stream to the registered IP address due to the need to support the SIP phones behind NAT. All these enable an MITM to divert any chosen RTP voice stream through any remote device on the Internet.

We have explored the transparent VoIP call detour in four scenarios: (1) a PSTN phone calls AT&T SIP phone; (2) an AT&T SIP phone calls a PSTN phone; (3) a PSTN phone calls a Vonage SIP phone; and (4) a Vonage SIP phone calls a PSTN phone. We assume there is an MITM between the SIP phone and the SIP signaling server and the MITM is collaborating with a remote device. By careful manipulation of the SDP part of the INVITE and 200 OK messages, we are able to divert the RTP voice streams in all of the four abovementioned scenarios through arbitrary node on the Internet.

Figure 6 shows the SIP message flows of the transparent detour of calls between an AT&T SIP phone and a PSTN phone. Note these SIP message flows differ from that of normal calls. First, the MITM intercepts the (1) INVITE message toward either SIP phone or the SIP server and send a copy (message (2) INVITE) to the remote device. This is to inform the remote device about the IP address and port number of the upcoming RTP stream selected by the caller side so that it can forward the RTP stream to the caller side. Second, the MITM modifies the SDP part of the intercepted INVITE message such that the IP address and port number for RTP will be that of the remote device. This essentially tells the callee side to send the RTP voice stream to the remote device. Then the MITM sends the modified (3) INVITE message to its original destination. The MITM will not intercept any 100 TRYING or 180 RINGING message. When the callee side accepts the call and sends the (6) 200 OK message to the caller side, the MITM intercept it and send a copy (message (7) 200 OK) to the remote device. This would informs the remote device about the IP address and port number of the upcoming RTP stream selected by the callee side.

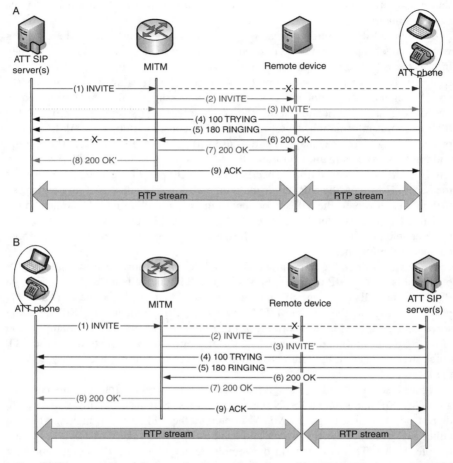

Fig. 6. SIP message flows of transparent detour of calls between an AT&T SIP phone and a PSTN phone. (A) PSTN phone calls AT&T SIP phone. (B) AT&T SIP phone calls PSTN phone.

Such an information is necessary for the remote device to relay the received RTP voice stream. Then the MITM changes the IP address and port number in the SDP part of the intercepted (6) 200 OK message to that of the remote device and sends the modified (8) 200 OK message to its original destination. This would trick the caller into sending the RTP voice stream to the remote device. Once the caller side responds with the (9) ACK message, the SIP call setup completes. Now the caller and callee will send their RTP voice streams to the remote device, which will relay the received RTP streams to their original destination and function as a transparent proxy between the caller and the callee.

When we apply the above attack procedures to the calls to and from our Vonage SIP phone, we see mixed result. While the Vonage SIP phone is tricked into sending its RTP voice stream to the remote device, the Vonage RTP server does not send out any RTP voice stream at all. It appears that the Vonage server checks the RTP stream IP address in the SDP part of the received INVITE or 200 OK messages and refuses to send out any RTP stream if it is different from the registered IP address of the SIP phone. This means that the MITM could not change the RTP stream IP address in the SDP part of the INVITE or 200 OK message if he wants the call established. It seems that Vonage's SIP-based VoIP service is robust against the transparent detour attack.

After further research, we have found that while the Vonage SIP server validates the RTP stream IP address in the SDP part of the received INVITE or 200 OK messages, it actually sends the RTP stream to the source IP address of the INVITE or 200 OK message. This is necessary for the Vonage VoIP service to support SIP phone behind NAT where the registered SIP IP address is private.

Based on this finding, we have changed the SIP message flow of our transparent detour exploit. Figure 7 shows the SIP message flows of the transparent detour of calls between a Vonage phone and a PSTN phone. The key difference between Figures 6A and 7A is who will send message 200 OK to the SIP server. When MITM intercepts message (6) 200 OK from the Vonage SIP phone, it sends a copy (message (7) 200 OK) to the remote device. Instead of letting the MITM to send out the modified 200 OK message to the Vonage SIP server, the remote device will modify the RTP stream port number only and send the modified message (8) 200 OK' to the Vonage server. Since the RTP stream IP address in the SDP part of message (8) 200 OK' is not changed, message (8) 200 OK' will pass the check by the Vonage server. The Vonage server will send the RTP stream to the source IP address of message (8) 200 OK', which is the remote device. The key difference between Figures 6B and 7B is who will send the INVITE message to the SIP server. For VoIP calls from Vonage SIP phone to PSTN phone, the MITM does not send the intercepted message (1) INVITE to the Vonage server, but rather let the remote device to modify the RTP port number in the SDP part of message (2) INVITE message and send the modified message (3) INVITE to the Vonage server. This would cause the Vonage SIP server to send its RTP stream to the remote device.

In summary, the MITM can transparently detour the RTP voice stream of any selected Vonage and AT&T SIP calls through any remote device on the Internet and let the original caller and callee establish the VoIP call. In this case, the remote device will have access to all the voice streams between the caller and the callee.

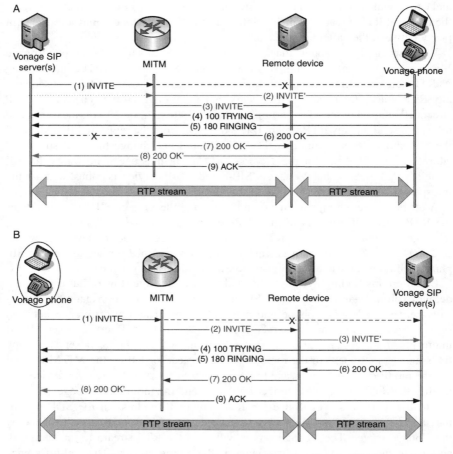

FIG. 7. SIP message flows of transparent detour of calls between a Vonage SIP phone and a PSTN phone. (A) PSTN phone calls Vonage SIP phone. (B) Vonage SIP phone calls PSTN phone.

6.3 Transparent Redirection of Selected VoIP Calls

In this section, we explore how the active adversary could transparently redirect any selected VoIP call to any third party chosen by the adversary. As a result, the caller will be connected to the third party rather than the original callee. However, the caller will think he has reached the original callee while he has actually reached the third party. On the other hand, the original callee has never received the call from the caller.

6.3.1 Callee Side Call Redirection

When a caller wants to initiate a call via SIP, he sends an INVITE message to the callee, who is identified by the request-URI field in the INVITE message. Although the request-URI field is part of the SIP digest authentication, the SIP digest authentication is only applied to those INVITE messages from the SIP phone to the SIP servers. In other words, any INVITE messages from the SIP proxy to the SIP phone are not authenticated with the digest. Therefore, the MITM in between the SIP proxy and the SIP phone could freely change the request-URI field and redirect the SIP calls to any other SIP phone.

We have explored such call redirection attack at the callee side of our Vonage and AT&T SIP phones, and we are able to transparently redirect calls between our Vonage phone and AT&T phone. Figure 8 illustrates SIP message flows of the call redirection attack. When someone wants to call phone A, he sends an INVITE message to phone A. The MITM at the callee side intercepts the INVITE message and modifies the request-URI, To fields, the IP address, and the port number of the INVITE message, and sends the modified INVITE message to phone B. When phone B responds with 100 TRYING, 180 RINGING, or 200 OK messages, the MITM intercepts them and modifies the To field, the IP address, and the port number. Then the MITM forward the modified SIP message from phone B to the SIP proxy of phone A—pretending that those messages were from phone A. When the SIP proxy acknowledges the receipt of 200 OK, it sends out ACK message to

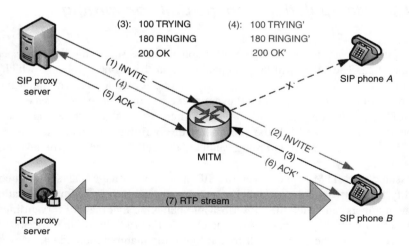

FIG. 8. Unauthorized call redirection via MITM.

phone A. The MITM intercepts it and sends the modified ACK message to phone B. This will establish the call between the caller and phone B instead of phone A.

6.3.2 Caller Side Call Redirection

We have also been able to transparently redirect selected VoIP calls at the caller side. In this case, the MITM intercepts the SIP messages from the victim caller, and pretends to be the SIP server by responding with spoofed SIP messages. The message flow in this case is similar to that shown in Figure 8.

6.4 Manipulating and Hijacking Call Forwarding Setup

Call forwarding is a feature that allows the telephone subscribers to specify where the incoming calls will be forwarded. For example, people can set up call forwarding so that they can receive calls to their office phones with their cell phones while they are away from their offices.

Now, we describe two attacks that would allow attacker to transparently manipulate the phone number to which the calls to the victim will be forwarded to. Unlike attacks described in Sections 6.2 and 6.3, the attacks on call forwarding setup exploit the vulnerabilities of the media stream (e.g., RTP) rather than that of the signaling protocols (e.g., SIP). Therefore, these attacks could work even if the VoIP signaling is fully protected.

6.4.1 Manipulating Vonage Call Forwarding Setup

The call forwarding of Vonage VoIP phones can be set up by dialing a special number *72. The caller will be prompted to input the phone number to which the incoming calls (to the subscriber's phone) will be forwarded to. The input phone number will be transferred via RTP event packets to the Vonage RTP server. After the RTP server receives the call forwarding number, it will acknowledge the call forwarding number and ask the subscriber for confirmation. Once the subscriber confirms the input call forwarding number, the call forwarding will take effect immediately.

Assume the MITM is in between the SIP phone and Vonage RTP server, it could modify the call forwarding number to any phone number (including international phone numbers) and trick the subscriber into believing that the call forwarding has been set up with the number he/she has chosen. Figure 9 shows the SIP and RTP message flows of the Vonage call forwarding setup manipulation attack. Messages (1) (2), and (3) show the authenticated call setup sequence for call to *72.

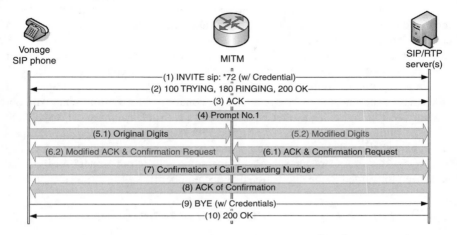

FIG. 9. Manipulating Vonage call forwarding setup via man-in-the-middle (MITM) in between SIP phones and Vonage servers.

Once the call to *72 has been established, the Vonage RTP server will send the caller voice prompt For in country call forwarding, please enter ..., for international call forwarding, please enter ... in RTP (represented by message (4) Prompt No. 1) and wait for caller's response.

Once the caller inputs the call forwarding number, the SIP phone will send the call forwarding number in RTP event packets (represented by message (5.1) Original Digits) to the Vonage RTP server. The MITM intercepts the RTP event packets, and sends the modified call forwarding number in the bogus RTP event packets (represented by message (5.2) Modified Digits) to the RTP server.

Note the MITM could change the number of digits of the call forwarding number. For example, the MITM could change the call forwarding number from an 11-digit domestic phone number (1-xxx-xxx-0416) to a 15-digit international phone number (011-44-xxx-xxx-3648). This means that the MITM needs to send more bogus RTP event packets than the original RTP event packets from the caller. To maintain the correct RTP seq# and extended seq# in the RTP stream, the MITM needs to drop some normal RTP packets, which essentially contains background noise in between the keystrokes by the caller. This will make sure the RTP server accepts the modified bogus RTP (event) packets.

The RTP server will acknowledge the bogus call forwarding number it received and ask the caller for confirmation: you have entered 011-44-xxx-xxx-3648, press 1 to ... (represented by message (6.1) ACK & Confirmation Request). To prevent the caller from knowing the bogus call forwarding number received by the RTP server, the MITM needs to intercept the original acknowledgment and confirmation request and send the caller the modify acknowledgment and confirmation request (message (6.2) Modified ACK & Confirmation Request) so that the caller will hear the original call forwarding number (1-xxx-xxx-0416) he/she entered. After that, the MITM could let the rest RTP stream pass without modification.

We have experimented the above attack on Vonage call forwarding setup with our Vonage VoIP account. The caller have chosen to forward incoming call to an U.S. domestic number (1-xxx-xxx-0416), the MITM have successfully and transparently changed the call forwarding number to an international phone number (011-44-xxx-xxx-3648). As a result, subsequent incoming calls to our Vonage VoIP phone have been forwarded to the international phone number 011-44-xxx-xxx-3648.

6.4.2 Hijacking Gizmo Call Forwarding Setup

In Section 6.4.1, we have shown that an MITM can transparently modify the call forwarding number to any preselected phone number while keeping the caller thinking that the call forwarding has been set up with the number he/she has chosen. In fact, the MITM can hijack the call forwarding setup session completely and let the attacker impersonate the VoIP subscriber and set up the call forwarding for the victim. We choose to use Gizmo, a popular SIP softphone system, to demonstrate such hijacking attack on call forwarding setup.

To set up the call forwarding for a Gizmo phone number, the Gizmo subscriber dials 611 from his/her Gizmo softphone to begin a call forwarding setup session. The Gizmo caller will be prompted to input the call forwarding number after the session to 611 has been established. Similar to the Vonage RTP server, the Gizmo RTP server will acknowledge the received call forwarding number and ask the Gizmo caller for confirmation. Once the Gizmo caller confirms the number, the call forwarding will take effect immediately.

Assume the victim uses Gizmo softphone 1, the attacker uses Gizmo softphone 2, and the MITM is in between the Gizmo softphone 1 and Gizmo SIP, RTP servers. The MITM could let the attacker at Gizmo softphone 2 hijack the call forwarding setup session between Gizmo softphone 1 and Gizmo RTP server and configure the call forwarding of Gizmo softphone 1. At the same time, the victim at Gizmo softphone 1 will hear a bogus voice message: the number you are trying to reach is busy. This would make the victim think that the call forwarding setup server is busy and the call forwarding has not been set up. Figure 10 shows the SIP and RTP message flows of the hijacking of the Gizmo call forwarding setup session.

First, the attacker at Gizmo phone 2 calls 611, and establishes a session with the Gizmo RTP server. Messages (1) (2), and (3) show the call setup sequence

VOIP SECURITY

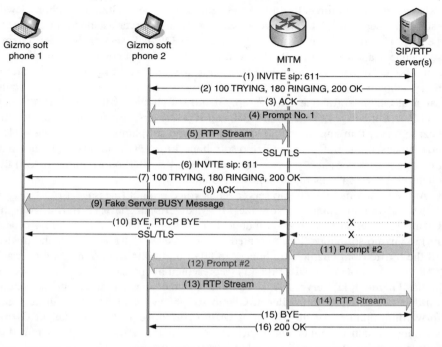

FIG. 10. Hijacking gizmo call forwarding setup via man-in-the-middle (MITM) in between Gizmo softphones and Gizmo servers.

between Gizmo softphone 2 and the Gizmo SIP server. Then the Gizmo RTP server will send Gizmo softphone 2 voice prompt in RTP (represented by message (4) Prompt No. 1). At the same time, the Gizmo softphone 2 will start sending RTP stream to the negotiated UDP port (6824) at the Gizmo RTP server. The MITM temporarily blocks the RTP stream from the Gizmo softphone 2 (represented by message (5) RTP Stream).

We notice that Gizmo softphone 2 and the Gizmo RTP server have established some SSL/TLS connection during the call establishment phase, which appears to be some secure out-of-band management channel. The MITM does not block the SSL/TLS connection between them. The purpose of establishing the session between Gizmo phone 2 and the Gizmo RTP server is to facilitate the quick hijacking of the 611 call session between Gizmo softphone 1 and the Gizmo RTP server. In theory, we can establish the session between Gizmo phone 2 and the Gizmo RTP server on the fly, but this will incur some extra delay in the call hijacking.

Now once the victim calls 611 from Gizmo softphone 1, Gizmo softphone 1 will establish a separate 611 call session with the Gizmo server as shown in messages (6), (7), and (8). To hijack the established 611 call session between Gizmo softphone 1 and the Gizmo RTP server, the MITM first sends Gizmo softphone 1 some bogus voice message in RTP: the number you are trying to reach is busy (shown as message (9) Fake Server BUSY Message). After the victim caller at Gizmo softphone 1 hangs up, Gizmo softphone 1 will send SIP BYE and RTCP BYE messages to the Gizmo SIP and RTP servers, respectively. To keep the Gizmo server thinking that its session with Gizmo softphone 1 is alive, the MITM now blocks all the traffic from Gizmo softphone 1 to the Gizmo server (as shown in message (10) BYE, RTCP BYE) and remembers the UDP port number (6454) the Gizmo RTP server uses for session with Gizmo softphone 1.

At the same time, the Gizmo server sends voice prompt (message (11) Prompt #2) to Gizmo softphone 1. Now the MITM diverts all the RTP traffic from the Gizmo RTP server to Gizmo softphone 1 (represented by message (11) Prompt #2) to Gizmo softphone 2 (represented by message (12) Prompt #2), and diverts all the traffic from Gizmo softphone 2 to UDP port 6824 (represented by message (13) RTP Stream) to UDP port 6454 (represented by message (14) RTP Stream) at the Gizmo RTP server. This would allow the attacker at Gizmo softphone 2 impersonate the victim caller at Gizmo softphone 1 and freely set up any call forwarding number for the victim at Gizmo softphone 1. The attacker at Gizmo softphone 2 can terminate the hijacked 611 call forwarding setup session after setting up any call forwarding number he/she has chosen.

We have experimented the above hijacking attack on the Gizmo call forwarding setup with our Gizmo VoIP accounts, and we have been able to hijack the call forwarding setup session for our Gizmo phone number 1 and configure the call forwarding to an international phone number 011-44-xxx-xxx-1284 from our Gizmo phone number 2. As a result, subsequent incoming calls to our Gizmo phone number 1 have been transparently forwarded to the international phone number 011-44-xxx-xxx-1284.

6.5 Voice Pharming

In Sections 6.2–6.4, we have empirically demonstrated that an MITM could detour or redirect any selected Vonage and AT&T VoIP calls via or to anywhere on the Internet. In addition, the MITM could manipulate and hijack the call forwarding setup of selected Vonage and Gizmo SIP subscribers such that the attacker can control where the calls to the victims will be forwarded to. All these call diversion attacks essentially violate the VoIP users' basic trust that their calls will reach the intended callees only. Furthermore, such a call diversion capability enables the attacker to launch the voice pharming attack against targeted VoIP callers, where the selected VoIP calls are transparently transferred to the bogus IVR or representative even if the callers have dialed correct phone numbers. In this case, the victim callers have no easy way to tell if they have reached the bogus IVR or representative. Therefore, even the most cautious callers could be tricked into giving out their credentials (e.g., SSN, credit card number, PIN) to the adversary. Such a voice pharming attack, enabled by the unauthorized call diversion, could indeed shake the long time trust that the general public have in voice communication.

6.5.1 Hypothetical Voice Pharming Attack

Citibank provides a phone banking service which allows its customers to have checks issued and paid to anyone by calling Citibank. Specifically, when a customer dials the Citibank phone banking phone number (1-800-374-9700), the IVR will prompt the caller to speak or enter his/her 9-digit SSN or personal taxpayer identification number. Then the IVR will ask for the telephone access code before allowing the caller to choose the available service options. After choosing the bill payment option, the caller will be connected to a Citibank service representative. To authenticate the caller, the service representative usually asks a few questions about the following information: (1) *debit card number*, (2) *checking account number*, (3) *phone PIN*, (4) *mother's maiden name*, (5) *the state on which the account was opened*, and (6) *personal full name*. If the caller correctly answers the questions, the service representative would issue checks paying to anyone at any address the caller wants. One coauthor of this chapter has successfully had one check issued, mailed, and paid to another coauthor via Citibank phone banking.

Figure 11 illustrates a hypothetical voice pharming attack against Citibank phone banking. In order to launch voice pharming attack, the attacker needs to (1) set up a bogus IVR that sounds exactly the same as the real IVR; (2) redirect the calls toward Citibank phone banking to the bogus IVR and/or a phone the attacker uses. Setting up a bogus IVR is quite straight forward with VoIP technology. For example, the attacker could simply call the real Citibank IVR via VoIP and record all the prompts as RTP traces. Then the attacker could construct the bogus IVR by replaying the collected RTP traces. Such a bogus IVR would have exactly the same voice as that of the real IVR. Now suppose the MITM is in a place (e.g., gateway, wireless router, firewall) that can intercept VoIP traffic, then it can check if there is any call toward any number of targeted financial institutions (e.g., 1-800-374-9700 of Citibank phone banking). By using the real-time call redirection attack described in Section 6.3, the attacker could transparently divert the call to his bogus Citibank IVR. In addition, the attacker could pretend to be a Citibank service representative asking the caller the same questions (e.g., debit card number, mother's maiden

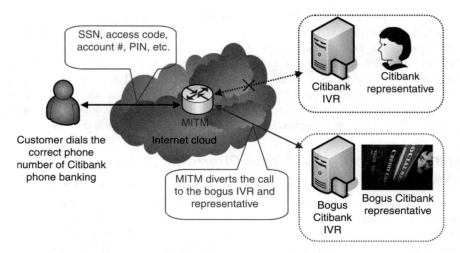

FIG. 11. Hypothetical voice pharming attack.

name). Since the caller has dialed the correct phone number and has heard exactly the same voice menu, he/she simply has no way to tell if he/she is talking to a bogus IVR. Given that a bank customer usually does not know the bank representative personally, he/she cannot tell if he/she is talking to a real bank representative or a bogus one. Therefore, voice pharming could let the attacker obtain all the information needed to impersonate the victim caller and gain financially (e.g., pay bill at the victim's expense).

6.6 Remote MITM

In previous sections, we have shown that a MITM could launch various attacks (e.g., transparent VoIP diversion, hijacking and wiretap, voice pharming) against the targeted VoIP users. Intuitively, people think those MITM attacks require the attacker to be in the path of VoIP traffic and the MITM attacks are infeasible if the attacker is not initially in the path of VoIP traffic.

In this section, we demonstrate that a remote attacker who is not initially in the path of VoIP traffic can indeed launch all kinds of MITM attacks on VoIP by exploiting DNS and VoIP implementation vulnerabilities. Our case study of Vonage VoIP, the No. 1 residential VoIP service in the U.S. market, shows that a remote attacker from anywhere on the Internet can stealthily become a remote MITM through DNS spoofing attack on a Vonage phone, as long as the remote attacker knows the phone number and the IP address of the Vonage phone.

First, we describe our testbed setup and message flow of the normal startup or reboot of the Vonage SIP phone. Then we present the identified DNS implementation weaknesses of the Vonage phone and its vulnerability in handling the malformed INVITE message. Next, we illustrate the message flow of the DNS spoofing attack and describe our experimental results.

6.6.1 Network Setup

Figure 12 illustrates the network setup of our testbed. The remote attacker runs Red Hat Linux on a Dell D610 laptop computer. NAT router 1 is a FreeBSD machine running on a virtual machine and NAT router 2 is a Linksys router.

Figure 12A illustrates the network setup where the SIP phone is directly connected to the Internet. We use SIP/RTP server(s) to denote the SIP server and the RTP server which handle the signaling messages and the RTP stream, respectively. The remote attacker could be anywhere on the Internet. In our experiment, we use a wiretap device to capture live network traffic transited from/to the SIP phone. The wiretap device and the SIP phone connect to a four port 10BASE-T Ethernet hub.

Figure 12B illustrates the network setup where the SIP phone is behind NATs. Note this setup is different from the most popular settings where the SIP phone is behind only one NAT router. We notice that the SIP phone will send some destination unreachable ICMP packets to the Vonage DNS server when receiving spoofed DNS responses with unmatched port numbers. We use the NAT router2 to block these unwanted traffic from reaching the Vonage DNS server.

As a result, the SIP phone is behind 2 NAT routers. For convenience, we placed the remote attacker outside NAT router2 but inside the private network of NAT router1. From the remote attacker's perspective, the targeted SIP phone is behind one NAT router, which is the most likely configuration for residential VoIP phones. In this configuration, the wiretap device and NAT router2 connect to a four port 10BASE-T Ethernet hub. We notice that none of the NAT router will change the source port number of the passing packet, this enables the remote attacker to become the remote MITM via the identified exploit even if the targeted Vonage phone is behind two levels of NAT routers.

6.6.2 Message Flow of Normal Startup or Reboot

Figure 13 depicts the message flow of normal startup or reboot of a Vonage phone. At the beginning, the SIP phone sends a DNS query to the Vonage DNS server to ask for SIP servers' IP addresses in step (1). All DNS queries from the

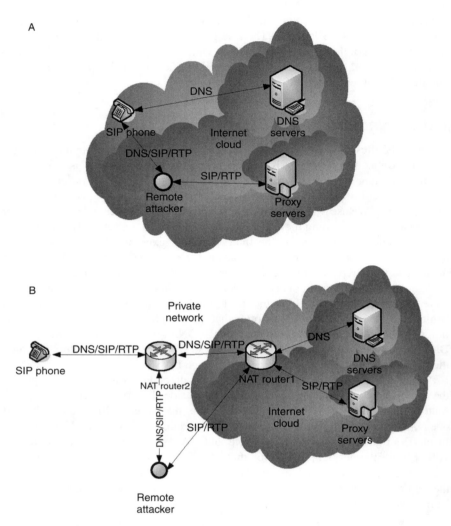

FIG. 12. Remote MIMT testbed setup. (A) SIP phone directly connected to the Internet. (B) SIP phone behind NATs.

Vonage SIP phone go to the Vonage DNS server at IP address 216.115.31.140. Then in step (2), the Vonage DNS server replies with a DNS response packet containing four IP addresses of Vonage SIP servers: 69.59.252.35, 69.59.232.42, 69.59.242.84, and 69.59.227.87. At step (3), the Vonage phone sends to one of four SIP servers a

VOIP SECURITY

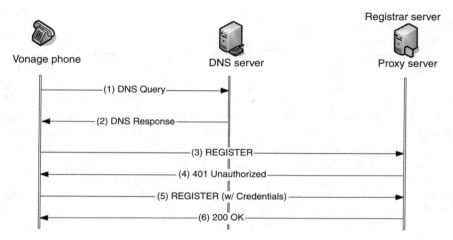

FIG. 13. Message flow of normal startup or reboot.

SIP REGISTER message. Then in step (4), the SIP server challenges the SIP phone with a 401 Unauthorized message. After receiving the 401 response, the SIP phone sends the SIP server a new SIP REGISTER message containing credentials. Note the "expires" field in the SIP REGISTER message specifies the duration for which this registration will be valid. So the SIP phone needs to refresh its registration from time to time.

6.6.3 Exploitable Vulnerabilities of Vonage SIP Phone

The implementation of DNS query/response in the Vonage phone has several weaknesses.

- The SIP phone always uses a static ID value, 0×0001, in all DNS queries.
- The source port number range of DNS queries is limited to 45,000–46,100.
- The question sections of all DNS queries are identical, and contain 11 bytes of string d.voncp.com.
- The SIP phone does not check the source IP address of a DNS response. Even if the source IP address is not that of the Vonage DNS server, the Vonage phone still accepts a spoofed DNS response.

Due to these vulnerabilities, the brute-force search space for forging a matching DNS response is no more than 1100.

We have found that our Vonage SIP phone fails to handle a malformed INVITE message correctly and it will reboot when receives a malformed INVITE message with an over length phone number in the From field. This allows the remote attacker to crash and reboot the targeted Vonage phone by sending it one malformed INVITE message. To launch such an attack, the remote attacker needs to spoof the source IP address as that of one of Vonage SIP servers. Otherwise, the Vonage phone will discard the INVITE message. Our experiments have shown that the Vonage phone does not ring but replies with a Trying message after receiving the malformed INVITE messages. Then the phone crashes and reboots almost immediately. After a few seconds (e.g., 13 s), the Vonage phone sends a DNS query to the Vonage DNS sever. Note the SIP phone crash attack is stealthy in that the SIP phone does not ring at all when receives the malformed INVITE message.

6.6.4 Message Flow of DNS Spoofing Attack

Figure 14 shows the SIP message flow of the DNS spoofing attack on the Vonage SIP phone. At the beginning, the remote attacker sends a malformed INVITE message to the SIP phone with a spoofed source IP in step (1). In response, the SIP phone sends a Trying message to the real SIP server in step (2). Then the SIP phone crashes and reboots. Several seconds later, the SIP phone sends a DNS query to the Vonage DNS server asking for the SIP servers' IP addresses in step (3). Within several milliseconds, the legitimate DNS response from the Vonage DNS server reaches the SIP phone in step (6).

If the remote attacker sends the spoofed DNS response packets to the Vonage phone within the time window from step (3) to (6), the Vonage phone will receive the spoofed DNS response before the legitimate DNS response arrives. This process is represented at step (4). Since the remote attacker does not have access to the original DNS query from the Vonage phone, he has to try each of the 1100 possible port numbers in the spoofed DNS response packets. If the spoofed DNS response packet contains the wrong port number, the Vonage phone sends a port unreachable ICMP packet to the DNS server at step (5). If the spoofed DNS response packet contains the matching port number, the Vonage phone accepts the spoofed DNS response packet and sends out REGISTER message to the remote attacker at step (7) as it now thinks the remote attacker is the Vonage SIP server. Therefore, the remote attacker can determine the success of the DNS spoofing by checking if he receives the expected REGISTER from the targeted Vonage phone within a predefined period of time.

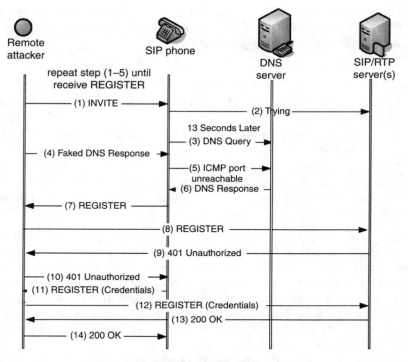

FIG. 14. Message flow of DNS spoofing attack.

If the remote attacker does not receive the expected REGISTER from the targeted Vonage phone within predefined period of time, he knows that the Vonage phone has accepted the authentic DNS response from the Vonage DNS server. The remote attacker needs to start a new round of attack by repeating steps (1–6) until he receives a REGISTER message from the SIP phone in step (7). We define steps from (1) to (6) as a round of the attack. Normally it will take several rounds before the SIP phone finally sends the REGISTER message to the remote attacker.

After receiving the REGISTER message at step (7) or (11), the remote attacker forward them to the real SIP server in step (8) or (12). Meanwhile the remote attacker forward the 401 Unauthorized message at step (9) and the 200 OK message at step (13) from the SIP server to the SIP phone in steps (10) and (14). Now the remote attacker becomes the MITM in that (1) the SIP phone thinks the remote attacker is the SIP server and (2) the SIP server thinks the remote attacker is the SIP phone.

To launch the DNS spoofing attack, the remote attacker only needs to construct 1000 fake DNS response packets with 1000 different destination port numbers. Specifically, the remote attacker just needs to

- Fill 0×0001 into the ID field of all spoofed DNS responses.
- Fill `d.voncp.com` into the question section of all DNS responses.
- Fill the IP address of the remote attacker into the answer section of all spoofed DNS responses.
- Set the destination port number of 1st, 2nd, ..., 1000th packet as 45000, 45001, ..., 45999.
- The SIP phone does not check source IP address. So we set it to the IP address of the remote attacker when the victim phone is on the Internet. When the phone is behind NATs, the source IP address of spoofed DNS packets is set to that of Vonage SIP server to pass through NAT router2.

Figure 15 illustrates the timeline of a round of the attack. T_0 is the time when the remote attacker sends a malformed `INVITE`. T_2 and T_3 are the times when the SIP phone sends a DNS query and receives the legitimate response from the DNS server, respectively. We refer to the time interval from T_2 to T_3 as the Vulnerable Window (VM). T_1 and T_4 denote the start time and end time, respectively, of sending spoofed DNS response packets. We refer to the time interval from T_1 to T_4 as an Attack Window (AW). Apparently, the larger the attack window is, the fewer rounds the remote attacker needs in order to succeed.

Our experiments show that the Vonage phone actually accepts spoofed DNS response before it sends out the DNS query. In addition, if the remote attacker keeps sending many spoofed DNS response packets with very shot interpacket arrival time, it will have a good chance to block the targeted SIP phone from

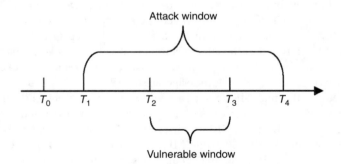

FIG. 15. Timeline of a round of attack.

receiving the authentic DNS response. Therefore, the attack window could start earlier and end later than the vulnerable window.

6.6.5 Experimental Results and Analysis of DNS Spoofing Attack

Ideally, we want T_1 to be earlier but not too much earlier than T_2. We have measured the time interval from the moment the remote attacker sends the malformed `INVITE` message to the moment the crashed and reboot SIP phone sends the first DNS query. Table III shows the measured time intervals for 10 runs of crashing the SIP phones. It shows that it takes 12.9–15.5 s for the SIP phone to send the first DNS query after receiving the malformed `INVITE` packet. Therefore, we set T_1 at 12 s after T_0. We have set transmission rate of the spoofed DNS response packets at 1000 pkt/s. To maximize the chance of hitting the correct port number while keeping the duration of each round short, we set the duration of attack window to be 8 s. Therefore, T_4 is 20 s after T_0. At each round, the remote attacker sends the 1000 spoofed DNS response packets for maximum eight times, and the duration of one round of attack is 20 s. As shown in Table IV, the average number of rounds and the required time of 10 instances of DNS spoofing attack against the SIP phone on the Internet is 39.8 and 789 s (about 13 min).

When the SIP phone is behind NATs, the attack is similar except that the IP address of fake DNS responses should be spoofed as that of the Vonage DNS server to pass through NAT router2. The result of one test showed that the number of rounds is eight and the required time is 169 s.

TABLE III
Measured Time Interval from INVITE to DNS Query Without Spoofed DNS

10 Times	1	2	3	4	5	6	7	8	9	10	Range	Average
Time (s)	14.9	13.8	13.0	18.8	14.6	12.9	15.5	12.8	15.5	14.1	12.9–15.5	14.9

TABLE IV
Number of Rounds and Time Needed to Become the Remote MITM

10 Instances	1	2	3	4	5	6	7	8	9	10	Sum	Average
#round	7	9	11	15	22	28	41	54	105	106	398	39.8
Time (s)	135	175	213	296	437	556	800	1080	2081	2117	7890	789

Our preliminary investigation shows that port numbers of DNS queries are all in the range 45,000–45,999, so that the range 45,000–45,999 is applied.

The packet size of a spoofed DNS response is 87 bytes, including 14 bytes of Ethernet header, 20 bytes of IP header, 8 bytes of UDP header, and 45 bytes of UDP payload. Given that the DNS spoofed packets are transmitted at 1000 pkt/s, the transmission rate is about 700 kbps. Since most household broadband Internet access has at least than 2 Mbps downstream rate, our DNS spoofing is practically applicable to household broadband VoIP.

7. Related Works in VoIP Security Research

Besides the VoIP built-in security mechanisms, a number of VoIP defense mechanisms have been proposed. Arkko et al. [20] proposed a new way for negotiating the security mechanisms (e.g., IPSec [10], TLS [21], HTTP authentication [8]) used between a SIP UA and its next-hop SIP entity. Reynolds and Ghosal [22] proposed a multilayered protection against flooding type of DoS attack on VoIP network. They described a DoS detection method based on measuring the difference between the number of attempted connection establishments and the number of completed handshakes. Wu et al. [23] proposed a stateful, cross protocol VoIP intrusion detection system called SCIDIVE, which detects BYE-attack via identification of the orphan RTP flows. If the attacker sends a BYE message to only one end of an established VoIP call, SCIDIVE is able to detect it since there is some orphan RTP flow left alive. However, SCIDIVE is not able to detect the case where the attacker sends BYE messages to both ends of the SIP session. Sengar et al. [24,25] extended the cross protocol VoIP intrusion detection method by using Hellinger distance to detect flooding DoS attacks that may use a combination of SIP, RTP, and IP streams. Specifically, the learning phase of their detection method learns the normal traffic pattern and the detection phase uses the Hellinger distance to detect abnormal deviations from the normal behaviors. However, none of these VoIP defense mechanisms is able to detect or prevent the unauthorized call diversion attacks we have presented in Section 6.

Salsano et al. [26] evaluated the performance of SIP digest authentication and showed that the processing overhead for implementing SIP digest authentication ranges from 77% to 156%. McGann and Sicker [27] analyzed several VoIP security tools and they showed that there exists large gap between known VoIP security vulnerabilities and the tool's detection capability. Geneiatakis et al. [28] looked the several potential security problems in SIP and listed several potential threats (e.g., DoS attack) to SIP and their remedies. However, they have not considered any of the transparent call diversion attacks we have demonstrated in this chapter.

Zhang et al. [29] recently demonstrated that current VoIP users are vulnerable to billing attacks, which would allow the attack to incur overcharges to the victims on calls they have made. Enck et al. [30] studied the security ramification of SMS (Short Messaging Services) of cell phone, and they showed that SMS of cell phone could be exploited to launch DoS attack on cellular networks. They suggested a number of methods to avoid such attacks, such as limiting message acceptance rates per phone number, separating voice and text data streams, resource provisioning, and making active phone lists difficult to obtain freely on the Internet. Similarly, Racic et al. [31] showed that Multimedia Messaging Services (MMS) of cell phones can be exploited to surreptitiously drain cell phone's battery and they suggested using message and server authentication, information hiding at WAP gateway, MMS message filtering, and improved PDP context management as mitigating techniques.

8. Technical Challenges in Securing VoIP

In Section 6, we have shown a number of attacks on currently deployed VoIP systems. While some of these attacks can be defeated by the full integrity protection of the VoIP signaling as well as VoIP voice stream, it is very hard to secure VoIP. In this section, we discuss the technical challenges in securing VoIP.

8.1 Open Architecture of VoIP

The VoIP network is as open as the Internet. First, the VoIP endpoints (e.g., VoIP phone) could be anywhere on the Internet. Second, VoIP endpoints could be highly mobile in that people frequently carry (especially soft) VoIP phone to different places. Because VoIP is intended to be a public phone service, it is likely that we will interact with some endpoint that has no prior established trust at all. Specifically, we have to allow an unknown person to call us from an unknown phone at a previously unknown IP address, and we need to be able to call an unknown person at a previously unknown IP address. These requirements raise the question about how we can secure VoIP with such an open architecture. Specifically, how do we authenticate a caller/callee we do not know at all?

8.2 Real-time Constraints of VoIP

Unlike other Internet applications such as Web, VoIP has stringent real-time constraints. Specifically, the end-to-end delay of a VoIP call should be no more than 150 ms. VoIP voice stream packets should be delivered at constant rate (e.g., once every 20 or 30 ms). This means the processing of each VoIP voice stream packet should never be more than 20 or 30 ms. Furthermore, the call setup time cannot be long since the callers expect to hear the ring tone or voice with seconds after dialing. All these put an upper bound on the total time that all the security mechanisms in all VoIP components (e.g., SIP proxy, SIP phone) can use. Such real-time constraints made it exceedingly hard, if possible at all, to do real-time voice content filtering.

8.3 Multiple Protocols

VoIP involves a numbers of complicated protocols. For example, VoIP requires signaling protocol (e.g., SIP, MGCP, H.323) and voice stream protocol (e.g., RTP). VoIP security uses different protocols such as HTTP digest, TLS, IPsec, and so on. In addition, VoIP interacts with a number of general networking protocols such as DNS, and NAT. Such a suite of diversified protocols make it very difficult to secure VoIP as the whole VoIP system is as strong as its weakest point. For example, we have shown in Section 6.6 that attacker can launch remote MITM attack against Vonage VoIP phones by exploiting it DNS implementation vulnerabilities.

8.4 E-911

E-911 is perhaps one of the most difficult features for VoIP. 911 in traditional PSTN is implemented in the digital switch with reserved trunk. When someone dials 911 from a traditional landline phone, it will be connected to the appropriate local PSAP (Public Safety Answering Point). Because every landline phone number corresponds to a fixed physical address, the PSAP can get the accurate physical address and the phone number of the caller immediately. Furthermore, the special implementation at the digital switch ensures that only the PSAP is able to terminate the 911 call. In other words, if the 911 caller hangs up, the 911 call will not be terminated, and 911 caller cannot make or receive any new calls unless the PSAP hangs up. This is to help the PSAP to make sure the 911 caller was not forced (by someone) to hang up.

In VoIP, the intelligence is at the endpoint (e.g., VoIP phone) rather than the network. Such a fundamental difference in design has made it hard for VoIP to provide the same 911 feature as traditional landline phone.

- *How to determine the authentic callerID of E-911 calls?* Because current SIP lacks full protection of its messages and fields, it is technically trivial for the attacker to spoof any callerID. In fact, there are companies (e.g., www.telespoofing.com, http://www.spooftel.com) that offer callerID spoofing service to the public.

- *How to determine the real location of the E-911 caller?* Unlike traditional landline phone, where each phone number corresponds a fixed physical address, VoIP phone can be used from anywhere on the Internet. This means the PSAP cannot determine the physical address of the caller even if he knows the real phone number of the E-911 caller. Using source IP address of the VoIP call might give approximate geolocation (e.g., the city), but it is not accurate enough for 911 purpose.
- *How to route an E-911 to the appropriate PSAP?* The E-911 call cannot be routed to the appropriate PSAP until the real location of the E-911 caller is known.
- *How to keep the E-911 connection with PSAP?* To prevent the E-911 caller to terminate the E-911 call before the PSAP does, the VoIP phone can be programmed such that it will not allow new call unless it receives BYE from the other side when it dialed 911. This, however, can be bypassed by power cycle of the VoIP phone. Once the VoIP phone is power cycled, it will have a fresh start, which allows new call.

8.5 Firewall Traversal

Firewall is perhaps the most widely deployed network security devices. The primary functionality of firewall is to filter incoming and/or outgoing traffic based on rules. For example, a firewall may block incoming traffic based on the source IP address, traffic type. Most deployed firewalls are configured to block unsolicited incoming traffic from the Internet. On the other hand, VoIP needs to allow unsolicited incoming calls from unknown and untrusted sources. Therefore, VoIP has conflicting requirement with typical firewalls.

To enable VoIP call pass through firewall, the firewall can actively scan and try to understand the passing VoIP signaling traffic. Such a SIP-aware firewall can dynamically set up pinholes in the firewall to allow the unsolicited VoIP traffic to pass. However, the SIP-aware firewall requires all the SIP traffic are in plaintext. In other words, if the VoIP clients use TLS or IPsec tunnel to protect their SIP traffic, the SIP-aware firewall will be able to see the SIP traffic thus it cannot set up the pinhole for the incoming VoIP traffic. In addition, the dynamical pinhole in the firewall may introduces new security vulnerabilities—giving the adversary a new way to penetrate firewall by exploiting VoIP vulnerabilities.

8.6 NAT Traversal

NAT is a technique to automatically map internal, private IP address and port number to external, public IP address and port number. Specifically, when a host in a private network wants to access the public Internet, it initiates the connection to the destination on the public Internet. When the NAT sees the outgoing packet with a private source IP address, it automatically replaces the private source IP address with the public IP address and change the source port number. The NAT creates a mapping between pair ⟨PrivateSrcIP, PrivateSrcPort⟩ and pair ⟨PublicSrcIP, PublicSrcPort⟩ so that it knows how to translate the destination IP address and the destination port number of the returning traffic. NAT allows multiple hosts in a private network to share one public IP address and it protects the hosts behind NAT by blocking unsolicited incoming traffic. Since most homes only have one public IP address, most residential VoIP phones are behind NAT. Because the automatic mapping of NAT is set up by the initiating traffic from the private network to the public Internet, any unsolicited incoming traffic from the Internet will be blocked due to lack of NAT translation mapping.

On the other hand, VoIP needs to support unsolicited incoming calls. Due to performance considerations, many VoIP service providers have separate servers for VoIP signaling and voice stream, respectively. This means that the incoming voice stream will be from a different IP address than that of the signaling traffic. In these cases, NAT does not have the translation mapping for the incoming traffic thus does not know how to translate the destination IP and port number.

In addition, NAT makes it hard to enforce the integrity from end-to-end and, specifically, validate or authenticate the location of the VoIP phone. When the VoIP phone is behind NAT, it only knows its private IP address unless it uses some protocol to learn its public IP address. This means the VoIP phone has to use its private IP address in whatever authentication scheme that includes the source IP address of the VoIP phone. Due to NAT, the party that is communicating with the VoIP phone only sees the public IP address. Therefore, the other party cannot use the VoIP phone's private IP address to authenticate. This opens door for various attacks on VoIP such as registration hijacking, call hijacking, and MITM attack.

A number of NAT traversal solutions (e.g., UPnP [32], STUN [33], TURN [34], ICE [35]) have been proposed to help VoIP phone to discover the NAT public IP address. However, they are not widely support by existing VoIP phones. In addition, NAT traversal does not automatically solve the problem with unsolicited incoming calls. To allow unsolicited incoming VoIP traffic, NAT has to be SIP aware. Similar to SIP-aware firewall, SIP-aware NAT may introduce new security vulnerabilities— allowing the remote attacker to penetrate NAT and attack the devices behind NAT.

9. Conclusions

VoIP is replacing traditional PSTN as the main underlying technology to carry voice calls. The shift from PSTN, where the intelligence is in the network, to VoIP, where the intelligence is pushed to the edge device, has made it (1) much harder for the service providers to protect their customers and (2) much easier for the attacker to temper with voice communication. We have systematically studied the security issues of current VoIP, and have shown why we cannot trust currently deployed VoIP services the same as traditional landline phone services. We have also discussed why it is so difficult to secure VoIP and we have presented a number of open problems to be addressed. We hope our work helps to raise the public awareness of the huge gap between the VoIP security that people have expected and what is currently provided, and motivate more researchers and vendors to look into the VoIP security problems for better solutions.

ACKNOWLEDGMENTS

Some preliminary results of this work have been presented in *the 4th International Conference on Security and Privacy in Communication Networks (SecureComm 2008)* in September 2008 [36] and *the 2009 ACM Symposium on Information, Computer & Communication Security (ASIACCS 2009)* in March 2009 [37].

REFERENCES

[1] P. Barnard, ABI Study Predicts 267 Million Residential VoIP Subscribers Worldwide by 2012. http://www.tmcnet.com/voip/ip-communications/articles/4824-abi-study-predicts-267-million-residential-voip-subscribers.htm.

[2] J. Rosenberg, H. Schulzrinne, G. Camarillo, A. Johnston, J. Peterson, R. Sparks, et al., SIP: Session Initiation Protocol, 2002 (RFC 3261, IETF, June).

[3] F. Andreasen, B. Foster, Media Gateway Control Protocol (MGCP), 2003 (RFC 3435, IETF, January).

[4] M. Handley, V. Jacobson, SDP: Session Description Protocol, 1998 (RFC 2327, IETF, April).

[5] H. Schulzrinne, S. Casner, R. Frederick, V. Jacobson, RTP: A Transport Protocol for Real-Time Applications, 1996 (RFC 1889, IETF, January).

[6] H. Schulzrinne, S. Casner, R. Frederick, V. Jacobson, RTP: A Transport Protocol for Real-Time Applications, 2003 (RFC 3550, IETF, July).

[7] J. Hollingworth, Top 10 Concerns of Buying a VoIP Business Phone System. http://whitepapers.zdnet.com/abstract.aspx?docid=1734213.

[8] J. Franks, P. Hallam-Baker, J. Hostetler, S. Lawrence, P. Leach, A. Luotonen, et al., HTTP Authentication: Basic and Digest Access Authentication, 1999 (RFC 2617, IETF, June).

[9] T. Dierks, E. Rescorla, The Transport Layer Security (TLS) Protocol Version 1.2, 2008 (RFC 5246, IETF, August).

[10] S. Kent, R. Atkinson, Security Architecture for the Internet Protocol, 1998 (RFC 2401, IETF, November).

[11] B. Ramsdell, S/MIME Version 3 Message Specification, 1999 (RFC 2633, IETF, June).

[12] M. Baugher, D. McGrew, M. Naslund, E. Carrara, K. Norrman, The Secure Real-time Transport Protocol (SRTP), 2004 (RFC 3711, IETF, March).

[13] A. Lavallee, Email Scammers Try New Bait in Voice 'Phishing'. http://www.post-gazette.com/pg/06198/706477-96.stm.

[14] R. Naraine, Voice Phishers Dialing for PayPal Dollars. http://www.eweek.com/article2/0,1895,1985966,00.asp.
[15] Skype—the Global Internet Telephony Company. http://www.skype.org.
[16] A. Nappa, A. Fattori, M. Balduzzi, M. Dell'Amico, L. Cavallaro, Take a Deep Breath: A Stealthy, Resilient and Cost-Effective Botnet Using Skype, in: Proceedings of the 7th International Conference on Detection of Intrusions and Malware, and Vulnerability Assessment (DIMVA 2010), July, 2010.
[17] Gizmo. http://www.gizmoproject.com/.
[18] Vonage Is Still #1 In VoIP Market Share. http://www.voipnow.org/2006/07/vonage_is_still.html.
[19] P.D. Kretkowski, VoIP: How Free Can It Be? http://www.voip-news.com/feature/voip-how-free-can-be-120307/.
[20] J. Arkko, V. Torvinen, G. Camarillo, A. Niemi, T. Haukka, Security Mechanism Agreement for the Session Initiation Protocol (SIP), 2003 (RFC 3329, IETF, January).
[21] T. Dierks, C. Allen, The TLS Protocol, 1999 (RFC 2246, IETF, January).
[22] B. Reynolds, D. Ghosal, Secure IP Telephony Using Multi-layered Protection, in: Proceedings of the 2003 Network and Distributed System Security Symposium (NDSS 2003), February 2003.
[23] Y.-S. Wu, S. Bagchi, S. Garg, N. Singh, T. Tsai, SCIDIVE: A stateful and cross protocol intrusion detection architecture for voice-over-IP environments, in: Proceedings of the 2004 International Conference on Dependable Systems and Networks (DSN 2004), July, 2004, pp. 433–442.
[24] H. Sengar, D. Wijesekera, H. Wang, S. Jajodia, VoIP Intrusion detection through interacting protocol state machines, in: Proceedings of the 2006 International Conference on Dependable Systems and Networks (DSN 2006), June, 2006.
[25] H. Sengar, H. Wang, D. Wijesekera, S. Jajodia, Fast detection of denial of service attacks on IP Telephony, in: Proceedings of the 14th IEEE International Workshop on Quality of Service (IWQoS 2006), June, 2006, pp. 199–208.
[26] S. Salsano, L. Veltri, D. Papalilo, SIP security issues: The SIP authentication procedure and its processing load, IEEE Netw. 16 (6) (2002) 38–44.
[27] S. McGann, D.C. Sicker, An analysis of security threats and tools in SIP-based VoIP systems, in: The Second VoIP Security Workshop, 2005.
[28] T.D.C.L.D. Geneiatakis, G. Kambourakis, S. Gritzalis, SIP security mechanisms: a state-of-the-art review, in: Proceedings of the Fifth International Network Conference (INC 2005), Samos, Greece, July 2005, pp. 147–155.
[29] R. Zhang, X. Wang, X. Yang, X. Jiang, Billing attacks on SIP-based VoIP systems, in: The First USENIX Workshop on Offensive Technologies (WOOT 2007), August 2007.
[30] W. Enck, P. Traynor, P. McDaniel, T.L. Porta, Exploiting open functionality in SMS-capable cellular networks, in: Proceedings of the 12th ACM Conference on Computer and Communications Security (CCS 2005), ACM, November 2005, pp. 393–404.
[31] R. Racic, D. Ma, H. Chen, Exploiting MMS vulnerabilities to stealthily exhaust mobile phone's battery, in: Proceedings of the Second International Conference on Security and Privacy in Communication Networks (SecureComm 2006), August, 2006, pp. 1–10.
[32] UPnP Forum. http://www.UPnP.org.
[33] J. Rosenberg, R. Mahy, P. Matthews, D. Wing, Session Traversal Utilities for NAT (STUN), 2008 (RFC 5389, IETF, October).
[34] R. Mahy, P. Matthews, J. Rosenberg, Traversal Using Relays around NAT (TURN): Relay Extensions to Session Traversal Utilities for NAT (STUN), 2010 (RFC 5766, IETF, April).

[35] J. Rosenberg, Interactive Connectivity Establishment (ICE): A Protocol for Network Address Translator (NAT) Traversal for Offer/Answer Protocols, 2010 (RFC 5245, IETF, April).
[36] X. Wang, R. Zhang, X. Yang, X. Jiang, D. Wijesekera, X. Wang, et al., Voice pharming attack and the trust of VoIP, in: Proceedings of the 4th International Conference on Security and Privacy in Communication Networks (SecureComm 2008), September 2008.
[37] R. Zhang, X. Wang, R. Farley, X. Yang, X. Jiang, R. Zhang, et al., On the feasibility of launching the man-in-the-middle attacks on VoIP from remote attackers, in: Proceedings of the 2009 ACM Symposium on Information, Computer & Communication Security (ASIACCS 2009), ACM, March 2009, pp. 61–69.

Phone-to-Phone Configuration for Internet Telephony

YIU-WING LEUNG

Department of Computer Science, Hong Kong Baptist University, Kowloon Tong, Hong Kong

Abstract

Internet telephony is promising for long-distance calls because of its low service charge and value-added functions. To use Internet telephony, a direct method is to use a computer connected to the Internet. However, non-Internet users constitute a significant portion of the general public, and Internet users may not be able to access the Internet at a certain time (e.g., when they are walking in the street or traveling in a bus without Internet access). To serve all users, a service provider can adopt the following *phone-to-phone configuration*: in each servicing city, a telephone gateway is used to bridge the local telephone network and the Internet, so that users can use telephones or mobile phones to access this telephone gateway for long-distance calls through the Internet. The phone-to-phone configuration involves two important issues:

   (i) *Service coverage*: The service provider should provide service to many cities to attain good service coverage. However, it is costly to operate telephone gateways in many cities.
   (ii) *Voice quality*: Voice quality depends on various factors (e.g., coding method, available bandwidth, packet loss in the Internet, etc.). In the phone-to-phone configuration, *multiple* voice streams are sent from a source gateway to a destination gateway. This property can be exploited to tackle the packet loss problem for better voice quality.

In this chapter, we describe the current methods for tackling the above two issues. Specifically, we describe the *sparse telephone gateway configuration* [Y.W. Leung, Sparse telephone gateway for Internet telephony, Comput. Netw. 54(1) (2010) 150–164.] which can serve many cities at lower cost, and describe the *shared packet loss recovery* method [Y.W. Leung, Shared packet loss recovery for Internet telephony, IEEE Commun. Lett. 9 (1) (2005) 84–86.] and the *lightweight piggybacking* method [W.Y. Chow, Y.W. Leung, Lightweight piggybacking for packet loss recovery in Internet telephony, Proceedings of the IEEE International Conference on Communications, Glasgow, UK, June 2007, pp. 1809–1814 (Revised version is under review by a journal.)], which exploit the property of the phone-to-phone configuration for effective packet loss recovery.

1. Introduction 52
2. Sparse Telephone Gateway for Internet Telephony 55
    2.1. Main Idea 56
    2.2. Optimizing Sparse Telephone Gateway Configuration 56
    2.3. Numerical Results 75
3. Packet Loss Recovery 81
    3.1. Piggybacking 82
    3.2. Shared Packet Loss Recovery 84
    3.3. Lightweight Piggybacking 87
4. Summary 100
    References 100

1. Introduction

Telephony has been the most popular communication service. In the early days, voice signals were transmitted in analog form through telephone networks. As the digital technology advanced, voice signals have been transmitted in digital form through telephone networks. This can realize the advantages of digital transmission such as better voice quality via coding and error control, lower bandwidth requirement via compression, and lower cost. In recent years, there is an increasing trend that digital voice signals are transmitted through the Internet. Internet telephony, which is a telephone service over the Internet [1–5], has two major advantages: (i) it makes use of the existing and public Internet infrastructure for transmitting the digital voice signals, so its service charge is low and hence it is particularly promising for long-distance calls and (ii) it can potentially include many value-added services [4,6]. In fact, Internet telephony has huge market potential as revealed by some market analyses and predictions [7–9].

To use Internet telephony, a direct method is to use a computer connected to the Internet. As both users of a call session use computers, the resulting configuration is called *computer-to-computer*. Several software applications are adopting the computer-to-computer configuration, and Skype [10] is probably the most well-known one (e.g., it has over 500 million registered users and 20 million concurrent users [11]). Since Skype is very successful but its protocols are proprietary, some researchers have conducted various experiments to find out the features of Skype in order to understand its successful factors (e.g., see Refs. [11–17]). In particular, it was found that Skype uses the peer-to-peer paradigm to organize the users into an overlay peer-to-peer network for two purposes: (i) maintaining the Skype usernames and the corresponding IP addresses in a distributed manner (ii) NAT traversal such that users who are using private IP address can still talk to each other.

The computer-to-computer configuration is suitable for the Internet users when they can access the Internet. However, a significant portion of the general public are not Internet users (e.g., see the statistics in Table I [18]), and some Internet users may not be able to access the Internet at a certain time (e.g., when they are walking in the street or traveling in a bus without Internet access). To serve all users, a service provider can adopt the *phone-to-phone configuration* shown in Fig. 1 [1,19–33]. In each servicing city, the service provider operates a *telephone gateway* to bridge the local telephone network and the Internet, so that users can use telephones or mobile phones to access this telephone gateway for long-distance calls through the Internet. The phone-to-phone configuration is promising because it is suitable for non-Internet users as well as Internet users when they cannot access the Internet. In fact, it is being used by some Internet telephone applications (e.g., Skype [10] and Gizmo5 [34] allow users to use either computers or phones/mobile phones for long-distance calls).

In the phone-to-phone configuration, the telephone gateway provides several functions to support Internet telephony. Specifically, when it receives a request for setting up a telephone session, it performs admission control [27] to determine whether it can accommodate this new telephone session. If it can admit this session, it executes the necessary signaling protocols [35,36] to establish the telephone call to the destination gateway and the remote user. For the ongoing telephone sessions, the telephone gateway collects the voice streams from the users in its servicing city through the local telephone network, packetizes them, and transmits the resulting

TABLE I
STATISTICS OF INTERNET USAGE IN JUNE 2010 [18]

Region	Percentage of Internet users	Percentage of non-Internet users
Africa	10.9	89.1
Asia	21.5	78.5
Europe	58.4	41.6
Middle East	29.8	70.2
North America	77.4	22.6
Latin America/Caribbean	34.5	65.5
Oceania/Australia	61.3	38.7
World	*28.7*	*71.3*

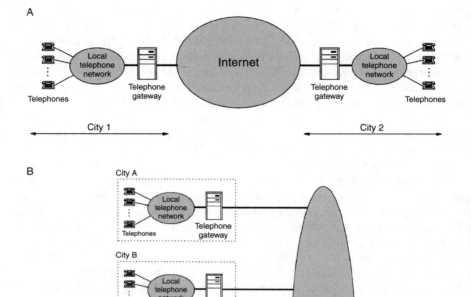

FIG. 1. Phone-to-phone configuration for long-distance Internet telephony [1,19–33]. In every city, a telephone gateway is used to bridge the local telephone network and the Internet, so that users can use telephones or mobile phones to access this gateway for long-distance calls. (A) Phone-to-phone configuration for two cities. (B) Phone-to-phone configuration for multiple cities.

voice packets to the destination gateway through the Internet. At the same time, it receives the voice packets from the remote telephone gateways, performs packet loss recovery to possibly recover the lost voice packets (this point will be further explained in the following), and extracts the voice bits from the packets and sends them to the users through the local telephone network.

The phone-to-phone configuration involves two important issues:

1. *Service coverage*: The service provider should provide service to many cities to attain good service coverage. Using the phone-to-phone configuration, it is necessary to operate a telephone gateway in each of these cities. When the number of cities is large, the total operating cost is high. It is desirable to serve many cities at lower cost so that the resulting service is attractive (because of good service coverage) and competitive (because of lower operating cost).
2. *Voice quality*: Voice quality depends on various factors (e.g., voice coding method, available bandwidth, packet loss, etc.). Since the Internet only provides the best effort delivery, the voice packets may be lost or erroneous. The lost or erroneous packets cannot be retransmitted because of the stringent delay requirement for interactive telephone voice communication. Consequently, the resulting voice would have lower quality. *Packet loss recovery* is a promising approach to tackle this problem [37]. Using this approach, the source transmits redundant information by which the destination can possibly recover the lost packets. In the phone-to-phone configuration, *multiple* voice streams are sent from a source gateway to a destination gateway. It is desirable to exploit this property for more effective packet loss recovery for better voice quality.

In this chapter, we describe the current methods for tackling the above two issues. In Section 2, we consider the first issue, describe the *sparse telephone gateway configuration* [38] which can serve many cities at lower cost, and describe how to optimize this configuration to achieve the best cost-effectiveness. In Section 3, we consider the second issue, describe a well-known packet loss recovery method called *piggybacking* [37,39,40], and describe two methods (called *shared packet loss recovery* [19], and *lightweight piggybacking* [20]) which exploit the property of the phone-to-phone configuration for more effective packet loss recovery.

2. Sparse Telephone Gateway for Internet Telephony

In this section, we describe the sparse telephone gateway configuration that we recently proposed in Ref. [38]. Using this configuration, the service provider can provide service to many cities at lower cost.

2.1 Main Idea

We use the following example to explain our main idea. Suppose we want to serve city A but this city has a low traffic volume and a high operating cost (e.g., high rental and manpower cost). In this case, we do not operate a telephone gateway in city A because of its low cost-effectiveness. Instead, we serve city A by using a telephone gateway in a nearby city (say, city B), such that users in city A access the telephone gateway in city B for long-distance calls. Figure 2 shows an example to illustrate this idea. In this manner, we need not operate a telephone gateway in city A but we can still serve this city.

The sparse telephone gateway configuration is based on the above idea. When we want to serve N cities, we operate telephone gateways in only M cities (where $M < N$ and M is to be optimized). If a telephone gateway is available in a city, the users in this city access this telephone gateway through local (intracity) calls for service; otherwise, they access a telephone gateway in a nearby city through intercity calls for service. In the example shown in Fig. 2, we use two telephone gateways to serve five cities.

When we do not operate a telephone gateway in city A but we serve city A by using the telephone gateway in city B, the cost is changed as follows:

- *Saved cost:* We can save the cost of operating a telephone gateway in city A, while the telephone gateway in city B can be shared by the users in both cities A and B for better resource utilization.
- *Additional cost:* Since the users in city A access the telephone gateway in city B for service, it is necessary to pay the tariff for intercity calls between cities A and B. Nevertheless, this tariff between two nearby cities is small compared with the service charge for long-distance calls, and it can be compensated by the saved cost through optimization.

From the above discussion, we see that it is necessary to optimize the sparse telephone gateway configuration in order to achieve the best cost-effectiveness. We formulate and solve this problem in Section 2.2.

2.2 Optimizing Sparse Telephone Gateway Configuration

2.2.1 Problem Outline and Related Existing Results

We optimize the sparse telephone gateway configuration to achieve the best cost-effectiveness. The cost-effectiveness depends on both income and cost. We measure the cost-effectiveness in terms of *revenue*, and our objective is to maximize the

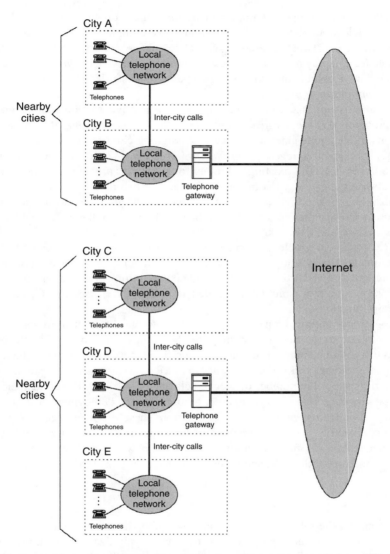

Fig. 2. Sparse telephone gateway configuration for long-distance Internet telephony [38]. In this configuration, nearby cities can share a telephone gateway for long-distance calls.

expected revenue. In optimization, we determine the following design choices for the sparse telephone gateway configuration: (1) the number and locations of telephone gateways, (2) the cities served by each telephone gateway, and (3) the service coverage (e.g., it may not be cost-effective to provide service between the cities that produce low income but involve high operating cost). This problem is called *gateway optimization problem*.

The gateway optimization problem is a mix of two types of operations research (OR) problems: (1) *facility location problem* [41] (for determining the number and locations of telephone gateways as well as assigning cities to telephone gateways), and (2) *nonlinear integer programming problem* [42] (for determining the service coverage). The existing solutions for these OR problems are not directly applicable to the gateway optimization problem for two reasons. First, the gateway optimization problem involves two *interdependent* OR problems, while the existing solutions were designed for each OR problem only. Second, these OR problems are general classes and the existing solutions were designed for specific problem instances (different problem instances involve different objectives and/or different constraints and hence they require different solutions). The gateway optimization problem has its own distinct objective and constraint functions (see Sections 2.2.2 and 2.2.3), so a new solution is needed to tackle these specific objective and constraint functions.

The gateway optimization problem is similar to but different from some placement problems in the networking literature. In wireless mesh networks, it is necessary to place a given number of gateways in order to maximize the throughput [43]. In cellular radio networks, it is necessary to place base stations in order to minimize the interference or the number of blocked channels [44]. In video-on-demand, it is necessary to place video servers in order to minimize the deployment cost [45]. The gateway optimization problem is different from these problems because it has a different objective (i.e., maximizing the revenue) and different constraints (as Internet telephony is fundamentally different from wireless mesh networks [43], cellular radio networks [44], and video-on-demand systems [45]).

2.2.2 Model

We consider the following model which specifies the Internet telephony system, cost components, and charging scheme. Our optimization methodology is also applicable to other models, and we will describe the generalizations in Section 2.2.8.

We use the sparse telephone gateway configuration (see Section 2.1 and Fig. 2) to provide service to N cities in one or more countries. The quality-of-service requirements can be transformed into cost requirements. For example, the per-call bandwidth requirement determines the necessary bandwidth of the Internet connection for each telephone gateway (i.e., the communication cost of each telephone gateway) and the residual packet loss rate requirement (i.e., the loss rate after performing packet loss recovery [20]) determines the processing cost of each telephone gateway. We will describe the cost components in the following.

The revenue is equal to income minus cost. We consider the revenue, income, and cost over one *servicing period* (e.g., 1 month). The income comes from the service charge paid by the users. Let R_{ij} be the service charge per unit time for a call from city i to city j. We define the *traffic load* to be the expected number of calls times the expected duration of each call over one servicing period. Let λ_{ij} be the traffic load from city i to city j. The income obtained from the calls from city i to city j in one servicing period is equal to $\lambda_{ij}R_{ij}$.

The cost has two components. The first one is the cost of operating telephone gateways. We consider a general operation cost model which can take various cost factors into account for wide applicability. In this model, the cost of operating a telephone gateway in city i is $C_i(\Lambda_i)$, where $C_i(\Lambda_i)$ depends on the traffic load Λ_i in this city. Using this model, we can include any cost components into $C_i(\Lambda_i)$, and the following are some examples:

- The operation cost may include the following cost components: bandwidth cost of the Internet connections, rental cost for physical space for locating the telephone gateway hardware, manpower cost for maintenance, and power and cooling cost. Each of these cost components can be either dependent on or independent of the traffic load Λ_i in this city. To take these cost components into account, we add them into $C_i(\Lambda_i)$.
- If we want to count the depreciation value for the initial setup and hardware cost (this is a common accounting practice for listed IT and telecommunication companies), we can add this cost component into $C_i(\Lambda_i)$.

The second cost component is the telephone tariff because users access telephone gateways through local or intercity telephone networks for long-distance calls (see Fig. 2). Let T_{ij} be the tariff per unit time for a call from city i to city j. If the distance between city i and city j is longer, T_{ij} is typically larger.

In the planning phase, the service provider may want to decide whether to provide service to some cities (e.g., if a city has a low traffic volume and a high operating cost, the service provider may decide not to provide service to this city). On the other hand, the service provider may be required to provide service to certain cities (e.g., to fulfill license requirements). The service provider can specify these requirements in terms of the set $\chi = \{(i,j):$ the service provider certainly provides service from city i to city $j\}$.

2.2.3 Problem Formulation

In this section, we formulate the problem of optimizing the sparse telephone gateway configuration. We define the following design variables:

$$x_i = \begin{cases} 1 & \text{a telephone gateway is operated in city } i, \\ 0 & \text{otherwise,} \end{cases} \quad (1)$$

$$y_{ij} = \begin{cases} 1 & \text{city } i \text{ is served by the telephone gateway in city } j, \\ 0 & \text{otherwise,} \end{cases} \quad (2)$$

$$z_{ij} = \begin{cases} 1 & \text{service is provided from city } i \text{ to city } j, \\ 0 & \text{otherwise.} \end{cases} \quad (3)$$

After optimization, if $x_i = 1$ and $y_{ji} = 1$, this means that cities i and j are nearby such that they share the telephone gateway in city i.

We express the expected revenue in terms of the above design variables. The expected revenue is equal to the expected income minus the expected cost. The expected income is the mean service charge received from the users and it is equal to

$$\sum_{i=1}^{N} \sum_{\substack{j=1 \\ i \neq j}}^{N} z_{ij} \lambda_{ij} R_{ij}. \quad (4)$$

The cost has two components. The first one is the expected operating cost of telephone gateways and it is equal to

$$\sum_{g=1}^{N} x_g C_g(\Lambda_g), \quad (5)$$

where the traffic load Λ_g handled by the telephone gateway in city g is

$$\Lambda_g = x_g \sum_{i=1}^{N} y_{ig} \left(\sum_{j=1}^{N} z_{ij} \lambda_{ij} + \sum_{j=1}^{N} z_{ji} \lambda_{ji} \right). \quad (6)$$

The second cost component is the expected tariff required. For city i, the mean outgoing and incoming traffic loads are equal to $\sum_{j=1}^{N} z_{ij} \lambda_{ij}$ and $\sum_{j=1}^{N} z_{ji} \lambda_{ji}$, respectively. If city i is served by the telephone gateway in city g, the expected tariff required is $T_{ig} \sum_{j=1}^{N} z_{ij} \lambda_{ij} + T_{gi} \sum_{j=1}^{N} z_{ji} \lambda_{ji}$. The total expected tariff required is equal to

$$\sum_{i=1}^{N} \sum_{g=1}^{N} x_g y_{ig} \left(T_{ig} \sum_{j=1}^{N} z_{ij} \lambda_{ij} + T_{gi} \sum_{j=1}^{N} z_{ji} \lambda_{ji} \right). \quad (7)$$

PHONE-TO-PHONE CONFIGURATION FOR INTERNET TELEPHONY

The expected revenue $\mathcal{R}$ is equal to the expected income minus the expected cost and it is equal to

$$\mathcal{R} = \sum_{i=1}^{N}\sum_{\substack{i=1 \\ i \neq j}}^{N} z_{ij}\lambda_{ij}R_{ij} - \sum_{g=1}^{N} x_g C_g(\Lambda_g) \qquad (8)$$

$$-\sum_{i=1}^{N}\sum_{g=1}^{N} x_g y_{ig} \left(T_{ig}\sum_{j=1}^{N} z_{ij}\lambda_{ij} + T_{gi}\sum_{j=1}^{N} z_{ji}\lambda_{ji} \right).$$

The gateway optimization problem is to maximize the expected revenue and it can be formulated as follows:

Gateway Optimization Problem

$$\text{Maximize} \quad \mathcal{R} = \sum_{i=1}^{N}\sum_{\substack{j=1 \\ i \neq j}}^{N} z_{ij}\lambda_{ij}R_{ij} - \sum_{g=1}^{N} x_g C_g(\Lambda_g)$$

$$-\sum_{i=1}^{N}\sum_{g=1}^{N} x_g y_{ig} \left(T_{ig}\sum_{j=1}^{N} z_{ij}\lambda_{ij} + T_{gi}\sum_{j=1}^{N} z_{ij}\lambda_{ji} \right).$$

Subject to

1. $x_i, y_{ig} \in \{0, 1\}$ for all $1 \leq i, g \leq N$
2. $z_{ij} = 1$ for all $(i,j) \in \chi$
3. $z_{ij} \in \{0, 1\}$ for all $(i,j) \notin \chi$ \qquad (9)
4. $$\sum_{g=1}^{N} x_g y_{ig} = \begin{cases} 1 & \text{if } \sum_{j=1}^{N}(z_{ij} + z_{ji}) \geq 1 \\ 0 & \text{otherwise,} \end{cases} \quad \text{for all } 1 \leq i \leq N$$

where the variables to be optimized are x_i for all $1 \leq i \leq N$, y_{ig} for all $1 \leq i, g \leq N$, and z_{ij} for all $1 \leq i, j \leq N$ and $(i,j) \notin \chi$. The objective function is the expected revenue. The first constraint ensures that the design variables x_i and y_{ig} are either 0 or 1. The second constraint ensures that the service is certainly provided to the city pairs specified by the service provider, while the third constraint allows the service provider to decide whether to provide service to other city pairs based on the cost-effectiveness. The fourth constraint ensures that, if service is provided between city i and at least one other city, a telephone gateway must be allocated to serve city i.

2.2.4 Problem Complexity

We prove that the gateway optimization problem is NP-hard through three main steps:

1. Define a special instance (called *Problem 1*) of the gateway optimization problem and a closely related decision problem (called *Problem 2*).
2. Make use of an NP-complete problem (namely, *vertex cover problem* [46]) to prove that Problem 2 is NP-complete and Problem 1 is NP-hard.
3. Make use of Problem 1 to prove that the gateway optimization problem is NP-hard.

Consider the first step. We define the following special instance of the gateway optimization problem. We provide service between city 0 in country A and N cities in country B (i.e., we only provide international calls between countries A and B). The operating cost of a telephone gateway in city i is fixed at C_i, and service charge and tariff are symmetric (i.e., $R_{ij} = R_{ji}$ and $T_{ij} = T_{ji}$). We operate a telephone gateway in city 0 of country A, and determine the number and locations of telephone gateways in country B. In this problem instance, the income is fixed, so the problem of maximizing the expected revenue is equivalent to the problem of minimizing the expected cost. The latter problem can be formulated as follows:

$$\text{Minimize} \quad \left(C_0 + \sum_{j=1}^{N} x_j C_j\right) + \sum_{j=1}^{N} (\lambda_{0j} + \lambda_{j0})$$
$$\times [T_{00} + \min(x'_1 T_{1j}, x'_2 T_{2j}, \ldots, x'_N T_{Nj})],$$

$$\text{Subject to} \quad x_i \in \{0, 1\} \text{ for all } 1 \leq i \leq N \qquad (10)$$

where $x'_i = 1$ if $x_i = 1$ and $x'_i = \infty$ if $x_i = 0$. We note that $(\lambda_{0j} + \lambda_{j0})T_{00}$ and C_{00} are independent of these decision variables, and let I be the index set $\{i : x_i = 1\}$. Then, the above problem is equivalent to the following problem:

Problem 1

$$\text{Minimize} \quad \sum_{j=1}^{N} \min_{i \in I} \{a_{ij}\} + \sum_{i \in I} C_i, \tag{11}$$

$$\text{Subject to} \quad I \subseteq \{1, 2, \ldots, N\},$$

where $a_{ij} = (\lambda_{0j} + \lambda_{j0}) T_{ij}$ and the decision variable is the index set I. We define the following decision problem that is closely related to Problem 1.

Problem 2

Let K be a positive integer ($1 \leq K \leq N$) and U be a positive real number. Is there an index set I with at most K elements such that $\sum_{j=1}^{N} \min_{i \in I} \{a_{ij}\} \leq U$?

In the second step, we first prove that Problem 2 is NP-complete. For this purpose, we make use of the following NP-complete problem called *vertex cover problem* [46].

Vertex Cover Problem

Let (V, E) be a graph where V and E are sets of vertexes and edges, respectively. Is there a vertex cover (i.e., a subset V' of V) with at most K vertexes such that at least one of the two vertexes of every edge in E belongs to this vertex cover V'?

Lemma 1

Problem 2 is NP-complete.

Proof

We prove by contradiction. Assume that Problem 2 is not NP-complete. We consider the following problem instance: $a_{ij} \in \{0, 1\}$ and $U = 0$. We transform the solutions of this problem instance to give the solutions to the vertex cover problem in polynomial time as follows: Let (V, E) be a graph where V is a set of N vertexes and E is a set of N edges, and define the following polynomial-time transformation:

$$a_{ij} = \begin{cases} 0 & \text{if vertex } i \text{ is one of the two vertexes of edge } j, \\ 1 & \text{otherwise.} \end{cases} \quad (12)$$

If the answer to Problem 2 is "yes" (i.e., there is an index set with at most K elements such that $\sum_{j=1}^{N}\sum_{i\in I}\min\{a_{ij}\} \leq 0$), the answer to the vertex cover problem is "yes" (i.e., there is vertex cover I with at most K vertexes); otherwise, the answer to the vertex cover problem is "no."

If Problem 2 is not NP-complete, we can find its solutions in polynomial time and transform them to give the solutions to the vertex cover problem using the polynomial transformation in Eq. (12) (i.e., we can solve the vertex cover problem in polynomial time). This leads to a contradiction because the vertex cover problem is NP-complete [46]. Therefore, Problem 2 is NP-complete. This completes the proof. □

Based on Lemma 1, we prove that Problem 1 is NP-hard in the following.

Lemma 2

Problem 1 is NP-hard.

Proof

We prove by contradiction. Assume that Problem 1 is not NP-hard. We consider the following problem instance: $a_{ij} \in \{0,1\}$ and $C_1 = C_2 = \cdots = C_N = C$. Let $I(n)$ be an index set with n elements where $I(n) \subseteq \{1, 2, \ldots, N\}$. We denote the objective function of Problem 1 by $F(n)$ where $F(n) = F_1(n) + F_2(n)$ and

$$\begin{cases} F_1(n) = \sum_{j=1}^{N} \min_{i \in I(n)} \{a_{ij}\}, \\ F_2(n) = \sum_{i \in I(n)} C_i = nC. \end{cases} \quad (13)$$

To minimize $F(n)$ for any given C, let the optimal n be $n^*(C)$ (when there are more than one optimal n, $n^*(C)$ denotes the largest one), $I^*(C)$ be the corresponding optimal index set, and $F^*(n^*(C))$ be the minimal value of $F(n)$. Since $n^*(C)$ is optimal, we have

$$\begin{cases} F(n^*(C)) < F(n^*(C) + 1), \\ F(n^*(C)) \leq F(n^*(C) - 1), \end{cases} \quad (14)$$

which is equivalent to

$$\begin{cases} F_1(n^*(C) - 1) - F_1(n^*(C)) \geq C, \\ F_1(n^*(C)) - F_1(n^*(C) + 1) < C. \end{cases} \quad (15)$$

Since $F_2(n)$ is independent of $I(n)$, the index set with $n^*(C)$ elements that minimizes $F_1(n^*(C))$ is also given by $I^*(C)$. Therefore, if we solve Problem 1 for any given C to get $n^*(C)$ and $I^*(C)$, we also solve Problem 2 for $K = n^*(C)$. We solve Problem 1 for $C = 0, 1, 2, \ldots, N$ to get $n^*(C)$ and $I^*(C)$ where $n^*(C)$ has the following characteristics:

- $C = 0$: Since $F(n) = F_1(n)$ and $F_1(n)$ is a decreasing function of n, $n^*(0) = N$.
- $C = N$: Since $F(n) = F_1(n) + nN$ and $0 \leq F_1(n) \leq N$, $n^*(N) = 1$.
- $0 \leq C \leq N$: Since $0 \leq F_1(n) \leq N$ and $F_1(n)$ is a convex decreasing function, the conditions in Eq. (15) ensure that $n^*(0), n^*(1), \ldots, n^*(N)$ span the range between 1 and N.

To solve Problem 2, we solve Problem 1 for $C = 0, 1, 2, \ldots, N$, select $n^*(c)$ such that $n^*(c-1) < K \leq n^*(c)$, and check whether $F_1(n^*(c)) = F^*(n^*(c)) - n^*(c)c$ is at most U to get an answer to Problem 2. In other words, if we can solve Problem 1 in polynomial time, we can also solve Problem 2 in polynomial time. This leads to a contradiction because Problem 2 is NP-complete (see Lemma 1). Therefore, Problem 1 is NP-hard. This completes the proof. □

In the third step, we apply Lemma 2 to prove that the gateway optimization problem is NP-hard:

Theorem

The gateway optimization problem is NP-hard.

Proof

Problem 1 is NP-hard (as proved in Lemma 2), and it is a special instance of the gateway optimization problem. Therefore, the latter problem is NP-hard. This completes the proof. □

2.2.5 Two-Stage Heuristic Algorithm: First Stage

We design an efficient two-stage heuristic algorithm to solve the NP-hard gateway optimization problem. In the first stage, we relax the third constraint so that the relaxed problem is simpler and solve this simpler problem to find a potential configuration.

In the second stage, we restore the third constraint and iteratively refine the configuration such that the resulting configuration is optimal or close-to-optimal and fulfills all the constraints of the gateway optimization problem. We describe the first stage in this section while we will describe the second stage in Section 2.2.6.

In the first stage, we relax the third constraint of the gateway optimization problem to $z_{ij}=1$ for all $(i,j) \notin \chi$. (We will restore and handle the third constraint in the second stage in Section 2.2.6.) After relaxation, the income becomes fixed at $\sum_{i \neq j} \lambda_{ij} R_{ij}$ and hence, the problem of maximizing the expected revenue is equivalent to the problem of minimizing the expected cost. To minimize the expected cost, we determine a good initial configuration and then improve this configuration iteratively. Before describing the details, we need the following terms:

- If a gateway is operated in city j, this gateway is called *gateway j*.
- If city i is served by gateway j, we say that city i is *assigned* to *gateway j*.

As we will explain shortly, we will often assign the N cities to some given gateways. Therefore, we first describe how to perform this assignment as follows: Let τ_i be the total traffic load of city i (i.e., $\tau_i = \sum_{j=1}^{N}(z_{ij}\lambda_{ij} + z_{ji}\lambda_{ji})$) and Γ_g be the current traffic load handled by gateway g. We assign the cities iteratively in the following order: the city with the largest traffic load, then the city with the next largest traffic load, etc. In each iteration, if we assign city i to gateway g, we evaluate the resulting increase in cost as follows:

$$\Delta_{i,g} = \left(\sum_{j=1}^{N} z_{ij}\lambda_{ij}\right)T_{ig} + \left(\sum_{j=1}^{N} z_{ji}\lambda_{ji}\right)T_{gi} + \left(C_g(\Gamma_g + \tau_i) - C_g(\Gamma_g)\right). \quad (16)$$

We assign city i to the gateway such that the resulting increase in cost is the smallest. The details of assignment are given in the following algorithm:

Algorithm 1

Assignment of cities to given gateways

1. Initialize $y_{ij}=0$ for all $1 \leq i, j \leq N$.
2. Sort $\tau_1, \tau_2, \ldots, \tau_N$ in the order $\tau_{(1)} \geq \tau_{(2)} \geq \cdots \geq \tau_{(N)}$ (where (i) represents an integer between 1 and N and $\{(1),(2),\ldots,(N)\}=\{1,2,\ldots,N\}$.
3. For all $1 \leq i \leq N$, do the following: If $\sum_{j=1}^{N}(z_{(i)j} + z_{j(i)}) \geq 1$, assign city (i) to a telephone gateway as follows: When this city is assigned to each of the given gateways, evaluate the resulting increase in cost based on Eq.(16). Select and perform the assignment (say, assign city (i) to gateway j) such that the increase in cost is the smallest and set $y_{(i)j}=1$.

Now, we determine a good initial configuration as follows: First, we locate two gateways and execute Algorithm 1 to assign the N cities to these two gateways such that the resulting cost is the smallest. In each subsequent iteration, we locate one additional gateway and execute Algorithm 1 to assign the N cities to the resulting gateways to result in the smallest cost. We repeat this iterative process until the cost cannot be further reduced. The framework is shown in the flowchart in Fig. 3 and the details are given in the following algorithm:

Algorithm 2

Determination of an Initial Configuration

1. Initialize $x_i = 0$ for all $1 \leq i \leq N$.
2. Locate two telephone gateways in two respective cities as follows: For all $1 \leq i \leq N$ and $1 \leq j \leq N$, if two telephone gateways are operated in city i and city j, respectively, execute Algorithm 1 to assign the N cities to these two gateways and evaluate the resulting cost. Select a pair of cities that result in the smallest cost (say, city c_1 and city c_2), operate a telephone gateway in each of these two cities, and set $x_{c_1} = 1$ and $x_{c_2} = 1$.
3. Locate an additional telephone gateway in a city as follows: For all $1 \leq i \leq N$ and $x_i = 0$, if a telephone gateway is operated in city i, execute Algorithm 1 to assign the N cities to the gateways and evaluate the resulting cost. Select a city (say, city c) that results in the smallest cost. If this cost is smaller than the original cost, operate a telephone gateway in city c, set $x_c = 1$, and repeat step 3; otherwise, STOP.

After determining an initial configuration, we improve this configuration iteratively. In each iteration, we add, remove, and/or relocate telephone gateways and perform reassignment in order to further reduce the cost. Specifically, when we attempt to add or remove a gateway, we execute Algorithm 1 to reassign the cities to the resulting gateways. If each city can still be assigned to one gateway, we evaluate the resulting cost. We select and perform an addition or a removal that can further reduce the cost, and the cost reduction is the largest. Moreover, when we attempt to relocate a gateway from one city to another city, we execute Algorithm 1 to reassign the cities to the resulting gateways. If each city can still be assigned to one gateway, we evaluate the resulting cost. We select and perform a relocation that can further reduce the cost, and the cost reduction is the largest. Furthermore, when we attempt

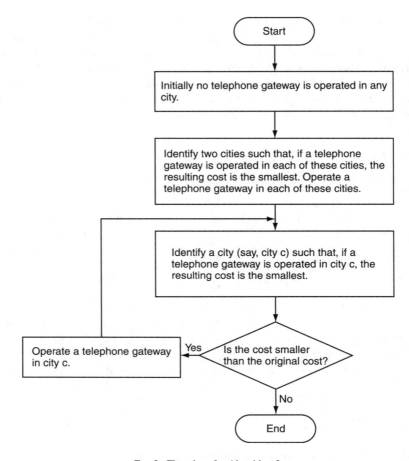

Fig. 3. Flowchart for Algorithm 2.

to add or remove two gateways, we execute Algorithm 1 to reassign the cities to the resulting gateways. If each city can still be assigned to one gateway, we evaluate the resulting cost. We select and perform the addition or removal that can further reduce the cost, and the cost reduction is the largest. We perform the above addition, removal, and relocation operations iteratively until the cost cannot be further reduced. The framework is shown in the flowchart in Fig. 4 and the details are given in the following algorithm:

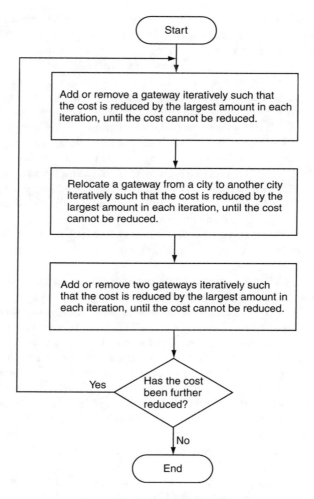

FIG. 4. Flowchart for Algorithm 3.

Algorithm 3

Iterative Refinement

1. For all $1 \leq i \leq N$ and $x_i = 0$, if a telephone gateway is added in city i, execute Algorithm 1 to assign the N cities to the resulting telephone gateways and evaluate the resulting cost.

2. If there exists an addition (say, adding a telephone gateway in city c) that results in the smallest cost and this cost is smaller than the original cost, add a telephone gateway in city c (i.e., set $x_c=1$) and go to Step 1.
3. For all $1 \leq i \leq N$ and $x_i=1$, if a telephone gateway in city i is removed, execute Algorithm 1 to assign the N cities to the remaining telephone gateways and evaluate the resulting cost if each city can still be assigned to one gateway.
4. If there exists a feasible removal (say, removing the telephone gateway in city c) that results in the smallest cost and this cost is smaller than the original cost, remove the telephone gateway in city c (i.e., set $x_c=0$) and go to Step 3.
5. For all $1 \leq i, j \leq N$, $i \neq j$, $x_i=1$ and $x_j=0$, when the telephone gateway in city i is relocated to city j, execute Algorithm 1 to assign the N cities to the resulting gateways and evaluate the resulting cost if each city can still be assigned to one gateway.
6. If there exists a feasible relocation (say, from city c_1 to city c_2) that results in the smallest cost and this cost is smaller than the original cost, relocate the telephone gateway from city c_1 to city c_2 (i.e., set $x_{c_1}=0$ and $x_{c_2}=1$) and go to Step 5.
7. For all $1 \leq i, j \leq N$, $i \neq j$, $x_i=0$ and $x_j=0$, if telephone gateways are added in city i and city j, execute Algorithm 1 to assign the N cities to the resulting telephone gateways and evaluate the resulting cost.
8. If there exists an addition (say, adding telephone gateways in city c_1 and city c_2) that results in the smallest cost and this cost is smaller than the original cost, add telephone gateways in city c_1 and city c_2 (i.e., set $x_{c_1}=1$ and $x_{c_2}=1$) and go to Step 7.
9. For all $1 \leq i, j \leq N$, $i \neq j$, $x_i=1$ and $x_j=1$, if the telephone gateways in city i and city j are removed, execute Algorithm 1 to assign the N cities to the remaining telephone gateways and evaluate the resulting cost if each city can still be assigned to one gateway.
10. If there exists a feasible removal (say, removing the telephone gateways in city c_1 and city c_2) that results in the smallest cost and this cost is smaller than the original cost, remove the telephone gateways in city c_1 and city c_2 (i.e., set $x_{c_1}=0$ and $x_{c_2}=0$) and go to Step 9.
11. If the cost has been reduced in any of the above steps, go to Step 1; otherwise, STOP.

2.2.6 Two-Stage Heuristic Algorithm: Second Stage

In the second stage, we restore the third constraint of the gateway optimization problem (i.e., z_{ij} can be 0 or 1 for all $(i,j) \notin \chi$) and further optimize the configuration determined in the first stage while fulfilling all the constraints. After restoring the third constraint, it is necessary to determine whether to provide service to the city pairs not specified in χ. We do not provide service to the city pairs of low cost-effectiveness (e.g., the service charge received is smaller than the cost involved). Specifically, if two cities are served by the same gateway, we do not provide service between them. For any two cities which are served by different gateways, if we do not provide service between them, we determine the change in the expected revenue. We select a pair of cities such that the change in the expected revenue is positive and the largest, and do not provide service between them. We repeat this step until the expected revenue cannot be further increased.

After we decide not to provide service to the city pairs of low cost-effectiveness, it may be possible to remove some telephone gateways for further cost reduction. In particular, when we attempt to remove a gateway, we execute Algorithm 1 to reassign the cities to the resulting gateways. If each city can still be assigned to one gateway, this removal is feasible and we evaluate the resulting expected revenue. We select and perform a removal that can further increase the expected revenue by the largest amount. We remove gateways iteratively until the expected revenue cannot be further increased. The framework is shown in the flowchart in Fig. 5 and the details are given in the following algorithm:

Algorithm 4

Further Refinement Subject to All Constraints

1. For all $1 \leq g \leq N$, if city i and city j are served by gateway g, $(i,j) \notin \chi$ and $T_{ig} + T_{gj} \geq R_{ij}$, then set $z_{ij}=0$ and $z_{ji}=0$.
2. For all $1 \leq i \leq N$, $i+1 \leq j \leq N$, $(i,j) \notin \chi$, and $z_{ij}=1$ or $z_{ji}=1$, if service is not provided between city i and city j, compute the change in the expected revenue. If there exist two cities (say, city m and city n) such that the resulting change in the expected revenue is positive and the largest, set $z_{mn}=0$ and $z_{nm}=0$ and repeat Step 2.
3. For all $1 \leq g \leq N$ and $x_g=1$, if gateway g is removed, execute Algorithm 1 to assign the N cities to the remaining gateways and compute the change in the expected revenue if each city can still be assigned to one gateway. If there exists a removal such that the resulting change in the expected revenue is positive and the largest, remove this gateway and repeat Step 3.
4. Repeat Steps 1–3 until the expected revenue cannot be further increased.

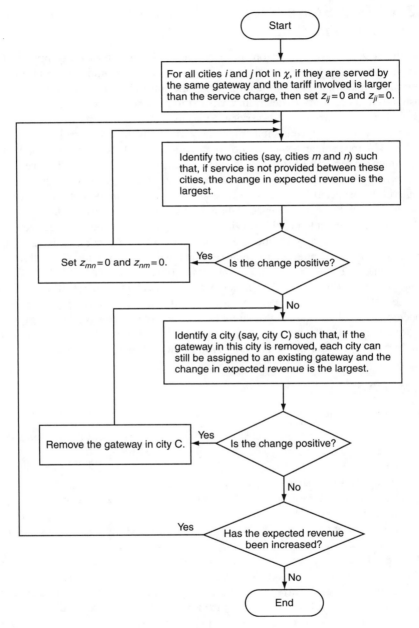

FIG. 5. Flowchart for Algorithm 4.

2.2.7 Overall Two-Stage Heuristic Algorithm

The overall two-stage heuristic algorithm is as follows: In the first stage, it relaxes the third constraint of the gateway optimization problem to a looser one (this constraint will be restored in the second stage), executes Algorithm 2 to determine an initial configuration, and then executes Algorithm 3 to refine the configuration iteratively. In the second stage, it restores the third constraint to the original one and executes Algorithm 4 to further refine the sparse configuration subject to all the constraints. The framework is shown in the flowchart in Fig. 6 and the details are given below.

Two-stage heuristic algorithm

1. *First Stage*
   1.1 Relax the third constraint of the gateway optimization problem to $z_{ij}=1$ for all $(i,j) \notin \chi$.
   1.2 Execute Algorithm 2 to determine an initial configuration.
   1.3 Execute Algorithm 3 to refine the configuration iteratively.
2. *Second Stage*
   2.1 Restore the third constraint to the original one $z_{ij} \in \{0,1\}$ for all $(i,j) \notin \chi$.
   2.2 Execute Algorithm 4 to further refine the sparse telephone gateway configuration subject to all the constraints of the Gateway Optimization Problem.

2.2.8 Generalization

In Sections 2.2.3, 2.2.4, and 2.2.5, we consider the model described in Section 2.2.2. The proposed two-stage heuristic algorithm is also applicable to other models for two reasons:

1. The proposed algorithm computes the expected cost and the expected revenue for iterative optimization. When we adopt another model, the proposed algorithm is still applicable by just using the new cost and revenue functions.

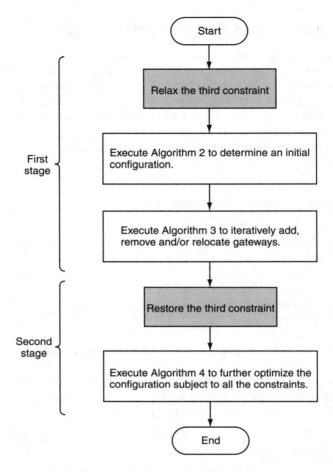

FIG. 6. Flowchart for the two-stage heuristic algorithm.

2. In each iteration of the optimization process, the proposed algorithm can take additional factors or constraints into account before making a choice in this iteration.

To illustrate, we give two examples in the following:

Example 1: Suppose the tariff is a fixed amount for one servicing period (e.g., fixed monthly charge). Let T'_{ij} be the tariff for the calls from city i to city j for one servicing period. The total tariff required becomes $\sum_{i=1}^{N}\sum_{g=1}^{N} x_g y_{ig}(T'_{ig} + T'_{gi})$ and hence, the expected revenue $\mathcal{R}$ becomes

$$\mathcal{R} = \sum_{\substack{i=1 \\ i \neq j}}^{N} \sum_{j=1}^{N} z_{ij} \lambda_{ij} R_{ij} - \sum_{g=1}^{N} x_g C_g(\Lambda_g) - \sum_{i=1}^{N} \sum_{g=1}^{N} x_g y_{ig}(T'_{ig} + T'_{gi}). \qquad (17)$$

The two-stage heuristic algorithm can tackle the resulting gateway optimization problem by using the above revenue function.

Example 2: Suppose each telephone gateway can handle a maximum traffic load of L. With minor modifications, the two-stage heuristic algorithm is applicable to handle this constraint as follows:

- Algorithm 1: In Step 3, we only consider the assignment that will not overload the telephone gateways.
- Algorithm 2: In Step 2, we assign the cities to the two telephone gateways without overloading them. In Step 3, we proceed until all the N cities have been assigned to the telephone gateways without overloading them and the cost cannot be further reduced.
- Algorithm 3: We only consider the removal of gateway(s) such that the N cities can still be assigned to the remaining telephone gateways without overloading them.

2.3 Numerical Results

We conduct two numerical experiments to study (i) the effectiveness of the sparse telephone gateway configuration (hereinafter referred to as *sparse configuration*) and (ii) the performance of the proposed two-stage heuristic algorithm.

2.3.1 First Numerical Experiment

The gateway optimization problem is NP-hard but we can still get its optimal solutions for small N by exhaustive search. In this experiment, we select a small N and compare the expected revenue of the following three cases for performance evaluation: (1) the optimal sparse configuration found by exhaustive search; (2) the sparse configuration found by the two-stage heuristic algorithm; and (3) the existing configuration.

We adopt the following parameter values: There are $N=8$ cities with 4 cities in one country and another 4 cities in another country. The system supports long-distance intercountry calls only. The service charge R_{ij} is 10 and the tariff T_{ij} is

$$T_{ij} = \begin{cases} |i-j| & \text{if city } i \text{ and city } j \text{ are in the same country} \\ 10 & \text{otherwise} \end{cases} \quad (18)$$

The operating cost $C_i(\Lambda_i)$ of a telephone gateway in city i is composed of a fixed cost and a load-dependent cost, where the fixed cost is randomly generated from a uniform distribution between 0 and 10,000 and the load-dependent cost is $2000\lceil\Lambda_i/1000\rceil$. We consider two traffic patterns:

1. *Uniformly generated:* The outgoing traffic load from each city is randomly generated from a uniform distribution with mean $\bar{\lambda}$. Each outgoing call is connected to any city in the other country with equal probability.
2. *Nonuniformly generated:* Each country has a *hot spot city* with a heavier traffic load. The outgoing traffic load from a hot spot city is randomly generated from a uniform distribution with mean $2\bar{\lambda}$, while that from an ordinary city has a uniform distribution with mean $\bar{\lambda}$. An outgoing call is connected to a hot spot city and an ordinary city with probabilities $2p$ and p, respectively.

For each value of $\bar{\lambda}$, we generate 1000 random cases and compute the average results as well as the 95% confidence intervals.

Figure 7A and Table II show the performance of the sparse configuration and the two-stage heuristic algorithm for uniformly generated traffic. The sparse configuration gives significantly higher revenue than the existing configuration. For example, when $\bar{\lambda} = 500$, the sparse configuration and the existing configuration give expected revenues of 8830 and $-24{,}259$, respectively (the revenue is negative when the income is smaller than the cost); when $\bar{\lambda} = 1000$, their expected revenues are 25,196 and -102, respectively. It is because the sparse configuration does not operate telephone gateways in the cities of low cost-effectiveness while it can still serve these cities. This can reduce the cost of operating telephone gateways, thereby increasing the expected revenue. These results show that the sparse configuration is more cost-effective than the existing configuration. Figure 7A and Table III also show that the two-stage heuristic algorithm can find close-to-optimal solutions. The average percentage difference from the optimal is only around 1%. These results show that the proposed algorithm is very effective.

Figure 7B and Table III show the performance of the sparse configuration and the two-stage heuristic algorithm for nonuniformly generated traffic. Again, the sparse configuration is more cost-effective than the existing configuration, and the proposed algorithm can find close-to-optimal solutions.

Table IV shows the mean execution time required by the two-stage heuristic algorithm on a personal computer using Intel P4-3.0G. The mean execution time required is only several milliseconds. Therefore, the proposed algorithm is fast

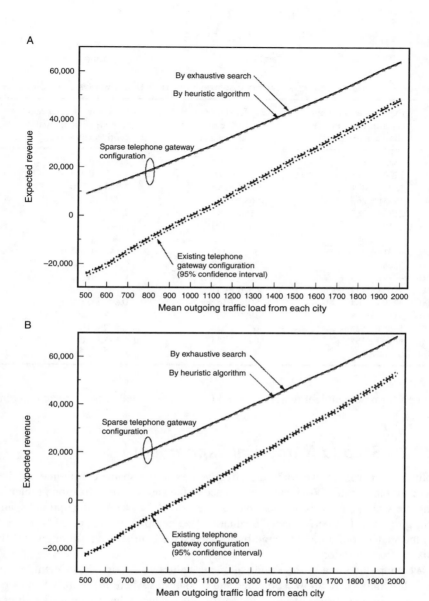

FIG. 7. Performance of the sparse telephone gateway configuration and the two-stage heuristic algorithm in the first numerical experiment; $N=8$. (For the sparse telephone gateway configuration, the exhaustive search algorithm and the heuristic algorithm give overlapping 95% confidence intervals, so these confidence intervals are given in Tables II and III for clarity.) (A) Uniformly generated traffic. (B) Nonuniformly generated traffic.

TABLE II
PERFORMANCE OF THE TWO-STAGE HEURISTIC ALGORITHM; TRAFFIC IS UNIFORMLY GENERATED AND $N=8$

$\bar{\lambda}$	Expected revenue by heuristic algorithm (95% confidence interval)	Expected revenue by exhaustive search (95% confidence interval)	Percentage difference in revenue from optimal (95% confidence interval)
500	8830 ± 194	8961 ± 197	1.343 ± 0.232
600	12003 ± 213	12192 ± 218	1.405 ± 0.223
700	15095 ± 244	15420 ± 252	1.928 ± 0.250
800	18351 ± 269	18682 ± 272	1.723 ± 0.214
900	21758 ± 285	22158 ± 287	1.790 ± 0.215
1000	25197 ± 317	25570 ± 320	1.445 ± 0.177
1100	28699 ± 30	29052 ± 342	1.205 ± 0.166
1200	32716 ± 73	33089 ± 372	1.149 ± 0.156
1300	36668 ± 399	37044 ± 400	1.021 ± 0.138
1400	40291 ± 429	40646 ± 428	0.895 ± 0.123
1500	44058 ± 437	44396 ± 434	0.783 ± 0.112
1600	47694 ± 464	48018 ± 462	0.691 ± 0.102
1700	51612 ± 508	51990 ± 506	0.742 ± 0.101
1800	55751 ± 526	56057 ± 524	0.562 ± 0.080
1900	60109 ± 557	60444 ± 554	0.575 ± 0.081
2000	63868 ± 597	64266 ± 592	0.642 ± 0.087

enough for practical deployment (remind that it is executed in the service planning phase).

2.3.2 Second Numerical Experiment

In this experiment, we study the effectiveness of the sparse configuration when there are many cities. Now, the search space of the gateway optimization problem is extremely large, and it is not possible to exhaustively search for the optimal sparse configurations. For performance evaluation, we compare the expected revenue of (1) the sparse configuration found by the two-stage heuristic algorithm and (2) the existing configuration.

We adopt the same settings and parameter values as the first experiment, except the following: There are $N=50$ cities and 5 countries where there are 10 cities per country. For nonuniformly generated traffic, each country has two hot spot cities.

Figure 8 compares the expected revenue of the sparse configuration and the existing configuration. We see that the sparse configuration gives significantly higher expected revenue than does the existing configuration. For example, when $\bar{\lambda} = 500$ and the traffic is uniformly generated, the sparse configuration and the existing configuration give expected revenues of 61,228 and −149,822,

TABLE III
PERFORMANCE OF THE TWO-STAGE HEURISTIC ALGORITHM; TRAFFIC IS NONUNIFORMLY GENERATED AND $N=8$

$\bar{\lambda}$	Expected revenue by heuristic algorithm (95% confidence interval)	Expected revenue by exhaustive search (95% confidence interval)	Percentage difference in revenue from optimal (95% confidence interval)
500	9817 ± 209	9954 ± 212	1.389 ± 0.287
600	13040 ± 236	13239 ± 242	1.348 ± 0.219
700	16529 ± 264	16863 ± 270	1.852 ± 0.233
800	19975 ± 280	20298 ± 284	1.539 ± 0.198
900	23859 ± 326	24215 ± 329	1.454 ± 0.185
1000	27292 ± 352	27669 ± 353	1.367 ± 0.165
1100	31360 ± 379	31739 ± 376	1.238 ± 0.159
1200	35233 ± 406	35590 ± 404	1.031 ± 0.137
1300	39298 ± 456	39701 ± 450	1.072 ± 0.140
1400	43008 ± 476	43368 ± 472	0.860 ± 0.127
1500	47240 ± 492	47592 ± 489	0.767 ± 0.108
1600	51386 ± 547	51695 ± 543	0.629 ± 0.089
1700	55141 ± 547	55474 ± 545	0.615 ± 0.082
1800	59510 ± 589	59887 ± 589	0.635 ± 0.083
1900	63565 ± 648	63910 ± 646	0.557 ± 0.082
2000	68391 ± 621	68732 ± 620	0.506 ± 0.071

TABLE IV
MEAN EXECUTION TIME REQUIRED BY THE TWO-STAGE HEURISTIC ALGORITHM ON A PERSONAL COMPUTER USING INTEL P4-3.0G IN THE FIRST EXPERIMENT FOR $N=8$

$\bar{\lambda}$	Mean execution time required (ms)	
	Uniformly generated traffic	Nonuniformly generated traffic
500	1.71	3.78
600	3.30	2.66
700	4.09	2.98
800	2.96	2.65
900	2.82	2.19
1000	4.21	2.51
1100	3.08	3.90
1200	4.86	2.65
1300	4.20	2.97
1400	4.65	3.32
1500	5.20	2.31
1600	4.09	5.49
1700	4.84	2.79
1800	5.00	3.60
1900	3.60	3.14
2000	3.60	4.54

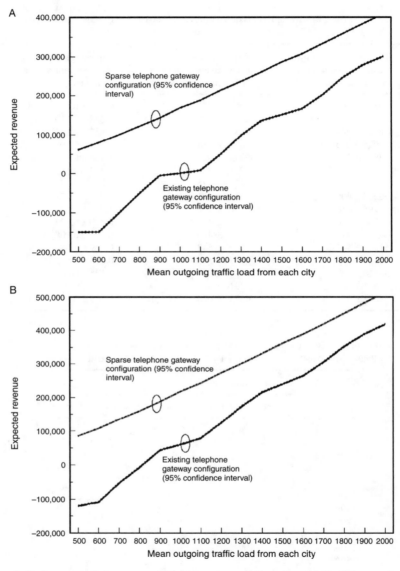

FIG. 8. Performance of the sparse telephone gateway configuration and the two-stage heuristic algorithm in the second numerical experiment; $N=50$. (A) Uniformly generated traffic. (B) Nonuniformly generated traffic.

TABLE V
MEAN EXECUTION TIME REQUIRED BY THE TWO-STAGE HEURISTIC ALGORITHM ON A PERSONAL COMPUTER USING INTEL P4-3.0G IN THE SECOND EXPERIMENT FOR $N=50$

	Mean execution time required (s)	
$\bar{\lambda}$	Uniformly generated traffic	Nonuniformly generated traffic
500	11.51	12.54
600	10.31	11.91
700	13.02	13.01
800	13.45	14.23
900	16.90	13.94
1000	15.22	16.34
1100	12.74	13.90
1200	14.65	15.55
1300	14.43	17.66
1400	15.43	17.92
1500	25.66	20.65
1600	21.77	20.89
1700	19.42	20.51
1800	18.83	21.88
1900	24.20	22.09
2000	21.11	23.34

respectively; when $\bar{\lambda} = 1000$, their expected revenues are 168,932 and 1171, respectively. These results show that the sparse configuration is significantly more cost-effective than the existing configuration.

Table V shows the mean execution time required by the two-stage heuristic algorithm on a personal computer using Intel P4-3.0G. When there are 50 cities, the mean execution time required is within half a minute. These results show that the proposed algorithm is fast enough for practical deployment (remind that it is executed in the service planning phase).

3. Packet Loss Recovery

When voice packets are transmitted through the Internet, they may be lost or become erroneous. The lost or erroneous packets cannot be retransmitted because of the stringent delay requirements for interactive voice, and this affects the resulting voice quality. *Packet loss recovery* is a promising approach to tackle this problem [37]. Its main idea is that the source adds redundant information by which the destination can possibly recover the lost packets. Some packet loss recovery methods have been proposed in the literature, and they produce and utilize the redundancy in different ways for packet loss recovery. In this section, we survey the packet loss recovery methods that are suitable for the phone-to-phone configuration shown in Fig. 1. In particular, we describe piggybacking [37,39,40] in Section 3.1, shared packet loss recovery [19] in Section 3.2, and lightweight piggybacking [20] in Section 3.3.

3.1 Piggybacking

Piggybacking [37,39,40] is a well-known packet loss recovery method for interactive voice communication. It is being adopted by some voice-over-IP applications (e.g., Free Phone [47] and Robust Audio Tools [48]). Piggybacking makes use of the following property of voice: different voice bits have different importance. For example, for each voice sample in a PCM voice stream, the most significant bits are more important than the least significant bits; in the frequency domain, some low-frequency components are more important than some high-frequency components, and so on.

Figure 9A shows an example to illustrate this method. For voice packet i, the source discards the least important bits to produce a small redundant packet i and attaches it to voice packet $i+1$. The resulting packet (i.e., voice packet $i+1$ with redundant packet i) is transmitted to the destination through the Internet. When voice packet i is lost, the destination extracts redundant packet i from voice packet $i+1$ and uses redundant packet i as a substitution of voice packet i. In the example shown in Fig. 9A, voice packet 3 is lost, so the destination extracts redundant packet 3 from voice packet 4 and uses redundant packet 3 as a substitution of voice packet 3.

Piggybacking is relatively simple. Besides, it involves a small recovery delay because the destination just waits for the voice packet which carries the voice bits of the next packetization period. After excluding the communication delay, this waiting time is only equal to two packetization periods. On the other hand, if two or more consecutive packets are lost, piggybacking cannot recover all the lost packets. Figure 9B shows an example in which two consecutive packets are lost and one cannot be recovered.

Skype is a well-known voice-over-IP application [10]. Although Skype is using proprietary protocols, some experimental studies have been carried out in the literature to find out its packet loss recovery scheme [11,17]. It was deduced that Skype is using a variant of piggybacking to send redundant packets for packet loss recovery [11,17].

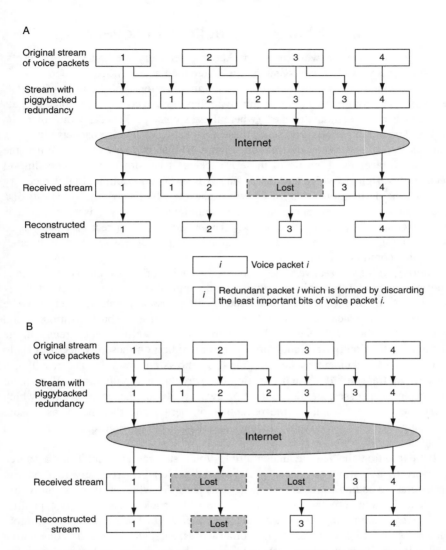

FIG. 9. Piggybacking for packet loss recovery for one voice stream [37,39]. (A) When one voice packet is lost, its low-quality version can be recovered. (B) When two consecutive voice packets are lost, one cannot be recovered.

3.2 Shared Packet Loss Recovery

Erasure coding is a well-known packet loss recovery method (e.g., see Refs. [49–53]). It is very effective for streaming stored media and real-time video, but it is not directly suitable for interactive voice communication. Specifically, when it is used for voice communication, the source applies erasure coding on k voice packets to produce $n-k$ redundant packets (where these k voice packets carry the voice bits of k packetization periods) and sends these n packets (i.e., k voice packets plus $n-k$ redundant packets) to the destination. Figure 10A shows an example. When the destination receives at least k of these n packets, it performs erasure decoding to recover the original k voice packets. Figure 10B shows an example. In this recovery process, the destination waits for the voice packets which carry the voice bits of k packetization periods. After excluding the communication delay, this waiting time is already equal to k packetization periods. This waiting time or recovery delay is so long that erasure coding is not directly suitable for packet loss recovery in interactive voice communication.

Shared packet loss recovery [19] exploits the characteristic of the phone-to-phone configuration such that erasure coding can be applied while the recovery delay is small enough for interactive voice. Specifically, a source gateway gets k voice packets from k voice streams of k telephone calls in each packetization period (i.e., these k voice packets carry the voice bits of k respective voice streams in the same packetization period) and applies erasure coding on these k voice packets to produce $n-k$ redundant packets. Figure 11A shows an example. The resulting n packets are transmitted to the destination gateway through the Internet. When there are any packet losses, the destination gateway waits for the voice packets which carry the voice bits of the same packetization period and then performs erasure decoding in order to possibly recover the lost packets. Figure 11B shows an example.

By exploiting the feature that multiple voice streams are sent from a source gateway to a destination gateway, the shared packet loss recovery method can apply erasure coding for effective packet loss recovery while ensuring small recovery delay for interactive voice. It has more powerful packet loss recovery capability than piggybacking. Specifically, it cannot recover the lost packets only when more than $n-k$ out of n packets are lost. In other words, even when there are consecutive packet losses, it can still recover them as long as the number of the lost packets is not more than $n-k$. On the other hand, its implementation is slightly more complicated than the implementation of piggybacking because it involves erasure coding at the source and erasure decoding at the destination.

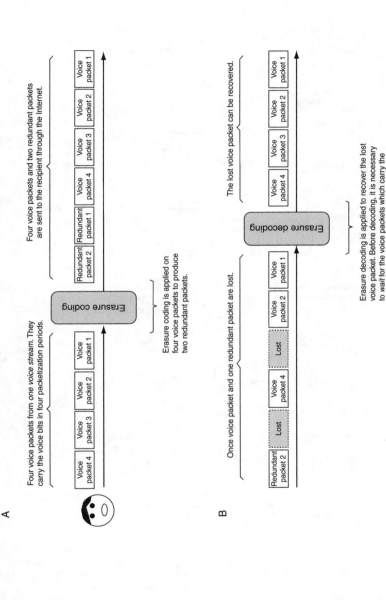

FIG. 10. Erasure coding and decoding for packet loss recovery for one voice stream [53]. (A) At the source, erasure coding is applied on one voice stream. (B) At the destination, erasure decoding is applied to recover the lost packet. It is necessary to wait for the voice packets which carry the voice bits in four packetization periods, so the recovery delay is long.

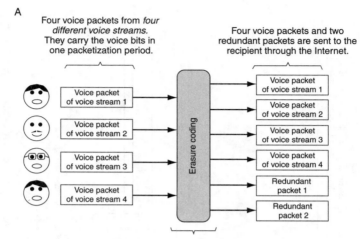

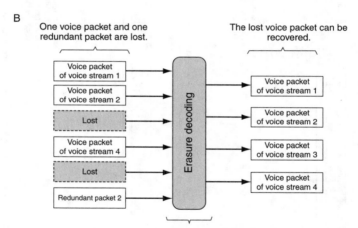

FIG. 11. The shared packet loss recovery method [19]. Using this method, *multiple* voice streams *share* the redundancy for more powerful packet loss recovery, and the recovery delay is small. (A) At the source, erasure coding is applied on the multiple voice packets from *multiple voice streams*. The resulting redundant packets can be shared by multiple voice streams. (B) At the destination, erasure coding is applied to recover the lost packets. It is only necessary to wait for the voice packets which carry the voice bits in one packetization periods, so the recovery delay is small.

3.3 Lightweight Piggybacking

Lightweight piggybacking [20] exploits two main properties: (i) in the phone-to-phone configuration, multiple voice streams are sent from a source gateway to a destination gateway and (ii) different voice bits have different importance. By exploiting these properties, lightweight piggybacking utilizes the resulting redundancy in the best possible way for packet loss recovery. Specifically, it applies erasure coding, fragmentation, and then erasure coding again, such that smaller redundancy is needed, and this redundancy can be better protected and fully shared by multiple voice streams for more effective packet loss recovery. We describe the details in the following subsections:

3.3.1 Lightweight Piggybacking Steps at the Source Gateway

At the source gateway, lightweight piggybacking has four steps: Steps 1 and 2 produce some small redundant packets via erasure coding. Step 3 is the most novel step and involves two operations: (i) fragments the redundant packets into smaller *pieces* in order to reduce the redundancy per voice packet and (ii) performs another stage of erasure coding in order to protect these pieces against loss. Step 4 piggybacks one piece to one voice packet for outgoing transmission. These four steps are described in the following and shown in Fig. 12. We let a source gateway send n voice streams of n respective telephone calls to a destination gateway.

- Step 1: *Discarding the least important bits.* The n voice packets from n respective voice streams in each packetization period form one group. The source gateway discards their least important bits to produce n respective *further-compressed voice packets* (FCVPs) (e.g., if voice is compressed by multilayer coding, the gateway discards the least-important enhancement layers). Let the packet size be reduced by a factor of c. Figure 12A shows the schematic.
- Step 2: *First stage of erasure coding.* The source gateway applies erasure coding on the n FCVPs to produce r *redundant packets* (RPs), where r is a design parameter. Figure 12B shows the schematic.
- Step 3: *Fragmentation and second stage of erasure coding.* The source gateway fragments each RP into p pieces called *lightweight pieces* (LPs), where p is a design parameter. Since r RPs are produced in Step 2, this fragmentation results in pr lightweight pieces. Then, the source gateway performs erasure coding on these pr lightweight pieces to produce $n - pr$

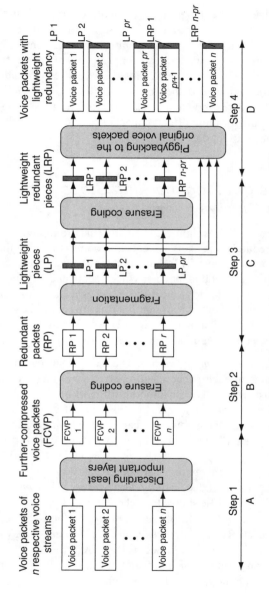

FIG. 12. Lightweight piggybacking steps at the source gateway [20].

PHONE-TO-PHONE CONFIGURATION FOR INTERNET TELEPHONY

lightweight redundant pieces (LRPs). Totally, there are n pieces. Figure 12C shows the schematic.

Step 4: *Piggybacking.* The source gateway piggybacks one piece (either lightweight piece or lightweight redundant piece) to one voice packet. Figure 12D shows the schematic.

After executing the above four steps, the source gateway transmits the resulting voice packets with lightweight redundancy to the destination gateway through the Internet.

3.3.2 Recovery Steps at the Destination Gateway

The destination gateway performs packet loss recovery as follows: It executes the reverse sequence of the corresponding operations performed by the source gateway. For convenience of explanation, we consider the example shown in Fig. 13, in which two out of six packets (of six respective voice streams) are lost, $p=2$, and $r=2$. The recovery steps are as follows:

Step 1: *Erasure Decoding and Reassembly.* The destination gateway extracts two lightweight pieces and two lightweight redundant pieces from the four received packets, performs erasure decoding on these pieces to get four lightweight pieces, and then reassembles these lightweight pieces to form two redundant packets. Figure 13A illustrates this step.

Step 2: *Creating further-compressed voice packets.* The destination gateway extracts four voice packets from the four received packets and discards their least important layers to create four respective further-compressed voice packets. Figure 13B illustrates this step.

Step 3: *Second stage of erasure decoding.* The destination gateway executes erasure decoding on the two redundant packets obtained in Step 1 and the four further-compressed voice packets obtained in Step 2, so that it recovers the two lost packets (further-compressed version). Figure 13C illustrates this step.

3.3.3 Properties of Lightweight Piggybacking

Conditions of successful recovery: The destination gateway can recover the lost packets (further-compressed version) if any one of the following conditions is satisfied:

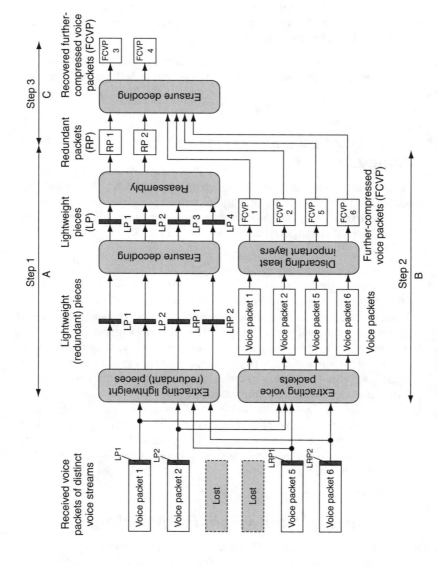

FIG. 13. An example to illustrate the packet loss recovery steps at the destination gateway [20].

(1) The destination gateway receives at least $n-r$ voice packets and pr pieces (lightweight pieces or lightweight redundant pieces). In this case, the destination gateway recovers pr lightweight pieces via erasure decoding, constructs r redundant packets by reassembling these pr lightweight pieces, and recovers all the lost packets (further-compressed version) via another stage of erasure decoding.

(2) For any $r' < r$, the destination gateway receives at least $n - r'$ voice packets and pr' lightweight pieces such that it can reassemble these lightweight pieces to produce r' redundant packets. In this case, the destination gateway recovers all the lost packets (further-compressed version) via erasure decoding on $n - r'$ further-compressed voice packets and r' redundant packets.

More powerful packet loss recovery using smaller redundancy: Lightweight piggybacking applies erasure coding, fragmentation, and then erasure coding. Through fragmentation, the redundancy per voice packet can be reduced (e.g., see Fig. 14). Through two stages of erasure coding, the redundancy can be better protected and fully shared by multiple voice streams of multiple telephone calls for more powerful packet loss recovery. In Section 3.3.4, we will present simulation results to demonstrate these properties.

Small delay in recovery: To perform packet loss recovery, the destination gateway waits for the voice packets which carry the voice bits of the same packetization period (remind that these voice packets come from distinct voice streams of distinct telephone calls, see Figs. 12 and 13). This waiting time is small.

Computation time required: In lightweight piggybacking, erasure coding and decoding are relatively time consuming, but they are still fast enough for practical deployment. Theoretically, for the erasure code described in Reg. [49], the coding and decoding time complexities are $O(nr)$ and $O(nl)$, respectively, where n, r, and l are the numbers of source, redundant, and lost packets, respectively [49]. Experimentally, we conducted an experiment in which the codec was implemented as software and run on a personal computer with an Intel® Xeon™ CPU 3.06 GHz. Table VI shows the coding and decoding time of software implementation of erasure

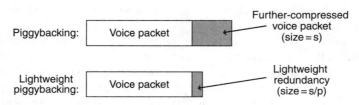

FIG. 14. Lightweight piggybacking needs smaller redundancy than piggybacking.

TABLE VI
CODING AND DECODING TIME OF SOFTWARE IMPLEMENTATION OF ERASURE CODES. THE SOFTWARE CODEC WAS RUN ON A PERSONAL COMPUTER WITH AN INTEL XEON CPU 3.06 GHZ; AND EACH PACKET HAS 256 BYTES

Number of source packets n	Number of redundant packets r	Number of lost packets l	Total encoding time (μs)	Total decoding time (μs)
8	1	1	13	16
	2	2	26	22
	4	4	58	38
	8	8	103	73
16	1	1	23	24
	2	2	46	40
	4	4	93	73
	8	8	187	148
	16	16	376	296
32	1	1	42	49
	2	2	91	81
	4	4	170	157
	8	8	337	291
	16	16	695	616
	32	32	1370	1395
64	1	1	83	97
	2	2	172	162
	4	4	341	294
	8	8	762	590
	16	16	1449	1247
	32	32	2711	2778
	64	64	5406	6798

code. It can be seen that the coding/decoding time has the order of microseconds. For example, when $n=32$, $r=16$, and $l=16$ and each packet has 256 bytes, the coding and decoding time are only 695 and 616 μs, respectively. If the codec is implemented as a hardware chip, the coding/decoding time is even smaller.

Uniform packet size: The voice packets with lightweight redundancy have the same size (see Fig. 12). As a result, the best packet size can be selected/used for the best transmission efficiency.

Erasure coding scheme: Lightweight piggybacking has the following mild requirement on the erasure coding scheme: producing a moderate number of redundant packets from a moderate number of source packets. We suggest adopting the linear block code [49] for lightweight piggybacking for several reasons: (i) it can already fulfill the above requirement, (ii) it is fast, and (iii) it can be easily implemented.

Guidelines for selecting parameter values: Lightweight piggybacking involves three parameters, n, r, and p. The optimal parameter values depend how the service provider would make tradeoffs among three conflicting factors: (i) recovery capability, (ii) amount of redundancy required, and (iii) computation complexity. These parameters have the following properties:

- If n is larger, the redundancy can be better utilized for packet loss recovery because it can be shared by more voice streams. However, the computation complexity is larger because more packets are involved in erasure coding and decoding.
- If r is larger, the recovery capability is better because more redundant packets are added to increase the chance of successful recovery. However, more redundancy is needed because more redundant packets are added and the computation complexity is larger because more packets are involved in erasure coding and decoding.
- If p is larger, the recovery capability is poorer because there are fewer lightweight redundant pieces. However, less redundancy is needed because each lightweight piece or lightweight redundant piece has smaller size.
- n, r, and p must fulfill the constraint $pr \leq n$, which ensures that the number of lightweight redundant pieces produced in Step 3 is nonnegative (see Fig. 12). In principle, pr can be equal to n. In practice, pr should be smaller than n in order to get the advantage of the second stage of erasure coding.

Based on the above properties, we suggest three guidelines for selecting the parameter values:

1. We select n such that n is as large as possible while the resulting computation complexity is acceptable.
2. For effective packet loss recovery, the number of redundant packets added should be larger than the average number of lost packets. Therefore, we suggest selecting r such that $r > n \times$ (packet loss rate).
3. If we want to have higher recovery capability at the expense of more redundancy, we select larger r and smaller p while fulfilling the constraint $pr < n$. On the other hand, if we want to reduce the redundancy required at the expense of lower recovery capability, we select smaller r and larger p while fulfilling $pr < n$.

3.3.4 Simulation Model and Results

Lightweight piggybacking involves interdependent steps, as shown in Figs. 12 and 13. Because of this interdependency, it is very difficult to mathematically analyze the performance of lightweight piggybacking. Therefore, we conduct computer simulation for performance evaluation.

We consider a dependent loss model in which packet loss is modeled by a two-state Markov chain [54]. This model is commonly adopted in the literature because packet losses in the Internet exhibit temporal dependency. We let the average burst length be 3, the number of voice streams n be 60 (i.e., n is large for effective utilization of the redundancy, while the resulting coding/decoding time is acceptably small, as shown in Table VI), the size of a voice packet be one unit, and the size of a FCVP be one third of the original size (e.g., this corresponds to multilayer coding with one base layer and two enhancement layers). For convenience, we let R be the ratio of redundant packets to source packets (i.e., $R = r/n$). For each set of parameter values, the simulation program runs continuously until the 99% confidence interval of the residual packet loss rate is sufficiently small (such that the upper endpoint and the lower endpoint of this 99% confidence interval are, after rounding to three significant digits, equal to each other).

For any packet loss recovery method in any packetization period, the source gateway computes the necessary redundancy for the n voice streams. Then, it multiplexes the resulting packets for transmission to the Internet as follows: it first sends the packet of the first voice stream, then it sends the packet of the second voice stream, and so on.

We adopt four performance metrics for comprehensive evaluation and comparisons:

1. The first metric is the *residual packet loss rate* [37,55,56]. Different packet loss recovery methods have different capabilities of recovering the lost packets. The first metric measures the recovery capability of any given method. It is equal to the packet loss rate after performing packet loss recovery. Mathematically, it is equal to $(Z - Y_1 - Y_2)/Z$, where Z is the number of voice packets sent, Y_1 is the number of voice packets received, and Y_2 is the number of recovered packets (including voice packets and further-compressed voice packets).

2. The second metric is the *total redundancy*. Redundancy is added to enable packet loss recovery. In general, there is a tradeoff between the recovery capability and the redundancy required. The second metric measures the amount of redundancy required. It is equal to the total number of redundant bits required by each group of n voice packets. For convenience, we adopt the following unit for the total redundancy: one unit of redundancy is equal to the number of bits of one FCVP. This unit is simple and convenient because it is independent of the coding scheme, the coding rate, and the duration of the packetization period. The total redundancies of piggybacking, shared packet loss recovery, and lightweight piggybacking can be found to be n, rc, and $r + (n - pr)/p = n/p$, respectively.

3. The third metric is the a*verage number of voice packets and average number of further-compressed voice packets after recovery*. We remind that piggybacking [37,39,40] and lightweight piggybacking can recover the FCVPs, while shared packet loss recovery can recover the voice packets. For a fair and comprehensive comparison, we measure the following two quantities after performing packet loss recovery: (i) the average number of voice packets per group and (ii) the average number of FVCPs per group.
4. The fourth metric is the *computation complexity*. Different packet loss recovery methods have different computation complexities. For comprehensive comparison, we measure the computation complexity of each method. We observe that coding and decoding are relatively time consuming, while the other steps are relatively simple and fast. Therefore, we measure the computational complexity of any packet loss recovery method in terms of its coding complexity and decoding complexity.

Figure 15 shows the performance of lightweight piggybacking and piggybacking where the packet loss rate is small to moderate (up to 0.10). We observe the following results:

1. Lightweight piggybacking gives a different tradeoff between recovery capability and redundancy requirement by using different r and p (as shown in Fig. 15A and B). For example, when ($R=0.15$, $p=5$) and the packet loss rate is 0.10, the residual packet loss rate is 4.64×10^{-2} and the total redundancy is 12; when ($R=0.30$, $p=2$) are used instead, the residual packet loss rate is reduced to 6.65×10^{-3}, but the total redundancy is increased to 30.
2. Lightweight piggybacking has better performance than piggybacking. Specifically, lightweight piggybacking gives lower residual packet loss rate (Fig. 15A) and requires smaller total redundancy (Fig. 15B). For example, when the packet loss rate is 0.10, lightweight piggybacking with ($R=0.30$, $p=2$) gives a residual packet loss rate of 6.65×10^{-3} and requires a total redundancy of 30, while piggybacking gives a residual packet loss rate of 6.67×10^{-2} and requires a total redundancy of 60. In addition, after packet loss recovery, lightweight piggybacking gives the same number of voice packets as but more FCVPs than piggybacking (Fig. 15C). These results show that lightweight piggybacking can better utilize the redundancy to combat losses.

However, Table VII shows that lightweight piggybacking involves higher computational complexity than piggybacking because lightweight piggybacking involves two stages of erasure coding and decoding. Nevertheless, as we discuss in

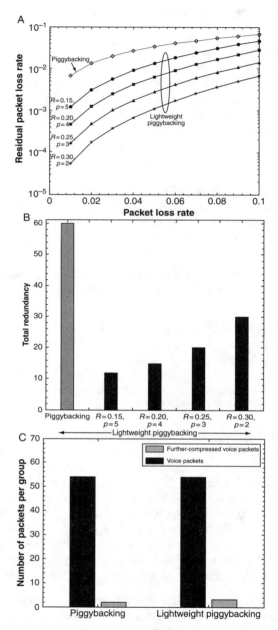

FIG. 15. Comparison between lightweight piggybacking and piggybacking. (A) Residual packet loss rate. (B) Total redundancy. (C) Average number of packets per group after packet loss recovery where the packet loss rate is 0.10 and ($R=0.15$, $p=5$).

TABLE VII
COMPARISON OF COMPUTATIONAL COMPLEXITY (IN TERMS OF THE NUMBER OF STAGES OF CODING AND DECODING). AS TABLE VI SHOWS, ERASURE CODING AND DECODING ARE VERY FAST IN PRACTICE

Packet loss recovery method	Coding complexity	Decoding complexity
Piggybacking [37,39]	Coding is not needed	Decoding is not needed
Shared packet loss recovery [19]	One stage of erasure coding	One stage of erasure decoding
Lightweight piggybacking [20]	Two stages of erasure coding	Two stages of erasure decoding

Section 3.3.3 and Table VI shows, erasure coding and decoding are very fast in practice. Therefore, the additional computational complexity incurred by lightweight piggybacking is small in practice.

Figure 16 shows the performance of lightweight piggybacking and shared packet loss recovery where the packet loss rate is small to moderate (up to 0.10). Compared with shared packet loss recovery, lightweight piggybacking offers favorable tradeoff: it gives smaller residual packet loss rate, requires smaller total redundancy, and gives more packets, but a few of these packets are FCVPs. For example, when the packet loss rate is 0.10 and ($R=0.15$, $p=5$), lightweight piggybacking gives 54.0 voice packets and 3.2 FCVPs in each group (i.e., the total number of packets is 57.2), while shared packet loss recovery gives 56.8 voice packets, but lightweight piggybacking reduces the total redundancy from 27 to 12 and the residual packet loss rate from 5.33 to 4.64×10^{-2}. Therefore, lightweight piggybacking offers favorable tradeoff and better utilizes the redundancy to combat losses. On the other hand, Table VII shows that lightweight piggybacking involves one additional stage of erasure coding/decoding than shared packet loss recovery. Since erasure coding and decoding are very fast in practice (see the discussion in Section 3.3.3 and the results in Table VI), the additional computational complexity incurred by lightweight piggybacking is small in practice

Figure 17 shows the performance of lightweight piggybacking, piggybacking, and shared packet loss recovery where the packet loss rate is high (up to 0.25, i.e., the loss is up to one quarter). We observe very similar properties under high packet loss rate: (i) lightweight piggybacking has better performance than piggybacking, and (ii) compared with shared packet loss recovery, lightweight piggybacking offers favorable tradeoff (i.e., it results in a small percentage of FCVPs, but it can significantly reduce the residual packet loss rate and/or the total redundancy).

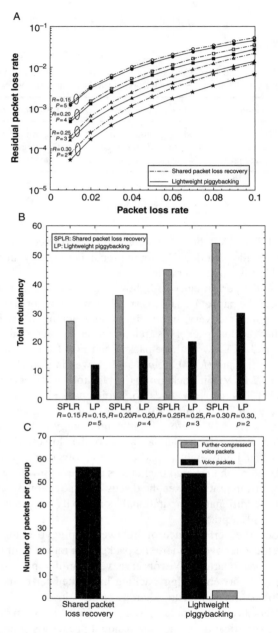

FIG. 16. Comparison between lightweight piggybacking and shared packet loss recovery. (A) Residual packet loss rate. (B) Total redundancy. (C) Average number of packets per group after packet loss recovery where the packet loss rate is 0.10 and ($R=0.15$, $p=5$).

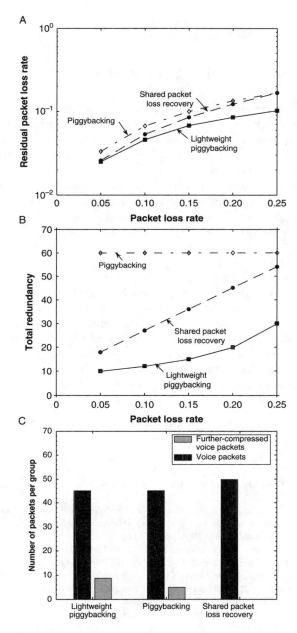

FIG. 17. Performance under high packet loss rate. The design parameters, r and p (where $r = nR$), are selected based on the guidelines in Section 3.3.3, such that (R, p) are (0.10, 6), (0.15, 5), (0.20, 4), (0.25, 3), and (0.30, 2) when packet loss rates are 0.05, 0.10, 0.15, 0.20, and 0.25, respectively. (A) Residual packet loss rate. (B) Total redundancy. (C) Average number of packets per group after packet loss recovery where the packet loss rate is 0.25.

4. Summary

The phone-to-phone configuration is promising for Internet telephony because users can use telephones or mobile phones to make long-distance calls through the Internet. We considered two major issues involved in this configuration: (i) how to cost-effectively achieve good service coverage and (ii) how to improve the voice quality via packet loss recovery.

To address the first issue, we described the sparse telephone gateway configuration. This configuration can provide service to many cities cost-effectively, so that the resulting service is attractive (because of good service coverage) and competitive (because of lower operating cost). We described how to optimize this configuration to achieve the best cost-effectiveness, and presented numerical results to demonstrate its cost-effectiveness.

To address the second issue, we described three packet loss recovery methods (namely, piggybacking, shared packet loss recovery, and lightweight piggybacking). In particular, the latter two methods exploit the following feature of the phone-to-phone configuration: multiple voice streams are sent from a source gateway to a destination gateway. As a result, the redundancy can be shared among multiple voice streams for more effective packet loss recovery. We presented numerical results to demonstrate that lightweight piggybacking is the most promising, because it uses smaller redundancy to give better recovery capability.

REFERENCES

[1] S. Karapantazis, F.N. Pavlidou, VoIP: a comprehensive survey on a promising technology, Comput. Netw. 53 (12) (2009) 2050–2090.
[2] M. Maresca, N. Zingirian, P. Baglietto, Internet protocol support for telephony, Proc. IEEE 92 (9) (2004) 1463–1477.
[3] H.M. Chong, H.S. Matthews, Comparative analysis of traditional telephone and voice-over-Internet protocol (VoIP) Systems, in: Proceedings of the IEEE International Symposium on Electronics and the Environment, Scottsdale, USA, 2004, pp. 106–111.
[4] IEEE Internet Computing, Special Issue on Internet Telephony 6(3) (May/June 2002).
[5] B. Goode, Voice over Internet protocol (VoIP), Proc. IEEE 90 (9) (2002) 1495–1517.
[6] R.H. Glitho, F. Khendek, A. De Marco, Creating value added services in Internet telephony: an overview and a case study on a high-level service creation environment, IEEE Trans. Syst. Man Cybern. C 33 (4) (2003) 446–457.
[7] Residential voice over IP subscribers to grow 7-fold to 267 million by 2012. Metrics 2.0—Business and Market Intelligence, February 2007. Available from: http://www.metrics2.com/blog/2007/02/02/residential_voice_over_ip_subscribers_to_grow_7fol.html.
[8] The mobile Internet report. Morgan Stanley Research Report, December 2009. Available from: http://www.morganstanley.com/institutional/techresearch/mobile_internet_report122009.html.

[9] Enterprise VoIP market trend 2009–2012. Osterman Research Report (2009). Available from: http://www.ostermanresearch.com/execsum/or_voip2009execsum.pdf.
[10] Skype. Available from: http://www.skype.com.
[11] T.Y. Huang, P. Huang, K.T. Chen, P.J. Wang, Could Skype be more satisfying? A QoE-centric study of the FEC mechanism in an Internet-scale VoIP system, IEEE Netw. 24 (2) (2010) 42–48.
[12] D. Rossi, M. Mellia, M. Meo, Understanding Skype signaling, Comput. Netw. 53 (2) (2009) 130–140.
[13] L.D. Cicco, S. Mascolo, A mathematical model of the Skype VoIP congestion control algorithm, IEEE Trans. Automatic Control 55 (3) (2010) 790–795.
[14] S.A. Baset, H.G. Schulzrinne, An analysis of the Skype peer-to-peer Internet telephony protocol, in: Proc. IEEE INFOCOM, Spain, 2006.
[15] D. Bonfiglio, M. Mellia, M. Meo, D. Rossi, Detailed analysis of Skype traffic, IEEE Trans. Multimedia 11 (1) (2009) 117–127.
[16] H. Sengar, Z. Ren, H. Wang, D. Wijesekera, S. Jajodia, Tracking Skype VoIP calls over the Internet, in: Proc. IEEE INFOCOM, California, 2010.
[17] B. Sat, B.W. Wah, Analysis and evaluation of the Skype and Google-Talk VoIP systems, in: Proc. IEEE Int. Conf. Multimedia and Expo, Toronto, 2006, pp. 2153–2156.
[18] Internet usage statistics: the Internet big picture. Available from: http://www.internetworldstats.com/stats.htm, 2003 (accessed 30-06-2010).
[19] Y.W. Leung, Shared packet loss recovery for Internet telephony, IEEE Commun. Lett. 9 (1) (2005) 84–86.
[20] W.Y. Chow, Y.W. Leung, Lightweight piggybacking for packet loss recovery in Internet telephony, in: Proceedings of the IEEE International Conference on Communications, Glasgow, UK, 2007, pp. 1809–1814 (Revised version is under review by a journal).
[21] T. Dagiuklas, P. Galiotos, Architecture and design of an enhanced H.323 VoIP gateway, in: Proceedings of the IEEE International Conference on Communications, New York, USA, April, 2002.
[22] A.S. Amin, H.M. El-Sheikh, Scalable VoIP gateway, in: Proceedings of the 9th International Conference on Advanced Communication Technology, 2007, pp. 1606–1612.
[23] H. Kim, M.J. Chae, I. Kang, The methods and the feasibility of frame grouping in Internet telephony, IEICE Trans. Commun. 85 (1) (2002) 173–182.
[24] H. Kim, I. Kang, E. Hwang, Measurement-based multi-call voice frame grouping in Internet telephony, IEEE Commun. Lett. 6 (5) (2002) 199–201.
[25] H.P. Sze, S.C. Liew, J.Y.B. Lee, D.C.S. Yip, A multiplexing scheme for H.323 voice-over-IP applications, IEEE J. Sel. Areas Commun. 20 (7) (2002) 1360–1368.
[26] J. Kim, H. Lee, W. Ryu, B. Lee, M. Hahn, An efficient shared adaptive packet loss concealment scheme through 1-port gateway system for Internet telephony service, IEICE Trans. Commun. 91 (5) (2008) 1370–1374.
[27] Y.W. Leung, Dynamic bandwidth allocation for Internet telephony, Comput. Commun. 29 (18) (2006) 3710–3717.
[28] M. Benini, S. Sicari, Assessing the risk of intercepting VoIP calls, Comput. Netw. 52 (12) (2008) 2432–2446.
[29] A. Ghafarian, R. Draughorne, S. Hargraves, S. Grainger, S. High, C. Jackson, Securing voice over Internet protocol, J. Inf. Assur. Secur. 2 (2007) 200–204.
[30] D.H. Lee, J.G. Kim, Voice over Internet protocol gateway and a method for controlling the same, U.S. Patent, Patent No: US 7, 376, 124 B2, 20 May 2008.
[31] H. Hagirahim, F. Waldman, Method and apparatus for providing efficient VoIP gateway-to-gateway communication, U.S. Patent, Patent No.: US 7, 330, 460 B1, 12 February 2008.

[32] A.S. Amin, H.M. El-Sheikh, Traffic control mechanism for VoIP gateway, in: Proceedings of the IEEE International Conference on Signal Processing and Communications, 2007, pp. 1235–1238.
[33] M. Jo, H. Kim, H. Kim, An adaptive routing method for VoIP gateways based on packet delay information, IEICE Trans. Commun. E88-B (2) (2005) 766–769.
[34] Gizmo5. Available from: http://gizmo5.com.
[35] N. Beijar, Signaling protocols for Internet telephony, Laboratory of Telecommunications Technology, Helsinki University of Technology, 1998. http://citeseerx.ist.psu.edu/viewdoc/download?doi=10.1.1.140.1115&rep=rep1&type=pdf.
[36] J. Glasmann, W. Kellerer, H. Muller, Service architectures in H.323 and SIP: a comparison, IEEE Commun. Surv. 5 (2) (2003) 32–47.
[37] C. Perkins, O. Hodson, V. Hardman, A survey of packet loss recovery techniques for streaming audio, IEEE Netw. 12 (5) (1998) 40–48.
[38] Y.W. Leung, Sparse telephone gateway for Internet telephony, Comput. Netw. 54 (1) (2010) 150–164.
[39] J.K. Kurose, K.W. Ross, Computer Networking: A Top-Down Approach, fourth ed., 2008, pp. 640-642.
[40] V. Hardman, A. Sasse, M. Handley, A. Watson, Reliable audio for use over the Internet, in: Proceedings of INET, Hawaii, 1995.
[41] Z. Drezner, H.W. Hamacher (Eds.), Facility Location: Applications and Theory, Springer, 2002.
[42] F.S. Hillier, G.J. Lieberman, Introduction to Operations Research, eighth ed., McGraw Hill, 2005.
[43] P. Zhou, B.S. Manoj, R. Rao, A gateway placement algorithm in wireless mesh networks, in: Proceedings of the 3rd International Conference on Wireless Internet, Texas, 2007.
[44] R. Mathar, T. Niessen, Optimum positioning of base stations for cellular radio networks, Wireless Netw. 6 (6) (2000) 421–428.
[45] F. Thouin, M. Coates, in: Video-on-demand server selection and placement, in: Proceedings of the 20th International Teletraffic Congress, Ottawa, 2007, pp. 18–29.
[46] M.H. Garey, D.S. Johnson, Computers and Intractability: A Guide to the Theory of NP-Completeness, W. H. Freeman & Company, 1979.
[47] Free Phone. Available from: http://www-sop.inria.fr/rodeo/fphone/.
[48] Robust Audio Tools. Available from: http://www-mice.cs.ucl.ac.uk/multimedia/software/rat.
[49] L. Rizzo, Effective erasure codes for reliable computer communication protocols, ACM Comput. Commun. Rev. 27 (2) (1997) 24–36.
[50] S.W. Yuk, M.G. Kang, B.C. Shin, D.H. Cho, An adaptive redundancy control method for erasure-code based real-time data transmission over the Internet, IEEE Trans. Multimedia 3 (3) (2001) 366–374.
[51] F. Zhai, Y. Eisenberg, T.N. Pappas, R. Berry, A.K. Katsaggelos, Rate-distortion optimized hybrid error control for packetized video communications, IEEE Trans. Image Process. 15 (2006) 40–53.
[52] S.H.G. Chan, X. Zheng, Q. Zhang, W.W. Zhu, Y.Q. Zhang, Video loss recovery with FEC and stream replication, IEEE Trans. Multimedia 8 (2) (2006) 370–381.
[53] G. M. Schuster, J. Mahler, I. Sidhu, M. Borella, Forward error correction system for packet based real time media, United States Patent, Patent No. 6, 145, 109, 7 November 2000.
[54] X. Yang, C. Zhu, Z.G. Li, X. Lin, N. Ling, An unequal packet loss resilience scheme for video over the Internet, IEEE Trans. Multimedia 7 (4) (2005) 753–765.
[55] V.R. Gandikota, B.R. Tamma, C.S.R. Murthy, Adaptive FEC-based packet loss resilience scheme for supporting voice communication over ad hoc wireless networks, IEEE Trans. Mobile Comput. 7 (10) (2008) 1184–1199.
[56] H.P. Sze, S.C. Liew, Y.B. Lee, A packet-loss-recovery scheme for continuous-media streaming over the Internet, IEEE Commun. Lett. 5 (3) (2001) 116–118.

SLAM for Pedestrians and Ultrasonic Landmarks in Emergency Response Scenarios[☆]

CARL FISCHER

School of Computing and Communications, Lancaster University, Lancaster, Lancashire, United Kingdom

KAVITHA MUTHUKRISHNAN

Department of Software Technology, Delft University of Technology, Delft, Netherlands

MIKE HAZAS

School of Computing and Communications, Lancaster University, Lancaster, Lancashire, United Kingdom

Abstract

Providing *ad hoc* solutions for positioning and tracking of emergency response teams is an important and safety-critical challenge. Although solutions based on inertial sensing systems are promising, they are subject to drift. We address the problem of positional drift by having the responders themselves deploy sensor nodes capable of sensing range and angle-of-arrival, as they progress into an unknown environment. Our research focuses on a sensor network approach that does not rely on preexisting infrastructure. This chapter targets two important aspects of such a solution: how to locate the deployed static sensor nodes, and

---

[☆] Portions of the state-of-the-art and literature review were previously published in a survey article [1] in *IEEE Pervasive Computing Magazine*, and the algorithms and analysis are based on initial research reported in Kavitha Muthukrishnan's Ph.D. dissertation [2, Chapter 6].

how to track the responders by using a combination of ultrasound and inertial measurements. The main contributions of this chapter are: (i) a characterization of the errors encountered in inertial-based pedestrian dead-reckoning as well as ultrasound range and bearing measurements in a *mobile* setting, (ii) the formulation of an extended Kalman filter for simultaneously locating sensor nodes and tracking a pedestrian using a combination of ultrasound range/bearing measurements and inertial measurements, and (iii) the validation of the presented algorithms using data collected from real deployments.

1. Localization for Emergency Responders: State of the Art 104
　1.1.　Application Overview 105
　1.2.　Existing Technologies and Systems 109
　1.3.　Simultaneous Localization and Mapping 111
　1.4.　Multimodal Pedestrian SLAM 115
2. Implementation: Multimodal Sensing and Algorithms 116
　2.1.　Sensing Technologies 116
　2.2.　Localization and Mapping Algorithms 121
3. System Evaluation 128
　3.1.　Description of Experiments 128
　3.2.　Results and Analysis 133
　3.3.　Conclusions 150
　3.4.　Improvements and Future Work 152
4. Conclusion 157
　References 158

1. Localization for Emergency Responders: State of the Art

Over the past decade, many researchers have attempted to use embedded sensing and computation to improve the safety and effectiveness of emergency responders. Despite recent progress in location systems and wearable computing, responders continue to use mostly traditional methods such as ropes and radios and suffer from their limitations. Wearable sensing and computing are promising technologies for providing location and navigation support but to date few systems are robust enough to be deployed in the field.

In this section, we look at some of the problems faced by firefighters and other search and rescue workers operating indoors, and we examine the current practices and their limitations. We give an overview of the proposed sensing and algorithms that have been explored in research projects in this area. Finally, we conclude this section with a brief review of simultaneous localization and mapping (SLAM), since the new method we present in Section 1.2.4 employs such techniques.

1.1 Application Overview

Although localization is becoming available for the general public and for businesses via widespread use of GPS receivers and commercial indoor location systems (e.g., Ubisense, Sonitor, and Ekahau), many solutions are not suitable for use by emergency responders such as firefighters. The conditions they work in are significantly more demanding than nonemergency environments. Darkness, smoke, fire, power cuts, water, and noise can all prevent a location system from working, and heavy protective clothing, gloves, and facemasks make using a standard mobile computer impossible. In the past decade, much research effort has been put into this challenging problem and a wide variety of ideas have been developed.

1.1.1 Scenarios

Location and navigation support is useful in many everyday situations but essential in emergency response scenarios. Teams need to be able to reach safety quickly if conditions become too dangerous, and the incident commander needs to keep track of where teams are. The simple task of finding one's way in a building becomes a challenge when there is little or no visibility due to smoke and power failure. The high levels of mental and physical stress add to the difficulty. Getting lost in a burning or collapsing building can have fatal consequences for both the rescue personnel and the evacuees as oxygen supplies run out and medical attention is delayed.

A report by the National Fire Protection Association (NFPA) [3] in the United States identifies "lost inside" as a major cause of traumatic injuries to firefighters. Reports by the National Institute for Occupational Safety and Health (NIOSH) [4] also reveal that disorientation and failure to locate victims are contributing factors to firefighter deaths.

In case of a sudden increase in temperature, a firefighter may only have seconds to reach safety. They need to find the exit as fast as possible. In some cases, they may not be able to retreat along the same path due to a collapsed ceiling or floor. Alternative exits may be available but not clearly visible. When a firefighter radios a distress call because they are trapped or when someone fails to report back to the command post, the rescue team must be able to locate them. Even when situations are not immediately life-threatening, precious time can be wasted by searching the same room twice or failing to search another. The incident commander also needs to know elements of the building layout, where the team members are, and which parts of the building have been searched.

Several recurring recommendations from the NIOSH reports [4] explicitly highlight the need for a navigation and tracking system, and suggest some solutions[1]:

- "train firefighters on actions to take if they become trapped or disoriented inside a burning structure";
- "consider using exit locators such as high intensity floodlights, flashing strobe lights, hose markings, or safety ropes to guide lost or disoriented firefighters to the exit";
- "ensure that the Incident Commander receives pertinent information (i.e., location of stairs, number of occupants in the structure, etc.) from occupants on scene and information is relayed to crews during size-up";
- "working in large structures (high rise buildings, warehouses, and supermarkets) requires that firefighters be cognizant of the distance traveled and the time required to reach the point of suppression activity from the point of entry."
- "conduct research into refining existing and developing new technology to track the movement of firefighters inside structures";

In addition to the location and navigation requirements, other reports emphasize the need for reliable communication of interior conditions to the incident commander and for monitoring building stability. Temperature, smoke, sounds, and vibrations are all indicators of the progression of the fire and the stability of the building.

1.1.2 Current Practices

Firefighters have developed their own specific navigation practices for use in poor visibility. Details vary but overall the same ideas are used worldwide. The methods tend to be simple and practical, and the equipment is seemingly low-tech and very robust.

Following a hose is a simple method to find the exit through a dark or smokey building. If no hose is available, firefighters may use dedicated ropes called *lifelines* which connect them to a point outside the dangerous area. The other end can be left

---

[1] See in particular reports F2007-18, F2006-19, and F2008-09.

attached deep inside the building if a new team comes in to continue the search [5]. Additional lines may be attached to rings on the main lifeline to allow several firefighters to branch off in different directions, while remaining physically linked to the rest of their team. A series of knots on the main lifeline helps firefighters determine the direction and distance to the exit and can be used as reference points when radioing positions to the commander [6]. A flashlight left in the doorway of a room helps locate the exit and indicates to colleagues that the room is currently being searched, and a chalk mark on the door indicates that a room has already been searched [5,6]. Teams returning from a search mission sketch the layout of the building to assist the commander and any further teams.

All firefighters entering hazardous areas wear a Personal Alert Safety System (PASS) device attached to their breathing apparatus [7] (as cited by Ref. [8]). The PASS device sounds an alarm, if the firefighter does not move for a short time. At a fire scene, the sound of a PASS alarm is a signal that a firefighter is in distress. By following the sound, the rescue team can locate that firefighter. While not strictly a navigation tool, thermal imaging cameras can also be used for finding people and seeing walls, doorways, and windows when vision is obscured.

Many firefighters are trained to search a dark room while keeping either their left or right hand in contact with the wall. This helps with orientation and provides a strategy for systematically exploring an unknown space [9].

Human contact and accountability are also essential. Searches are always performed in teams of at least two members who should avoid being separated [10]. During a lifeline search, one team member may remain at a fixed position to help with orientation and provide progress reports while their colleagues search further. Locations are reported as accurately as possible over the radio to the commander outside the building who keeps track of team locations on a whiteboard.

The Pathfinder system produced by SummitSafety[2] consists of a handheld tracker and beacons which transmit powerful ultrasound pulses. Firefighters can use the tracker to locate a beacon placed at the exit, while rescue teams can use it to locate a beacon with a different frequency worn by a firefighter in distress. Ultrasound waves are blocked by walls but will find a path around corners and under doors; this path can be followed by firefighters. Smoke, heat, humidity, and audible sounds from the fire do not interfere with the ultrasonic waves and a directional receiver for ultrasound is a lot smaller than for audible sound. The tracker displays the amplitude of the detected signal on a bar graph so a firefighter can locate the direction of a beacon by scanning a 360° circle.

---

[2] http://www.summitsafetyinc.com

1.1.3 Constraints and Limitations

Although these methods become more effective with training, they are practical and simple to understand. However, they sometimes fail. A lifeline may become tangled in furniture, a flashlight may be buried under debris, and the temperature of the environment may make a thermal imaging camera unusable. But the principles behind these methods are familiar—the physical properties of a rope, the propagation of light, even the principle of thermal imaging. Failure is understood and even expected in certain conditions. The left hand method for finding an exit can also be misleading and a person can find themselves walking in circles around a large pillar or repeatedly visiting two or three rooms connected by several doors. These techniques are used to aid and support navigation rather than impose an inflexible method. Human error can occur especially during complex and prolonged incidents. Simple techniques such as taking notes (for the commander) or following a rope (for the search teams) are designed to reduce the mental load. As pointed out in the NIOSH reports, many improvements can be made by following procedures and through adequate training. But localization, sensing, and communication are all areas where embedded computers, body-worn sensors, and wireless sensor nodes could play a role if they can be adapted to the harsh conditions.

Navigation by sight is impossible when darkness, smoke, or dust limit visibility to less than an arm's length. Persons or objects that are out of reach can be passed unnoticed. The environment can change as ceilings, floors, or shelves collapse, as furniture is moved and doors are opened or closed by people searching for an exit. The noise of the fire can mask PASS alarms, interfere with radio conversations, and make cries for help difficult to locate.

High-tech systems are generally not adapted to these conditions. Propagation of radio, ultrasound, and laser signals typically used for location is hindered by high temperatures, thick smoke, noise, gusts of air, obstacles, and falling debris. A report by the City of Phoenix Fire Department [11] analyses problems with radio communications inside buildings and identifies unreliable radio links as the cause of several injuries. Sensors deployed in the environment may be kicked, fall through the floor, or be buried. Firefighters may crawl or walk in unusual patterns, and body-worn sensors may lie at odd angles. In addition, there is the issue of presenting the right amount of information to the firefighter in an accessible way and ensuring that devices can be used in the dark with gloves. Finally, the casing and electronics of all devices must be made as robust as possible in the same way as PASS devices and radios to withstand rough handling and very high temperatures [8].

The FIRE project at UC Berkeley reports on some of the major difficulties in designing high-tech location systems for the emergency services [12]. Reliability is more important than high resolution or fast updates. Consistent room-level locations every 20 s are deemed more useful than finer resolution updates with higher probability of error. And the firefighters must be able to customize and service the equipment themselves to some extent. All this is key to acceptance of new technologies.

1.2 Existing Technologies and Systems

1.2.1 SmokeNet

Researchers at UC Berkeley have developed SmokeNet [13], a preinstalled sensor network which tracks firefighters in a multistorey building. Sensor nodes installed in each room and approximately every 10 m along corridors provide room-scale location accuracy. Additional sensor nodes monitor smoke and temperature, and relay data to the command post. Color-coded LEDs show occupants which escape routes are safe. The Fire-Eye display mounted inside each firefighter's face mask displays a floorplan and short text messages from the command post. The incident commander uses the electronic Incident Command System to see the locations and health status of firefighters, and the status of the smoke detectors. At Carnegie Mellon University, researchers have used robots to autonomously map the positions of radio [14] or ultrasound [15] beacons. This map can then be used to monitor the progress of a fire and to track firefighters in a similar way. The method developed by Djugash et al. [15] is very similar to the method we describe in this chapter.

1.2.2 LifeNet

The LifeNet concept developed by Klann [5] is designed to provide the functionality of the traditional lifeline. It consists of beacons and a wearable device that senses nearby beacons and shows navigational guidance on a head-mounted display. The beacons are dropped automatically at appropriate intervals from a device attached to the firefighter's breathing apparatus and form a trail of "breadcrumbs." Each beacon acts as a waypoint to guide the firefighter in either direction. Trails deployed by different firefighters combine to offer alternative escape routes, and loops create shortcuts instead of becoming a trap. The challenge is to present concise and clear information to the firefighters despite the inaccuracies in detecting the direction of the beacons.

1.2.3 Relate Trails

The Relate Trails project [16] provides navigation assistance by displaying an arrow on a head-mounted display to help a person retrace their path. Ultrasound beacons are dropped on the way in, and then used to correct pedestrian dead reckoning (PDR) position and direction estimates on the way out. Absolute positions may be inaccurate due to PDR drift over long distances but navigation only relies on the position of the user relative to the closest beacons. The use of PDR in addition to beacons allows the system to function to some extent even if beacons are destroyed or out of range.

1.2.4 Dräger Patent

Dräger Safety owns a patent covering a "device and process for guiding a person along a path traveled" [17]. Their system ejects transponders such as RFID tags as a person walks along a path. These transponders are detected by a portable receiver carried by the person and the direction to the nearest transponder is communicated via a visual or audio display. The system may use GPS to determine the absolute location of each transponder as it is dropped. The transponders can store an identification code and a sequence code that allow the system to cope with multiple users and loops. The transponders are carried in a container attached below the air canister. The ejection mechanism automatically drops a transponder at constant time intervals, at constant distance intervals, or based on the amplitude of the signal.

1.2.5 Precision Personnel Location System

The Precision Personnel Location (PPL) system [18] developed at the Worcester Polytechnic Institute uses RF receivers at fixed locations on emergency response vehicles outside the building to track the 3D position of personnel carrying a transmitter. The RF signals can be used alone to estimate location or they can be used to correct drift in dead-reckoned positions. The dead-reckoning is particularly useful in larger buildings where the RF position estimates are less accurate.

1.2.6 Thales Indoor Positioning System

The indoor positioning system [19] developed by Thales works similarly to GPS. Firetrucks parked around a building are located using standard GPS. Using ultrawide band (UWB) radio signals, firefighters inside the building can be located by means of time of arrival measurements. The system takes the GPS fixes of at least two firetrucks outside and their positions in the *ad hoc* network, and uses this information to calculate the location of the firefighters. The result is that the GPS signals aid position fixing, but do not need to penetrate the burning building.

1.2.7 Flipside RFID

A team from the National Institute of Standards and Technology (NIST) investigated how predeployed RFID tags embedded in the building could be used to correct PDR [20]. They call this the *flipside* of RFID because, unlike typical RFID systems, the tags are static and the mobile reader is worn by the firefighters. The range of the reader and the distance between tags are the key parameters. A long range will only give approximate locations but a short range will miss tags.

1.2.8 Map Matching with RFID

The drift in PDR position estimates can be corrected by using information from floorplans when these are available. A team from EPFL asks the first team of firefighters to identify doorways by placing an RFID tag on the frame as they pass through [21]. As each tag is placed, the location system adjusts the PDR position estimate based on the position of the nearest doorway on the floorplan. The system corrects the orientation estimate based on the direction in which the doorway must be crossed. Following teams wear an RFID reader which detects the tags deployed by the first team so that the system can correct the position and orientation estimates in the same way.

1.2.9 Map Matching with Particle Filters

Researchers from the WearIT@Work project also use floorplans to ensure that successive PDR position estimates do not pass through walls [22]. A particle filter keeps track of thousands of different position and orientation estimates (the particles) and each one is weighted according to how well it fits with the measurements from a wearable inertial sensor. Particles that pass through walls are eliminated and replaced by plausible ones. The map filtering method works with building outlines but benefits from more detailed floorplans. Woodman and Harle at the University of Cambridge use maps which also include vertical positions [23] to represent stairs. Their particle filter uses these 2.5-dimensional maps to track locations over several floors and improve estimates even further.

1.3 Simultaneous Localization and Mapping

The motivation for SLAM is to overcome the need for an *a priori* map and preinstalled infrastructure, and to enable mapping that is both extensible and adaptive to a changing environment. SLAM has been extensively used in the field of robotics [15,24] and has been specifically applied to a variety of environments such as indoor [25], outdoor [26], aerial [27], and undersea [28,29]. In this section, we report briefly some of the work developed within the context of robotic SLAM and how the interest over years has been shifted to SLAM for pedestrians. For a more comprehensive review of SLAM, the reader is referred to Thrun [30].

1.3.1 SLAM in Robotics

The basic setting for the SLAM problem is that a robot with a known kinematic model starts at an unknown location, and moves through an environment containing landmarks. The robot is equipped with sensors that can take measurements of the relative location between itself and any individual landmark. SLAM is performed by storing landmarks in a map, as they are observed by the sensors present in the robots, using the robot pose estimate (i.e., position and orientation) to determine the landmarks' locations (relative to the robot). At the same time, the algorithm utilizes these landmarks to improve the robot's pose estimate. As the landmarks are repeatedly observed, the confidence of their location estimates increases, and the map converges.

Some of the notable problems that are widely researched in SLAM community are: (i) *complexity*—of the SLAM methods, which is compounded by the number of landmarks in the map and (ii) *data association*—to correctly associate observations of landmarks with landmarks held in the map. This is particularly important for "loop closure," when a robot returns to a previously mapped region after traversing a long path. These problems are being addressed in detail within robotics research [31,32].

1.3.2 Approaches

SLAM methods fall mainly in three categories: (i) EKF-SLAM, which employs an extended Kalman filter (EKF) to represent the joint state space of robot pose and all landmarks that have been identified; (ii) FastSLAM, which uses a Rao-Blackwellized particle filter in which each particle effectively represents a pose and a set of independent compact EKFs for each landmark; and (iii) GraphSLAM, which models the landmarks and the successive positions of the target as nodes of a graph, and measurements as edges of this graph.

Although the EKF approach can be considered as the de facto standard in SLAM, it has two serious drawbacks that prevent it being applicable to large-scale real environments. First, quadratic complexity limits the number of landmarks that can be handled by this approach. Second, it relies heavily upon the assumption that the mapping between observations and landmarks is known. Associating a small number of observations with incorrect landmarks in the EKF can cause the filter to diverge. This shortcoming has been recognized and investigated by the community [33,34]. The computational effort has been reduced by using submapping methods—splitting the global map into a number of submaps [34] and by using sparse information matrices instead of covariance matrices [35].

FastSLAM was introduced by Montemerlo et al. [36] as an efficient SLAM algorithm. Basically, FastSLAM decomposes the SLAM problem into robot localization problem and landmark estimation problem that are conditioned on the robot's pose estimate. FastSLAM uses a particle filter for estimation. Each particle effectively represents a pose and a set of independent low-dimensional EKFs for every landmark. The conditioning on a pose allows the landmarks to be estimated independently, thus lowering the complexity. Further research by Montemerlo et al. [37] showed that FastSLAM is able to deal with ambiguous data association more reliably than EKF methods. This approach does have some drawbacks which are explored by Bailey et al. [38]. In particular, the uncertainty estimates inevitably become too optimistic and the filter is unable to explore the complete state space. A further publication by Brooks and Bailey [39] presents a hybrid approach combining the benefits of EKF-SLAM and FastSLAM.

Graph-based methods represent the positions of the landmarks as nodes in a graph. The series of target positions (robot or pedestrian) are also nodes in the graph. Odometry or dead-reckoning measurements become edges between target positions, and landmark observations (e.g., range measurements) are edges between target positions and landmark positions. The graph can be optimized, for instance' by using a method such as gradient descent or MDS-MAP. Kleiner and Sun [40] use a graph representation to build a map of landmarks (the nodes of the graph) after several pedestrians have walked between them (thus measuring the edges of the graph). Djugash et al. [15] use a similar representation to initialize landmark positions but then use EKF-SLAM to track their robot. Thrun and Montemerlo [41] give a formal description of GraphSLAM. They present it as a solution to the offline SLAM problem for large environments. In other words, it can only be used after all the data have been collected, not in real time. Offline solutions such as GraphSLAM are able to produce more accurate maps and traces than online methods such as EKF-SLAM and FastSLAM, because they have more data to work with. However, they rely on a good initial estimate of node positions which can be difficult to achieve, given the drift in dead-reckoning over long distances.

1.3.3 Common Sensing Modalities for SLAM

SLAM implementations for robot positioning typically build on sensors and odometry as these are often readily available on robot platforms. In most cases, sensors used in the robots include precise laser scanners, or a single (monocular) or a multiple (stereo) cameras to gather landmark observations. Stereo cameras provide both range and bearing estimates to landmarks in the environment, while monocular camera provides only bearing information. Little prior work has addressed the problem of building maps with sensors providing range-only information or bearing-only information.

Dead-reckoning traces its roots back to ship navigation. It is a common technique in robotics where it is often implemented using wheel odometry, but it is now becoming popular in the area of pedestrian tracking due to the miniaturization of inertial sensors. This method tracks the movements of the target (vehicle, robot, or person) and deduces their current position relative to their starting point and orientation. Not all SLAM methods rely on wheel-based odometry or inertial dead-reckoning. Camera-based methods often apply *visual odometry* [42] to estimate the trajectory of the camera using only a stream of images.

SLAM with range-only sensors [29,43] and bearing-only sensors [44] shows that a single measurement does not contain enough information to estimate the location of landmark. It must be observed from multiple vantage points. This essentially means that we collect measurements over a period of time, and initializing the landmark is performed using a batch update scheme [31,32].

Range-only SLAM utilizes the range measurement between the robot and the landmarks. The distinct advantage of using range-only SLAM is that it does not have any data-association problems, as one can infer the unique correspondences between the sensed ranges and particular landmarks. This is because in the existing range-only SLAM systems, the identity of the landmark is encoded in the radio trigger packet. Olson et al. present a range-only approach which is robust in identifying the outliers in the range measurements. The initial position of the landmarks is computed using a voting scheme [29]. Once the initial estimation converges, a standard EKF-SLAM implementation is performed. Undersea range-only SLAM has also been demonstrated to give good results despite unreliable information from odometry [28].

There is also work done as an extension to range-only SLAM with the landmarks themselves able to measure range to each other, in addition to robot-landmark measurements. This will essentially improve the robot pose estimation and the resulting map of the landmark. Djugash et al. distinguish both these approaches in their work [25], and illustrate that adding more observations to the SLAM algorithm yields better results.

1.3.4 Aerial SLAM

SLAM has also been applied to unmanned aerial vehicles (UAVs). Kim and Sukkarieh [27] utilize a single camera and inertial measurement unit (IMU) installed in a UAV. Their results show that both the map and the vehicle uncertainty are corrected even though the model of the system and observation is highly nonlinear. The results, however, also indicate that further work of observability and the relationship between vehicle model drift and the number and the location of landmarks need to be further analyzed, given the highly dynamic nature of the system.

1.3.5 Pedestrian SLAM

Some recent attention has been given to pedestrian SLAM systems. The primary goal is to automate the generation of maps that can be used later by others to navigate inside a building.

HeadSLAM [45] uses a combination of dead-reckoning and measurements from a laser scanner (for detecting direction and distance to obstacles) for building the map of the environment and for positioning purposes. The scanner detects the direction and distance of obstacles such as walls. The map produced resembles an actual floorplan showing corridors, rooms, and doorways. SLAM can be very effective when a robot is allowed to repeatedly scan the environment but it is unclear how well it would perform for a pedestrian in an emergency.

Recently, the SLAM problem has been addressed without using any visual or other type of sensors, thus leading to SLAM using only PDR measurements (known as *FootSLAM*). This is challenging because of the inherent error associated with the dead-reckoning measurements, despite mitigation of drift using periodic resetting of velocities. Robertson et al. [46] demonstrate that a pedestrian's location and the building layout can be jointly estimated by using the pedestrian's odometry alone, as measured by the foot-mounted IMU. When a pedestrian walks within a constrained area such as a building, then even drift-prone odometry measurements can give information about certain features like turns, doors, and walls, which can then be used to build a map of the explored area. The authors use the FastSLAM approach whereby each particle assumes a certain pose history and estimates the motion probabilities at each location. In order to compute and store the probability distribution of the motion as a function of location, they have chosen to partition the space into a regular grid of adjacent and uniform hexagons. A usable map emerges once the pedestrian has covered the main walkable areas of the map several times.

1.4 Multimodal Pedestrian SLAM

Considering the pros and cons of the various techniques described in the previous section for tracking pedestrians indoors, we have chosen to use two key technologies: inertial dead-reckoning and ultrasonic range and bearing measurements.

PDR using inertial sensors is one key sensing modality used in our work. The other modality is range and bearing estimation using ultrasound. By measuring the time of flight and amplitude of an ultrasonic pulse, we are able to estimate the range and bearing to a transmitter, relative to the receiver. By attaching such a sensor to our target (a pedestrian) and deploying other sensors on the ground, we obtain information that can be used to locate the target. This is achieved by combining these range and bearing measurements with the dead-reckoning estimates according to the methods described in the Section 2.

The benefits of inertial dead-reckoning are clear—it is able to track pedestrian motion using only a small foot-mounted sensor. We believe that the ultrasound sensors are a suitable technology for this application for three reasons: (1) each sensor has a unique identifier which is present in every measurement so landmark identification is not a problem, (2) each measurement contains explicit range and bearing information which does not need to be inferred from other types of data, and (3) sensors can be deployed as required, and their number and position can be chosen depending on the type of environment and the requirements of the situation. Visual landmarks satisfy none of these criteria. RFID tags provide unique identification and can be deployed as required, but they do not provide precise range and bearing information.

2. Implementation: Multimodal Sensing and Algorithms

In this section, we describe our implementation of a tracking system for emergency responders based on inertial PDR and ultrasonic sensors deployed by the responders as they explore a building. First, we implement an EKF which can track the pedestrian when the sensor positions are known. Second, we augment this EKF to track the pedestrian and simultaneously locate the sensors in the case where no prior knowledge is available.

2.1 Sensing Technologies

In this section, we start by describing PDR in detail, and discuss the inherent issue of drift. Then, we describe our ultrasound range and bearing sensors and characterize their error.

2.1.1 Inertial PDR with Zero-Velocity Updates

Dead-reckoning is a self-contained navigation technique in which measurements—typically from inertial sensors in the case of PDR—are used to update the position and orientation of an object, given an initial position, orientation, and velocity. No infrastructure is required but the position error can accumulate over time.

XSens's MTx[3] is an IMU comprising a tri-axis accelerometer, gyroscope, and magnetometer. The on-board processor computes 3D orientation. Our PDR algorithm is similar to other work [47,48] which uses shoe-mounted IMUs and applies periodic zero velocity updates (ZUPTs). In order to convert the MTx measurements into meaningful positions, the raw accelerations are rotated from the sensor coordinate system into the world coordinate system using the rotation matrix computed by the MTx as shown in Fig. 1. The transformed accelerations are then double integrated to yield position estimates in the world coordinate system.

In order to reduce the position error which increases quadratically with time, we reset the integrated velocities to zero at each footstep [47]. This has the effect of reducing the error, making it linear with distance covered. Two phases in walking are identified: the stance phase, when the foot is in contact with the ground, and the swing phase. During the stance phase, the velocity is reset and kept at zero; during the swing phase, the acceleration is double integrated. Our algorithm detects the stance phase of each step by comparing a threshold to the product of the norm of the acceleration and the norm of the rate of turn [48]. If this product is below an

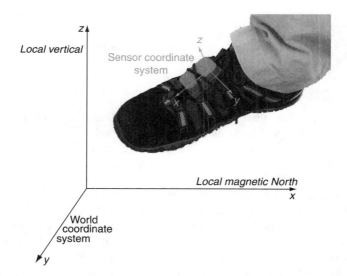

Fig. 1. Transformation from sensor to world coordinates via the direction cosine matrix: $\mathbf{pos}_{world} = R_{sensor \to world} \mathbf{pos}_{sensor}$.

[3] http://www.xsens.com

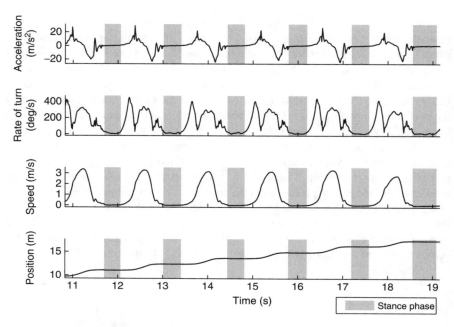

Fig. 2. PDR algorithm: each step has a stance phase (shaded) and a swing phase. Velocity is reset to zero during the stance phase, acceleration is double integrated during the swing phase.

empirically determined threshold for more than 200 ms, then a stance phase is detected. When the product rises above the threshold again, a swing phase is detected. This is illustrated in Fig. 2. If steps are taken at a significantly faster pace, then the stance phase may not always be detected and some opportunities for ZUPTs may be missed.

Although some distance drift is inevitable due to the integration of noise and small offsets in the raw sensor data, we also believe that most of the distance error is due to the MTx incorrectly estimating its orientation as explained by Foxlin [49]. Thus, we might interpret some of the forward motion as vertical motion, or vice versa. The traces appearing later in this chapter (Fig. 10) show the effects of drift on position estimates. The MTx internal algorithm which supplies the rotation matrix has not been disclosed by XSens, and is effectively a "black box." Thus, we have very little information about how the sensors (gyroscope, magnetometers, and accelerometers) are used in computing the MTx orientation, and little control over the unit's internal parameters. We assume that most of the heading errors are due to metallic objects or magnetic fields interfering with the MTx magnetometers, as we have observed that these errors occur systematically in the same locations. We also note that when using the system outdoors or in an open space, the results are much better and the orientation drift is often negligible. So it seems that magnetometers help in outdoor situations where they accurately determine magnetic North but that indoors they cause heading errors due to interference.

Our PDR implementation is a naive version—(1) rotate the accelerations from the sensor coordinate system into the world coordinate system, (2) integrate the accelerations to determine the speed, (3) integrate the speed to determine the position, (4) if the accelerations and the gyroscope readings are small then reset the speed to zero. More advanced algorithms (such as the one described by Foxlin) use the ZUPTs to correct not only the speed but also the position, and to estimate accelerometer and gyroscope biases which are responsible for much of the error accumulation. Such methods are able to achieve better accuracy over longer distances through tighter integration of the orientation computation, the dead-reckoning, and the ZUPTs. This improvement in performance comes at the expense of greater complexity in the design of the algorithm (typically an EKF). Beauregard gives more details on the challenges in his thesis [50]. Irrespective of the method used for PDR, the output will inevitably suffer from error accumulation. A more advanced method than ours could simply be used as a drop-in replacement in the algorithm presented above.[4]

2.1.2 Ultrasonic Ranging and Angulation

Our ultrasonic sensors [51] are able to determine range and bearing to each other. Each sensor contains a microcontroller, radio transceiver, and four ultrasonic transducers (Fig. 3). The transmitter broadcasts a trigger radio packet just before it emits a series of ultrasonic pulses. Upon receiving the radio trigger, the sensors listen for the ultrasonic pulses and determine the distance to the transmitter based on the time of flight of the pulse. The angle-of-arrival is estimated based on the measured pulse amplitudes on the four ultrasonic transducers. The receiver measures the local temperature in order to more accurately estimate the speed of sound but variations in the environment can introduce error into the measurements. The largest errors are due to reflections, so some simple heuristics are applied to detect and eliminate some of these bad measurements.

---

[4] Since writing this chapter, we have implemented a Kalman filter based PDR system. The main improvement has been in the altitude estimation.

FIG. 3. Ultrasonic sensor node—microcontroller and radio transceiver on the top board, transducers, and amplification circuit on the lower board.

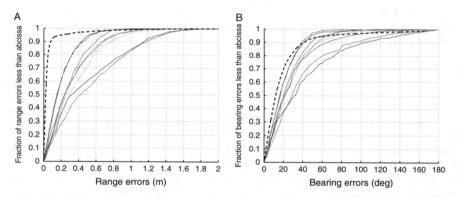

FIG. 4. Range and bearing measurement errors of the ultrasonic sensors. Each of the eight lines represents the error distribution for a different experiment. The thick dotted line is the error distribution measured in previous work using a more accurate visual fiducial-based tracking system to record the true positions of the mobile sensor. (A) Range errors. (B) Bearing errors.

The errors in the range and bearing measurements taken during our experiments are shown in Fig. 4. Each line represents the range or bearing error distributions for one of our eight selected experiments. The large error for the range and bearing measurements in our pedestrian tracking experiment can be explained by three factors. (1) The true position of the pedestrian is measured by the Ubisense UWB location system[5] with an accuracy of 30 cm according to the manufacturers; this accuracy is worse than the ranging accuracy of our ultrasonic sensors, and is further limited by a challenging environment in terms of RF propagation. (2) The UWB transmitter used to track the true position of the pedestrian was not colocated with the foot-mounted ultrasonic transmitter but attached to his cap. (3) The ultrasonic transmitter was modified to fit around the pedestrian's shoe but this causes the alignment of the transducers to be suboptimal; the other foot can also block the line of sight between the transmitter and the sensors lying on the ground. Thus, to give a better indication of our sensors' accuracy, we have also plotted the error from a separate set of experiments in which we deployed five static sensors and one mobile sensor on a Lego MindStorms robot in a 2.75×2.00 m arena. The thick dotted line in Fig. 4 represents the range and bearing errors for over 60,000 measurements involving the mobile sensor in that arena. In these previous experiments, the true sensor positions were measured using visual markers on top of each sensor and camera tracking software which estimates positions with an accuracy of the order of a centimeter but over a small area.

2.2 Localization and Mapping Algorithms

We use inertial PDR and ultrasonic sensor nodes to simultaneously locate a pedestrian and map the positions of the nodes (which effectively serve as landmarks). The SLAM techniques which we use have been developed by robotics researchers since the 1980s, but have only been applied to pedestrian localization in the past 5 years. We first describe an implementation which assumes that sensor positions are known; then we describe an implementation that also estimates the positions of the sensors, thus requiring no prior knowledge of the environment.

2.2.1 Kalman Filtering

Kalman filtering is a form of Bayesian estimation. Fox et al. [52] give a concise overview of Bayesian principles applied to location estimation and we also restrict our explanations to the context of location estimation. A Bayesian filter estimates the probability distribution of the target's location based on a probabilistic model of its movement and of the measurements. Using the movement model, the filter attempts to predict the location of the target; then using the measurement model, it

---

[5] http://www.ubisense.net

corrects the location estimate. The Kalman filter models all probabilities as Gaussian distributions and tracks their means and covariances noted state and P, respectively.

In our application, we track a pedestrian. Their state is modeled as state$_p = [x_p\ y_p\ \psi_p]^T$, the two Cartesian coordinates and the direction of travel, or heading. We only address the two-dimensional problem in this work. Tracking the z-dimension is more challenging because our ultrasound sensors only measure bearings in the horizontal plane and because in our experience PDR systems tend to drift more along the z-axis than in the horizontal plane. If necessary, stairs could be detected and flagged separately. Tracking in the horizontal plane may be sufficient for many scenarios, including emergency response. We have measurements from two sources—PDR which measures the movement of the pedestrian, and ultrasound sensor nodes which measure the range and bearing of the pedestrian. The PDR measurements are used in the prediction phase and the ultrasonic measurements are used in the correction phase. The movement model (or process model) takes advantage of the PDR measurements $[d\ \delta\psi]^T$ which tell us how the pedestrian has moved since the previous estimate. d is the distance traveled and $\delta\psi$ is the change in the direction of travel (heading) since the previous PDR measurement. We measure the change in heading because we know that the absolute heading measured inertially is prone to drift. These notations are presented in Eq. (1) and illustrated in Fig. 5.

$$\begin{aligned}\text{state}_p^{k+1} &= f\left(\text{state}_p^k, \text{meas}_{PDR}\right) \\ &= \begin{bmatrix} x_{k+1} \\ y_{k+1} \\ \psi_{k+1} \end{bmatrix} \\ &= \begin{bmatrix} x_k + d\cos(\psi_k + \delta\psi) \\ y_k + d\sin(\psi_k + \delta\psi) \\ \psi_k + \delta\psi \end{bmatrix}\end{aligned} \qquad (1)$$

We assume that both elements of the PDR measurement are subject to additive Gaussian noise. Measurement noise has been omitted from the above process equation where each occurrence of d is actually $d + d\varepsilon_d$, and each occurrence of $\delta\psi$ is actually $\delta\psi + d\varepsilon_{\delta\psi}$ so that the noise is applied proportionally to the distance moved since the previous measurement. The covariance of the noise $[\varepsilon_d\ \varepsilon_{\delta\psi}]^T$ is denoted by Q. By defining the process noise proportional to the distance traveled, we prevent the covariance (the position uncertainty) from increasing when the pedestrian is standing still. Unfortunately, this introduces additional nonlinearities which affect the pedestrian covariance as shown later in Fig. 11.

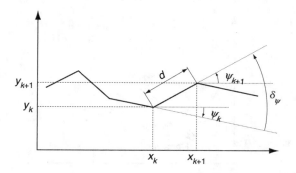

FIG. 5. Notations used for PDR measurements. d is the distance traveled and $\delta\psi$ is the change in heading since the previous PDR measurement.

The ultrasound sensor nodes are able to measure the range r and the bearing θ to a compatible device attached to the pedestrian's boot and colocated (as closely as possible) with the IMU used to perform PDR. The state of each sensor i is modeled as $\text{state}_{s_i} = \begin{bmatrix} x_{s_i} & y_{s_i} & \phi_{s_i} \end{bmatrix}^T$, the two Cartesian coordinates and the orientation. We can express the measurement as a function of the state of the pedestrian and of the relevant sensor as shown in Eq. (2) and illustrated in Fig. 6.

$$\text{meas}_{us} = h(\text{state}_p, \text{state}_{s_i})$$
$$= \begin{bmatrix} r \\ \theta \end{bmatrix}$$
$$= \begin{bmatrix} \sqrt{(x_p - x_{s_i})^2 + (y_p - y_{s_i})^2} \\ \arctan\left(\dfrac{y_p - y_{s_i}}{x_p - x_{s_i}}\right) - \phi_{s_i} \end{bmatrix} \quad (2)$$

We assume that the ultrasonic measurements are subject to additive Gaussian noise $[\varepsilon_r\ \varepsilon_\theta]^T$ with covariance R. This has been omitted from the above measurement equation. It is important to ensure that all angles remain between $-\pi$ and π, so the filter can determine whether two angles are similar or not.

The process function f (Eq. (1)) and the measurement function h (Eq. (2)) are both nonlinear functions of the state, measurements, and noise, so we will need to calculate their Jacobian matrices (Eqs. (3)–(5)). The Jacobian $W = \partial h / \partial \varepsilon_{us}$ is the identity matrix because additive noise is applied directly to the measurement, it can therefore be omitted.

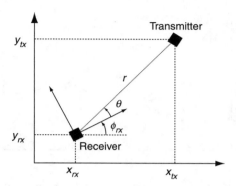

FIG. 6. Notations used for ultrasound sensor coordinates, orientations, and measurements. x_{rx}, y_{rx}, and ϕ_{rx} are the coordinates and orientation of the receiver (deployed sensor). x_{tx} and y_{tx} are the coordinates of the transmitter (attached to the pedestrian). r and θ are the range and bearing measurements.

$$A = \frac{\partial f}{\partial \text{state}_p}$$
$$= \begin{bmatrix} 1 & 0 & -d\sin(\psi_p) \\ 0 & 1 & d\cos(\psi_p) \\ 0 & 0 & 1 \end{bmatrix} \quad (3)$$

$$H = \frac{\partial h}{\partial \text{state}_p}$$
$$= \begin{bmatrix} \dfrac{x_p - x_{s_i}}{\sqrt{(x_p - x_{s_i})^2 + (y_p - y_{s_i})^2}} & \dfrac{y_p - y_{s_i}}{\sqrt{(x_p - x_{s_i})^2 + (y_p - y_{s_i})^2}} & 0 \\ -\dfrac{y_p - y_{s_i}}{\sqrt{(y_p - y_{s_i})^2 + (x_p - x_{s_i})^2}} & \dfrac{x_p - x_{s_i}}{\sqrt{(y_p - y_{s_i})^2 + (x_p - x_{s_i})^2}} & 0 \end{bmatrix} \quad (4)$$

$$V = \frac{\partial f}{\partial \varepsilon_{\text{PDR}}}$$
$$= \begin{bmatrix} d\cos(\psi_p) & -d^2\sin(\psi_p) \\ d\sin(\psi_p) & d^2\cos(\psi_p) \\ 0 & d \end{bmatrix} \quad (5)$$

We can apply the Kalman prediction equations whenever a PDR measurement is available (Eq. (6)).

$$\text{predicted\_state}_p = f(\text{state}_p, \text{measurement}_{PDR})$$
$$\text{predicted\_}P_p = AP_pA^T + VQV^T \qquad (6)$$

Similarly, when an ultrasonic measurement is available, we can apply the Kalman update equations (Eq. (7)).

$$K = PH^T(HPH^T + R)^{-1}$$
$$\text{updated\_state}_p^{k+1} = \text{predicted\_state}_p^k + K(\text{meas}_{us} - \text{predicted\_meas}_{us}) \qquad (7)$$
$$\text{updated\_}P_p^{k+1} = (I - KH)\text{predicted\_}P_p^k$$

2.2.2 Pedestrian SLAM

If the positions of the sensor nodes are unknown, then we have a SLAM problem where we must estimate the positions of the sensors (the map) in addition to the position of the pedestrian. This can be achieved by modifying the Kalman filter developed in the previous section. The positions of the sensors become variables instead of constants and the filter also tracks their covariances.

The prediction equations remain the same—the filter uses the PDR measurements to predict the position of the pedestrian. The sensors are assumed to be static, so their position estimates remain the same during the prediction phase. However, the correction phase is slightly different because each ultrasonic measurement can correct both the pedestrian position estimate and the position estimate of the sensor which took the measurement.

The state under consideration is now the concatenation of the position of the pedestrian and the position of the sensor i which took the measurement (Eq. (8)). The Jacobian matrix H is also different because now the elements of the sensor state state_{s_i} are also variable (Eq. (4)).

$$\text{state}_{p+s_i} = \begin{bmatrix} \text{state}_p \\ \text{state}_{s_i} \end{bmatrix}$$
$$P_{p+s_i} = \begin{bmatrix} P_p & 0 \\ 0 & P_{s_i} \end{bmatrix} \qquad (8)$$

$$H = \frac{\partial h}{\partial \text{state}_{p+s_i}}$$

$$= \begin{bmatrix} \frac{x_p - x_{s_i}}{\sqrt{(x_p - x_{s_i})^2 + (y_p - y_{s_i})^2}} & -\frac{y_p - y_{s_i}}{(x_p - x_{s_i})^2 + (y_p - y_{s_i})^2} \\ \frac{y_p - y_{s_i}}{\sqrt{(x_p - x_{s_i})^2 + (y_p - y_{s_i})^2}} & \frac{x_p - x_{s_i}}{(x_p - x_{s_i})^2 + (y_p - y_{s_i})^2} \\ 0 & 0 \\ -\frac{x_p - x_{s_i}}{\sqrt{(x_p - x_{s_i})^2 + (y_p - y_{s_i})^2}} & \frac{y_p - y_{s_i}}{(x_p - x_{s_i})^2 + (y_p - y_{s_i})^2} \\ -\frac{y_p - y_{s_i}}{\sqrt{(x_p - x_{s_i})^2 + (y_p - y_{s_i})^2}} & -\frac{x_p - x_{s_i}}{(x_p - x_{s_i})^2 + (y_p - y_{s_i})^2} \\ 0 & -1 \end{bmatrix}^T \quad (9)$$

The combined state state_{p+s_i} is created when each ultrasonic measurement is received, and when the correction phase is finished, the position of the pedestrian and the position of the sensor are stored separately. This allows us to process data only of dimension six for each measurement which scales better and is computationally much less expensive than considering all sensors when doing the update. However, there is a tradeoff because we lose the information about the cross-correlations between the different sensor positions.

In a typical Kalman filter, the covariance increases during the prediction phase due to the uncertainty of the process model and then decreases during the correction phase due to the information from the measurement. However, the covariances of the sensors never increase because they are modeled as static. To avoid the covariances decreasing too quickly, we use the fading memory technique described by Simon [53] which artificially increases the sensor covariances before each correction phase by multiplying P_{s_i} by a value slightly greater than one (Simon [53] suggests 1.01^2 and we also use this value).

2.2.3 Sensor Initialization

The first time a sensor reports a measurement, the filter has no information about it. A sensor cannot be initialized from a single measurement, so we delay initialization until we have at least three measurements taken from positions that are well spaced apart. Using these measurements, we can usually perform reliable trilateration using a nonlinear regression. However, if the points are nearly collinear there are two solutions. The bearing measurements can help us resolve such ambiguities, but we find that using them directly in the regression gives poor results. We adopt the following heuristic method for selecting the correct solution. First, we use only the range measurements in the regression. We know that the solution computed by the regression is either the true position of the sensor, or a position that is the symmetric of the true position with respect to the path traveled. Therefore, we estimate the least squares line which approximates the path traveled, and compute the symmetric of the estimated sensor position. One of these points should be close to the true sensor position. In other words, if there is indeed an ambiguity, we are likely to have found two local minima corresponding to the actual position of the sensor and its reflection. Using each of the bearing measurements, we estimate the orientation of the sensor for both positions. The correct position will yield similar orientation estimates for all of the bearing measurements. The incorrect symmetric position will yield inconsistent orientation estimates for each of the bearing measurements. We select the position which minimizes the bearing residuals (or the variance of the orientation estimates).

Figure 7 illustrates the sensor initialization process. In this example, the pedestrian walked along a straight path for approximately 2 m. An ultrasound sensor took range and bearing measurements to the pedestrian and we recorded those that were at least 50 cm apart. The circles represent the range measurements taken at these different points, and their intersections are the possible positions of the sensor. There are two solutions because the pedestrian path was almost collinear. Using a single bearing measurement is not enough to choose the correct solution, but the sequence of bearing measurements is only consistent with the correct solution. This heuristic initialization method works well but, because of the noise in our bearing measurements, there are times when it is unable to choose correctly between the true sensor position and the symmetric.

Djugash et al. [15] use a similar method for initializing sensor positions in their work on range-only SLAM for robots, but they explain it differently (and do not have the benefit of bearing measurements). They represent the sensors as nodes in a graph, and robot positions as *virtual nodes*. The edges are either sensor-to-sensor ranges (which we do not use) or sensor-to-robot ranges. Based on this graph representation, they are able to determine the locations of sensor nodes and of the robot by running a batch optimization. Deans and Hebert [44] also use a comparable approach in their Kalman filter SLAM implementation for bearing-only measurements. They initialize a landmark position by performing *bundle adjustment* (a nonlinear optimization) over a section of the robot trajectory and the first few bearing measurements in order to optimize both the landmark position and the robot trajectory (we only optimize the landmark positions).

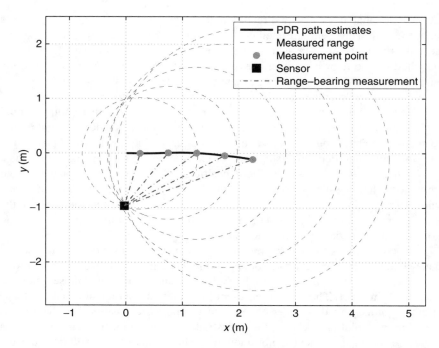

FIG. 7. Sensor initialization from a sequence of range/bearing measurements. First the ranges are used to determine the possible sensor locations. Then the bearings are used to select the correct one.

3. System Evaluation

We evaluate the EKF algorithm (with and without SLAM) described in the previous section using multiple sets of inertial and ultrasound measurements. These were recorded as a subject walked through an indoor office environment. In this section, we first describe the experimental setup, then we show the results achieved for different experiments before drawing conclusions about the suitability of this type of algorithm for emergency response scenarios.

3.1 Description of Experiments

Our experiments took place in an office building at the University of Twente in the Netherlands. The test area covered a large, mostly empty, office on one side of a corridor, and two smaller individual offices, each with a desk, on the other side of the corridor (Fig. 8).

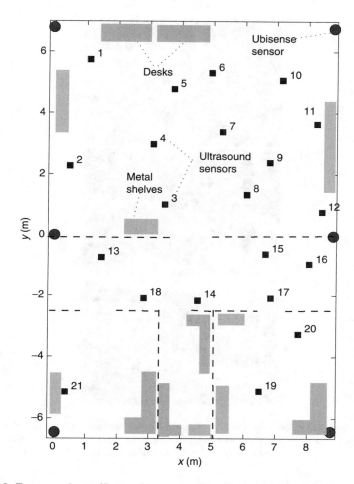

FIG. 8. Test area: a large office, corridor, two smaller offices covering a total of 9×13 m.

3.1.1 Inertial/Ultrasound Data Collection

In this section, we characterize our algorithms using data gathered during eight different collection sessions, each lasting from 3 to 8 min. We placed 21 ultrasound sensors on the floor (Fig. 9A), and surveyed their positions by hand. We also surveyed six additional reference points which the pedestrian walked to and from in some of the experiments.

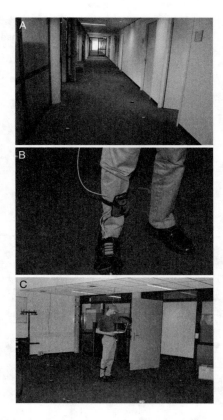

FIG. 9. Experiment and data collection. (A) Sensor nodes deployed along the corridor. (B) Ultrasound transmitter strapped to the leg, and Xsens inertial measurement unit attached to the pedestrian's foot. (C) Data collection performed in the large office. Ultrasound sensors are deployed on the floor.

During the data collection sessions, a pedestrian (one of the authors) walked within the test area at an average walking speed between 0.8 and 1.5 m/s. The pedestrian wore an MTx inertial sensor on his foot and an ultrasonic transmitter on his lower leg (Fig. 9B). The transducers of the transmitter were on flying leads and attached around the edge of the sole so they would be in the same plane as the sensors deployed on the ground. The paths are shown in Fig. 12 in Section 3.2. Paths (A), (G), and (H) were constrained to the large room. All other paths covered the large room, both smaller rooms, and the corridor. For path (A), the pedestrian walked around the room once and then to reference point 1 (not shown), then around the room again and to reference point 2, and so on. For paths (B), (C), (D), and (E), he walked continuously between the rooms following the same path each time; the path in the large room is a straight line across the middle of the room, hence the "T" shape. Path (F) is similar but he walked around the large room instead of across it. For paths (G) and (H), the pedestrian stopped at each reference point (not shown) for approximately 10 s after walking arbitrarily about the room for approximately a minute between each reference point.

3.1.2 Path Capture Using Ubisense

Ground truth was measured using a Ubisense commercial UWB location system.[6] We deployed six Ubisense receivers at the edges of the measurement volume. This system was calibrated using the Ubisense software's "full calibration" mode. The pedestrian wore a Ubisense tag on a cap in order to measure reference location data.

However, from our previous work, we know that the 90th percentile horizontal accuracy of this technology can vary from 11 cm to worse than 150 cm [54, Fig. 4 and Table 2], depending on calibration effort invested, sensor density, and environmental conditions affecting UWB line-of-sight and multipath. Thus, the Ubisense accuracy can be considerably worse than the ultrasonic sensor ranging accuracy, which we measured in a separate set of experiments to be about 7 cm at the 90th percentile confidence level (Fig. 4). It therefore cannot be used to characterize the lower bound of the accuracy for our pedestrian tracking and SLAM algorithms. However, it does give a reliable indication of the path traversed, and can to some extent be used to compare the relative accuracy of different types of path, side-by-side (assuming both types of path cover similar parts of the measurement volume, and thus will be subject to similar amounts of Ubisense tracking error).

3.1.3 Error Characterization

Typically, a location system is characterized by the error between the estimated positions of the pedestrian and of the sensors, and their true positions. For our system, this is not possible because the SLAM algorithm works in its own arbitrary coordinate system which can potentially change over time. In other words, the positions of the pedestrian and of the sensors can only be compared to each other, not to their true values. For this reason, we calculate the errors in the ranges and bearings between sensors and between sensors and pedestrian. We can then examine how these errors evolve over time and compare their distributions. The performance of the filter can

---

[6] http://www.ubisense.net

also be evaluated to some degree without any ground truth by examining the innovations (i.e., the differences between the predicted ranges and bearings, and the actual measurements). If the innovation remains small, then the filter is likely to have correctly estimated the positions of the pedestrian and of the sensor.

The estimated positions of the sensors may be aligned with their hand-surveyed positions either manually or by using a regression to determine the optimal translation and rotation to apply. This method can be useful for visualizing the output of the filter but it must be used with care because of the additional complexity introduced by the optimization and the fact that the coordinate system in which the filter locates the sensors and the pedestrian changes over time.

3.1.4 Algorithm Parameters and SLAM Initialization

The specific algorithm parameters we used are given in Table I. Note that the values we use for Q and R bear no clear relationship to measurable noise values, and were chosen empirically. This is a common situation when designing a Kalman filter for which the system model is not well known or not detailed enough; the modeled noise values need to be increased in order to account for the modeling "errors." In our system, this is emphasized by our defining the process noise proportional to the distance traveled (rather than the time elapsed) and the nonlinearities that this introduces in the model.

We use five ultrasonic measurements taken at least at 30 cm from each other to initialize the sensor positions. Measurements for which the range innovation divided

TABLE I
PARAMETERS USED IN OUR SLAM ALGORITHM

Name	Notation	Value
PDR noise covariance	Q	$\begin{bmatrix} 2 & 0 \\ 0 & 20 \end{bmatrix}^2$
Range/bearing measurement noise covariance	R	$\begin{bmatrix} 2 & 0 \\ 0 & 1 \end{bmatrix}^2$
Initial sensor position covariance	$P_{s_i}^0$	$\begin{bmatrix} 50 & 0 & 0 \\ 0 & 50 & 0 \\ 0 & 0 & 8.73 \end{bmatrix}^2$
Initial pedestrian position covariance	P_p^0	$[0]$
Fading memory factor		1.01^2

by the square root of the sum of x and y covariances for the pedestrian and the relevant sensor are greater than 0.5 are discarded as outliers.

In the case where we assume that the sensor positions are known in advance, we initialize them to their true values and set P_{s_i} to zero. The position of the pedestrian is set according to the Ubisense measurements. Their direction of travel is initialized by using the angular difference between the direction given by the first Ubisense measurements that are at least 3 m from each other, and the corresponding position estimates using PDR alone.

3.2 Results and Analysis

We now take a closer look at the different components of the system and how they perform on our datasets.

3.2.1 Inertial Pedestrian Dead-Reckoning

The inertial PDR measurements can be used for determining the path, but as explained in Section 2.1.1, the estimated position drifts after a few meters. Figure 10 shows how the position estimates suffer from drift when the PDR measurements are used alone. In data sets (C), (E), and (F), the drift manifests itself only as a translation. This is because the magnetometer (digital compass) included in the IMU prevents orientation error from accumulating on the yaw or heading angle (in the horizontal plane). Data sets (B) and (D) were recorded after configuring the IMU to use only the inertial sensors and discard the magnetometer readings. These data sets clearly show drift accumulates on the yaw. However, there are also disadvantages to using the magnetometers. In previous work [16], we observed that, in certain environments, the magnetometers suffer from interference due to metallic structures in the building fabric or heavy machinery. This causes drift that is unpredictable in the sense that it is dependent on the environment in which the system is used and not just on the characteristics of the sensor, the way it is attached, and the type of movement.

Good results can be achieved without using the magnetometers. Data set (A) for instance displays very little drift in the yaw angle despite being recorded without using the magnetometers. In data set (D), there are four iterations which are nearly superimposed.

Another cause for the variable quality of the dead-reckoning traces is the internal calibration algorithm of the IMU. The IMU we use is a generic sensor not designed specifically for PDR but for other forms of human motion capture or for augmenting machinery. The internal algorithm (an EKF according to the XSens documentation) adapts to the characteristics of the movement (amplitude of accelerations and rotations) and we have found that the results often improve during the course of the experiment.

This may explain why the latter part of data set (D) displays much less drift than the initial part. This also explains why the first 30 s (which we removed from our data) from most of the PDR traces contain artifacts which badly distort them. (This issue is specific to this particular IMU and does not invalidate our results.)

We know from observing these traces that the uncertainty in the PDR position estimates increases with distance traveled. The process model in our Kalman filter is designed to take this into account, and the value of the pedestrian position estimate covariance should therefore increase with distance traveled (and with time). Figure 11 shows how the position estimate covariance changes with time for experiment (A). We plot $\sqrt{P_p(1,1) + P_p(2,2)}$ (square root of the sum of the first two diagonal elements of the position estimate covariance) versus time which is an estimate of the pedestrian position uncertainty. The covariance starts at zero because we arbitrarily decide on the initial position of the pedestrian. We see that globally the uncertainty increases as expected but there are several instances where the

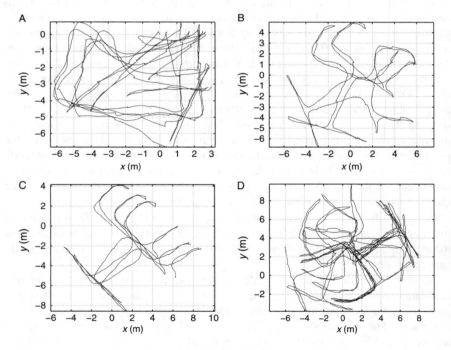

FIG. 10. (Continued)

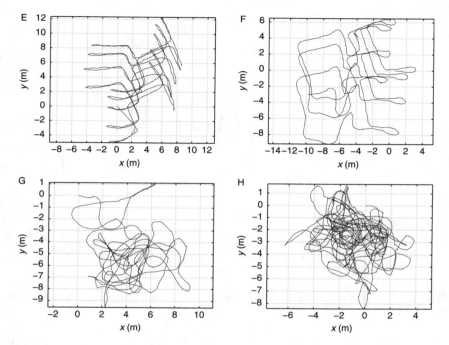

FIG. 10. Paths estimated from inertial pedestrian dead reckoning alone. (A) Large room. (B) All rooms (T) 1. (C) All rooms (T) 2. (D) All rooms (T) 3. (E) All rooms (T) 4. (F) All rooms (O). (G) Random 1. (H) Random 2.

uncertainty decreases. This occurs on corners when the change in heading is large. This decrease in uncertainty is normal, as position errors can cancel each other out for some time following a 180 degree turn. Wan and Foxlin give a more detailed analysis of PDR error [58].

3.2.2 Kalman Filter Using Known Sensor Positions

If the positions and orientations of the sensors are known in advance, we can use the first form of our Kalman filter which corrects the PDR location estimates. In our implementation of the Kalman filter, we set the sensor and pedestrian positions and orientations based on the surveyed positions. Setting the sensor position estimate

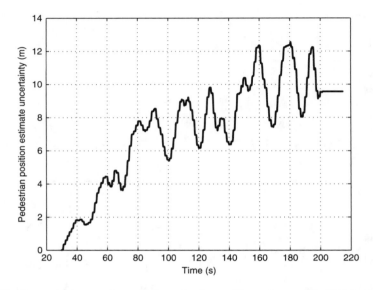

FIG. 11. Change in the covariance of the pedestrian position estimate when PDR is used alone (dataset (A)).

covariances to zero ensures that their position estimates do not change as range and bearing measurements are processed by the filter, only the pedestrian position is affected.

The results are shown in Fig. 12. In Fig. 12A, we see that all the *branches* of the path are clearly tracked and that the topology of the trace is close to the Ubisense groundtruth path. However, on the left part of the figure, there is an offset between the path estimated from our Kalman filter and the path estimated by Ubisense. This is also visible in other traces such as in Fig. 12E. Many of these offsets seem to be consistent across all traces, so they may be due to inaccuracies of Ubisense which can worsen when the tracked tag is placed close to the limits of the covered area.

Figure 13 illustrates how the covariance of the pedestrian position estimate evolves during these experiments when range and bearing measurements are taken into account. As they start to walk, the covariance increases due to the potential error in the PDR estimates, but when ultrasound measurements are received the covariance decreases due to the additional information. Eventually, a regular cycle of increases due to PDR and decreases due to ultrasonic measurements maintains the covariance around a constant value.

3.2.3 SLAM

Visualizing the output of the SLAM filter as we did when the sensor positions were known in advance can be misleading because the estimated coordinates of the sensors change during the course of the experiment as they are updated by the SLAM process. Figure 14 shows that in all the experiments, the estimated positions of the sensors change over time. Note that these plots are in an arbitrary coordinate

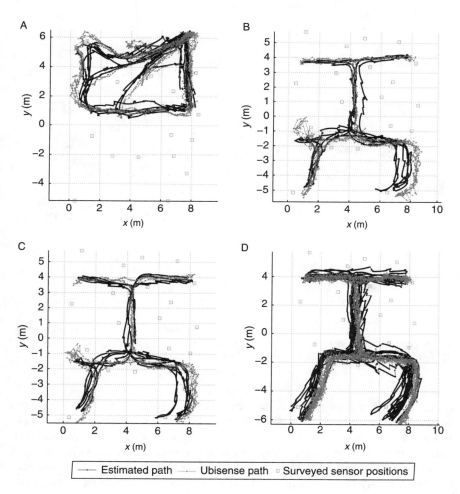

FIG. 12. (Continued)

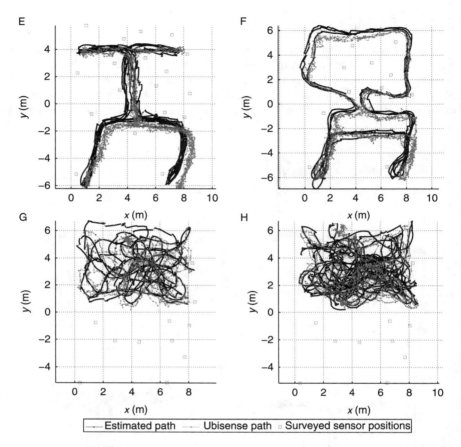

FIG. 12. Paths estimated from inertial pedestrian dead reckoning and ultrasonic measurements from sensors with known positions. (A) Large room. (B) All rooms (T) 1. (C) All rooms (T) 2. (D) All rooms (T) 3. (E) All rooms (T) 4. (F) All rooms (O). (G) Random 1. (H) Random 2.

system and do not directly map to the surveyed sensor positions without first finding and applying the most appropriate rotation and translation. Initially, they move a lot as more measurements are taken but even after they have stabilized they continue to drift slowly, and several sensors appear to move together in the same direction. Since our *map* is defined by the sensor positions, it also moves. In other words, this SLAM filter only gives positions of the sensors and of the pedestrian relative to the other sensors, not necessarily in an absolute coordinate system. This means that the

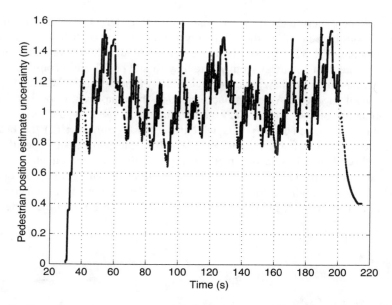

FIG. 13. Change in the covariance of the pedestrian position estimate—increases due to PDR error accumulation, decreases when ultrasonic measurements from known sensors are available.

filter might provide different coordinates for the pedestrian after he returns to a previous location but this could still be correct (in a relative sense) if the estimated positions of the sensors have also changed during that time. Conversely, if the estimated position of the pedestrian remains the same as previously but the estimated coordinates of the sensors have changed, the result could be incorrect (in a relative sense). This is because we are performing online SLAM. Offline methods such as GraphSLAM [41], which optimize the complete trajectory estimate and map after all data has been recorded, do not suffer from this limitation. They can be used to display the complete path and the landmarks on a single map.

Because of the shifting position estimates of the sensors, it is not necessarily helpful to plot the estimated path of the user. However, we can show snapshots of the sensor position estimates. In Fig. 15, we show the estimated positions of the sensors at the end of each experiment. These estimates were aligned to the surveyed sensor positions after running a nonlinear regression to determine the affine transformation (rotation and translation) that minimizes the sum of squared errors between surveyed and estimated positions. The line represents the Ubisense estimated path which is already in the same coordinate space as the surveyed sensor positions.

Three of the experiments give reasonable results in terms of sensor positions. Figure 15F was based on PDR data with only a small amount of drift so good results are expected, but Fig. 15E was similar in terms of drift but at least five of the sensors are placed more than 3 m away from where they should be. This is a consequence of the ambiguities in the initialization method—each of these sensors is placed on the wrong side of the path. As explained earlier in the chapter, this occurs due to the combination of two factors—(1) the measurements used for the initialization of the sensor are taken from points which are nearly collinear; (2) the bearing measurements are too noisy to determine which of the two possible positions is correct and our heuristic selection method (Fig. 7) fails. The consequences can be quite small if the misplaced sensor is only in range of the straight section of path with which it was initialized, for instance the two sensors in the upper right corner of Fig. 15E. But in general, these types of errors will create error in the pedestrian

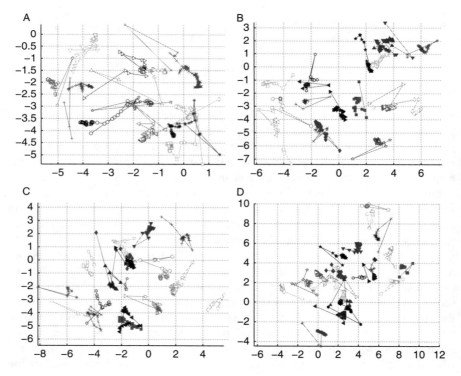

FIG. 14. (Continued)

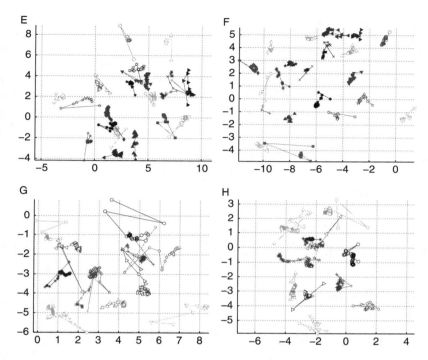

FIG. 14. Changes in the estimated positions of the sensors during the experiments. Initially the estimates change a lot, then they stabilize but continue to drift slowly. (A) Large room. (B) All rooms (T) 1. (C) All rooms (T) 2. (D) All rooms (T) 3. (E) All rooms (T) 4. (F) All rooms (O). (G) Random 1. (H) Random 2.

position estimates. This problem occurs for our data because the sensors were deployed in advance (although their positions were not made available to the SLAM algorithm). If the sensors were deployed by the pedestrian placing them at their feet as they walk around, then they could be directly initialized with the pedestrian's current position. The orientation would be trivial to initialize as the pedestrian moves on and a bearing measurement is taken. This type of manual deployment seems suitable for a real-world scenario if only a few sensors are required.

Figure 15G and H provides particularly good estimates. This is probably because these paths did not include any straight sections, therefore, the initialization was less likely to be ambiguous and sensors that were initialized incorrectly were adjusted thanks to the variety of range/bearing measurements taken from many different positions. In other words, the SLAM solution benefits from favorable geometric dilution of precision. This bears some similarity to situations where planes or ships are required to perform a particular maneuvre in order to improve tracking of a target [55].

In order to evaluate the performance of the filter in a more quantitative manner, we look at the range and bearing errors between the sensors, and between the sensors

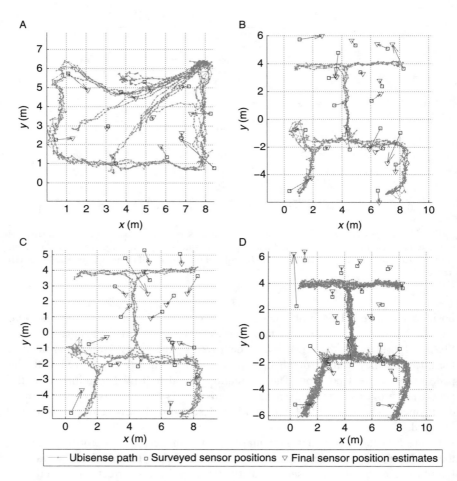

FIG. 15. (Continued)

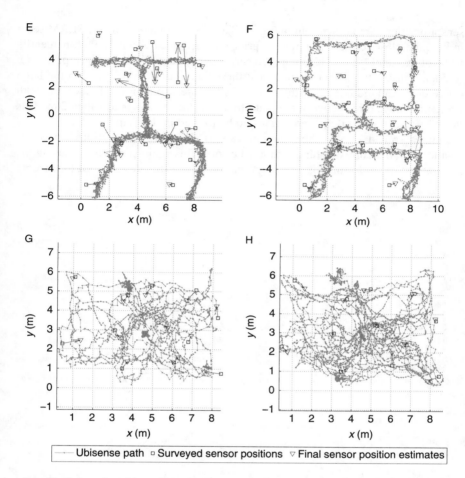

FIG. 15. Estimated positions of sensors at the end of each experiment rotated and translated to minimize the distance with the surveyed positions. (We have chosen not to show the estimated pedestrian path for reasons explained in Section 3.2.3). (A) Large room. (B) All rooms (T) 1. (C) All rooms (T) 2. (D) All rooms (T) 3. (E) All rooms (T) 4. (F) All rooms (O). (G) Random 1. (H) Random 2.

and the pedestrian for every update (Figs. 16 and 17). These errors reflect how accurately the sensors and the pedestrian are positioned relative to each other in the case where sensor positions are known in advance, and in the complete SLAM case with no prior knowledge. As expected, the errors for SLAM are higher than when the sensor positions are known. But in Fig. 16G and H, the range errors for SLAM seem to be quite close to the errors for the prior knowledge case, with a 90th percentile value of less than 1.2 m. This improved performance could be due to the type of path (random, unstructured) but we note that these data sets were recorded in the large office only, and did not cover the rest of the test area. In almost all cases, the 90th percentile range error between the pedestrian and the sensors is less than 2 m. This value reflects how well the pedestrian can be located in the map.

Figures 16 and 17 also show the innovations. The range innovations tend to be smaller than the corresponding estimated errors. This again suggests that using the position estimates from Ubisense as groundtruth overestimates some of the errors.

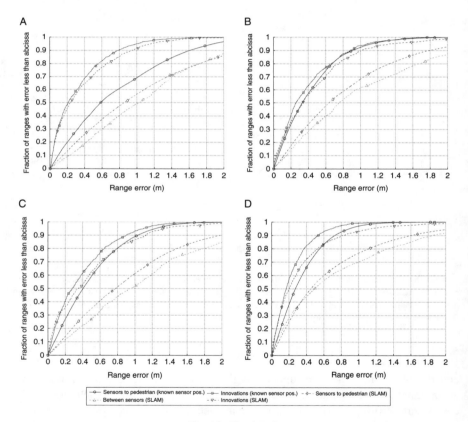

FIG. 16. (Continued)

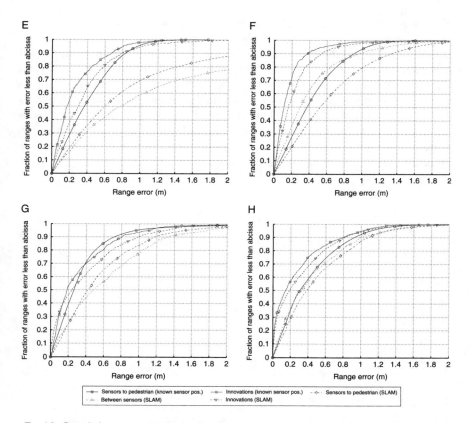

FIG. 16. Cumulative range error distributions for the full duration of each experiment. (A) Large room. (B) All rooms (T) 1. (C) All rooms (T) 2. (D) All rooms (T) 3. (E) All rooms (T) 4. (F) All rooms (O). (G) Random 1. (H) Random 2.

The bearing errors from Fig. 17 are more difficult to interpret because they depend on the position error of the pedestrian and sensors, and on the orientation error of the sensors. For instance, if a sensor's orientation is correct but the pedestrian is estimated to be a few centimeters in front of it, instead of a few centimeters behind it, the bearing error could be 180°.

In Figs. 16 and 17, we showed errors between all sensors, but it would also be reasonable to only take into account sensors that are either close to or far from the pedestrian depending on whether we are interested in local accuracy (position relative to nearby sensors, necessary for navigation) or global accuracy (position relative to sensors which are far away, necessary for mission planning). In Figs. 18

and 19, we have plotted separately the errors between sensors, and between pedestrian and sensors when they are less than 3 m apart, and those errors when they are more than 3 m apart. The 3 m limit is arbitrary, but corresponds to an area which could quickly be searched by a firefighter equipped with a long handled tool. These figures show that in many cases the local range error for SLAM is close to the error when the sensor positions are known. When the far range errors are larger than the local errors, this is due to large-scale distortion of the sensor positions. Large-scale distortion makes it difficult to overlay the estimated sensor positions onto a map or floorplan, but should not affect indoor navigation scenarios where a firefighter uses only nearby sensors as landmarks to progress toward a target in small steps.

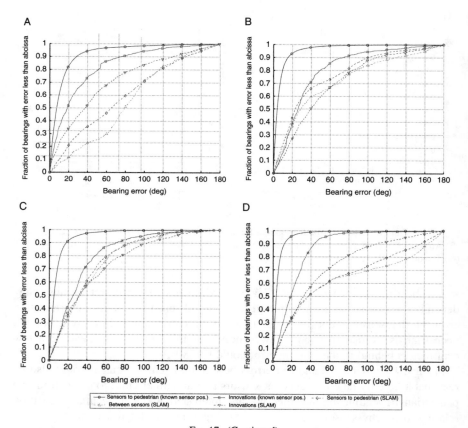

FIG. 17. (Continued)

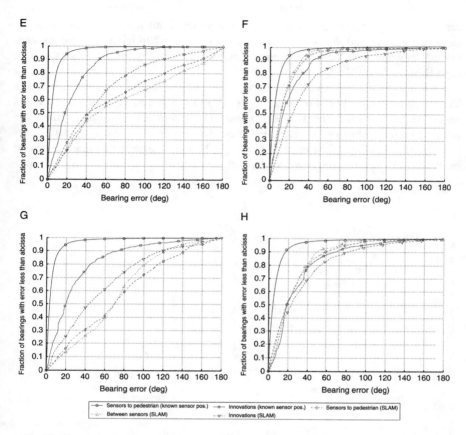

FIG. 17. Cumulative bearing error distributions for the full duration of each experiment. (A) Large room. (B) All rooms (T) 1. (C) All rooms (T) 2. (D) All rooms (T) 3. (E) All rooms (T) 4. (F) All rooms (O). (G) Random 1. (H) Random 2.

The bearing errors from far away sensors can be much smaller than the local bearing errors. Due to simple geometric properties, distance tends to dilute the effect of position error on the bearing. This is particularly visible in the case where sensor positions and orientations are known. SLAM bearing errors are generally large. Only Fig. 19F and H shows 90th percentile of bearing errors less than 60° for both short range and long range.

In some cases, despite our heuristic initialization method (Fig. 7), the SLAM algorithm sometimes places sensors on the wrong side of a straight section of path due to the symmetry ambiguity. This does not affect local range errors between the sensor and the pedestrian because of the symmetry but the errors for far away sensors are increased. Sensors in this situation are likely to have very inconsistent orientations and thus the bearing errors for both near and far away sensors will be high (Fig. 19C and G).

All our experiments can be split into sections of similar duration during which a similar path was walked. For experiments (B) to (F), the same path was repeated several times. For experiments (A), (G), and (H), the pedestrian returned to the same point at regular intervals. Whereas the previous figures show the aggregate errors over the full duration of the experiment, Figs. 20 and 21 give the median range and bearing errors for each section of the path. In a few cases, there is a noticeable

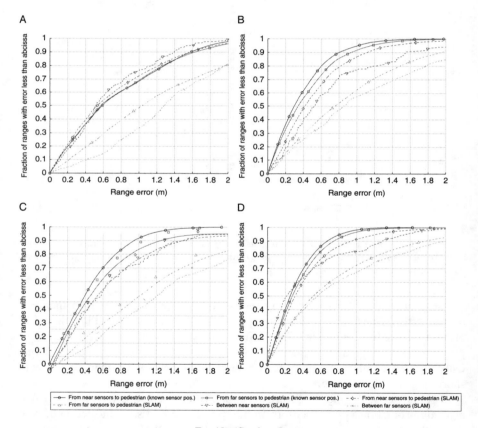

FIG. 18. (Continued)

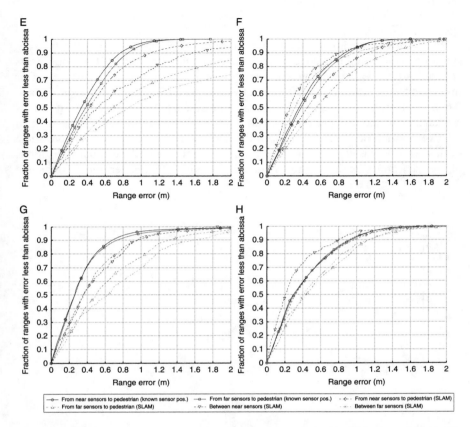

FIG. 18. Cumulative range error distributions for near sensors (≤ 3 m) and far sensors (> 3 m). (A) Large room. (B) All rooms (T) 1. (C) All rooms (T) 2. (D) All rooms (T) 3. (E) All rooms (T) 4. (F) All rooms (O). (G) Random 1. (H) Random 2.

improvement at each iteration but for the other cases the median error remains constant or even increases. For these latter cases, this could mean that sensor and pedestrian position estimates are as good as they are going to get after the first section and that there is no further improvement. After a firefighter has explored the building once and deployed the sensors, the system is immediately ready to help the following teams find their way.

3.3 Conclusions

We evaluated our implementation of an ultrasound-assisted pedestrian tracking system by comparing the ranges computed using the estimated positions of sensors and pedestrian with the ranges computed using the groundtruth positions. Our implementation achieves a median relative position accuracy of between 0.3 and 0.6 m when the sensor positions are known in advance. When the sensor positions are unknown and we perform SLAM, our implementation achieves a median relative position accuracy of between 0.4 and 0.8 m. In the SLAM version, some of this error is caused by large-scale distortions in the map (the estimated positions of the sensors). We have shown that the relative position errors between the

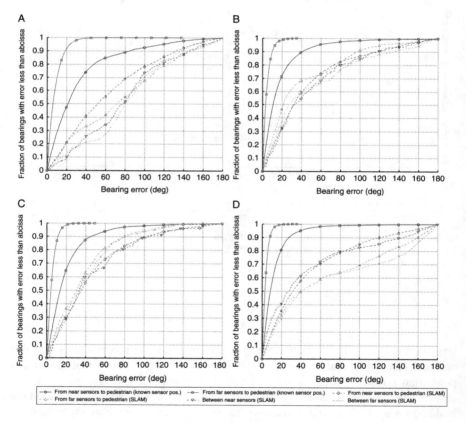

FIG. 19. (Continued)

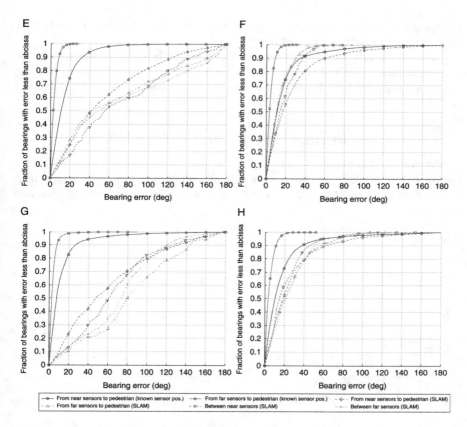

FIG. 19. Cumulative bearing error distributions for near sensors (≤3 m) and far sensors (>3 m). (A) Large room. (B) All rooms (T) 1. (C) All rooms (T) 2. (D) All rooms (T) 3. (E) All rooms (T) 4. (F) All rooms (O). (G) Random 1. (H) Random 2.

pedestrian and sensors closer than 3 m are comparable to the errors when the sensor positions are known.

Correct initialization of sensor positions is essential in order for the SLAM method to perform well. It appears to perform better when the paths are unstructured (not following a sequence of straight segments). We believe that this is primarily because sensor positions are more likely to be initialized correctly when the path contains many turns than when it is mostly straight.

For the SLAM method, there is no clear improvement of sensor position estimates as paths are repeated. However, the sensor orientations do improve with time. This result suggests that this type of SLAM system would work well if it was used in emergency response scenarios where the sensors are deployed on-the-fly, and where paths are not necessarily repeated many times.

Finally, our results suggest that some of the error which appears in our plots is due to inaccuracies in the UWB reference location system. This could mean that our system actually performs better than indicated by the results given in this chapter.

3.4 Improvements and Future Work

Although we believe the results we have presented confirm the viability of this type of hybrid tracking and navigation system, there are a number of areas which require further work. Here, we discuss a few avenues for further investigation and improvement.

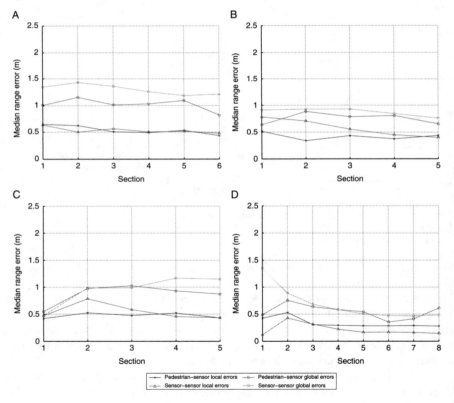

FIG. 20. (Continued)

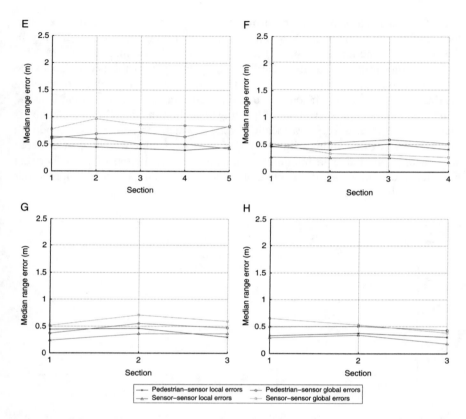

FIG. 20. Median range errors for each successive section of the path. (A) Large room. (B) All rooms (T) 1. (C) All rooms (T) 2. (D) All rooms (T) 3. (E) All rooms (T) 4. (F) All rooms (O). (G) Random 1. (H) Random 2.

3.4.1 Tracking Orientation

At the time of the experiment, our PDR system did not keep track of the orientation of the pedestrian, but only their direction of travel. Estimating the orientation of the pedestrian is a requirement if we are to provide them with navigation assistance, since they need to know what direction their target is in. There is a simple relationship between the direction of travel and the orientation of the inertial sensor, but a number of challenges remain. First, the orientation of the foot-mounted sensor does not accurately represent the orientation of the pedestrian or of the graphical interface that we are using to display our map. Second, in our current SLAM system, the orientation (and direction of travel) can only be corrected when the pedestrian is moving, by using the cross-correlation between the error in the estimated direction of travel and the error in the estimated position (an error in the estimated direction of travel will cause an error in the estimated position). This could be improved by directly measuring the bearing from the pedestrian to the landmarks (e.g., landmark X is behind the pedestrian). (Currently, we measure bearing but only from the landmarks to the pedestrian, e.g., the pedestrian is to the right of landmark Y.) Finally, it is challenging to evaluate how well the orientation is being tracked because very few location systems which can be used as groundtruth provide reliable orientation (a digital compass may work quite well in open areas but will become unreliable in buildings). Tracking the orientation of the pedestrian remains an essential part of a navigation system, so it is important to improve our system in this respect.

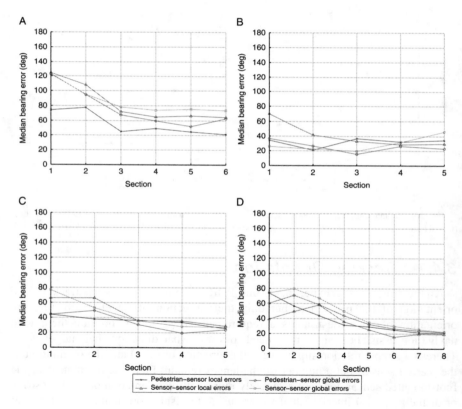

FIG. 21. (Continued)

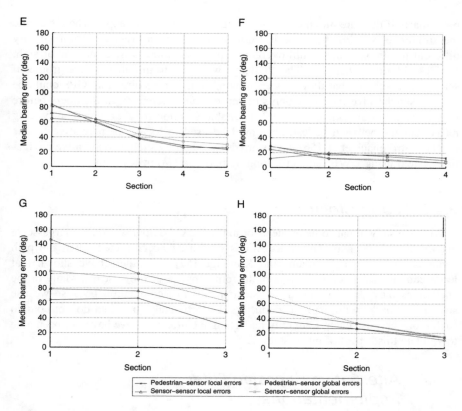

FIG. 21. Median bearing errors for each successive section of the path. (A) Large room. (B) All rooms (T) 1. (C) All rooms (T) 2. (D) All rooms (T) 3. (E) All rooms (T) 4. (F) All rooms (O). (G) Random 1. (H) Random 2.

3.4.2 Multiple Sensors

Groups of mobile sensors (both inertial and ultrasonic) could be used to great advantage. These sensors could be installed on both feet of the pedestrian or on multiple pedestrians traveling together. We anticipate that such scenarios would not only provide a lot more data than a single sensor in the same amount of time but would also benefit from the fact that the errors on each sensor are independent from the others. For instance, PDR errors are typically due to thermal noise and small vibrations in the MEMS components which will be different for each sensor particularly if they are mounted on different feet or different people. The ultrasonic measurements would also benefit from additional measurements from slightly different positions which could overcome many situations where the line of sight is blocked by the other foot. The challenge would be to initialize all the groups of sensors correctly so that they all start moving in the same direction. GPS could be used to synchronize the movements of sensors on different team members before they enter a building. If GPS is not a suitable solution maybe range and bearing measurements between pedestrians could be used to locate them relative to one another. Strömbäck et al. [56] have started investigating such a cooperative navigation system using UWB radios for ranging between two pedestrians.

3.4.3 Loop Closure

We have shown that some of the error in the SLAM output was due to large-scale distortions in the map. These distortions can often be corrected when the path contains large loops. Gutmann and Konolige [57] and others have shown that loops can be used to correct maps built from dense laser range scans. Many of the challenges they describe are nonexistent in our implementation because we use sparse landmarks with explicit identifiers. Unfortunately, our data do not contain such loops because each room only had a single entrance, so we have not been able to validate this aspect of our system.

3.4.4 Alternative Bayesian Filter Implementations

We have noted problems due to the nonlinearities in the process and measurement functions which cause the location estimate error to be underestimated. These could possibly be solved by using alternatives such as a particle filter or an unscented Kalman filter. The particle filter would also have the benefit of resolving the ambiguity in the initialization of sensor positions since it inherently allows multi-modal solutions. The initialization problem may also be solved more simply by having the pedestrian deploy the sensors just next to their feet so that the sensor positions can be directly initialized using the position of the pedestrian.

3.4.5 Evaluation Methods

Finally, the evaluation of such infrastructure-less tracking and navigation systems is problematic. First, it is difficult to obtain any accurate ground truth for large areas inside buildings. Second, even when accurate, high-update rate, wide area ground truth is available, it cannot easily be aligned to the estimates due to coordinate systems which change slightly over time. Some of the methods we have used in this chapter are useful evaluation tools, but we hope that future research will provide additional methods of comparing different navigation systems before passing them over to users for testing.

4. Conclusion

In this chapter, we have focused on the use of a sensor network to provide *positioning* and *tracking* capabilities that can directly support emergency responders. Although the application of sensor networks to support emergency response and in particular firefighting has been explored in a range of projects by predeploying positioning infrastructure, our research in contrast focuses on an *ad hoc* approach that does not require any predeployment.

Based on the understanding of the errors encountered in the PDR location estimates, we have looked into complementary technologies that can correct the drift. Specifically, we have used ultrasound sensors which have the capability to measure relative range and bearing.

We have used an EKF-based SLAM method to concurrently estimate the location of the deployed landmarks and the position of the pedestrian. Fusing ultrasound measurements with inertial tracking enables us to prevent the drift which makes inertial tracking alone unreliable. Ninety percent of range errors in our tests were less than 2 m even when no prior information about the environment was available. The deployment of small sensors during an intervention is consistent with practices in certain firefighting units where a first team searches a building before further teams enter to assist victims and attack the fire. Although there is plenty of room for improvement, we believe our EKF implementation provides location estimates suitable for indoor navigation of emergency responders and shows that the combination of ultrasound sensors deployed on-the-fly and inertial PDR is a viable solution.

We strongly believe that such a combination of modalities (inertial and ultrasound) can be extended to provide a fully functional *ad hoc* positioning system for tracking and guiding emergency responders. This chapter focused on the design of the system and we have shown using real-world data that this could be a viable solution; however, more work is required on both the sensing technology robustness and the algorithm to improve its characteristics.

References

[1] C. Fischer, H. Gellersen, Location and navigation support for emergency responders: a survey, IEEE Pervasive Comput. 9 (2010) 38–47.
[2] K. Muthukrishnan, Multimodal Localisation: Analysis, Algorithms and Experimental Evaluation, 2009, Ph.D. dissertation, University of Twente.
[3] R.F. Fahy, U.S. Fire Service Fatalities in Structure Fires, 1977–2000, NFPA, Quincy, MA, 2002, Technical report.
[4] National Institute for Occupational Safety and Health, Fire Fighter Fatality Investigation and Prevention Program, 2009, http://www.cdc.gov/niosh/fire/ (accessed 17.01.2011).
[5] M. Klann, Tactical navigation support for firefighters: the LifeNet Ad-Hoc Sensor-Network and Wearable System, in: J. Löffler, M. Klann (Eds). Mobile Response, Springer, Berlin, 2009, pp. 41–56.
[6] T.E. Sendelbach, Search Line Survival Training, TES2 Training, Savannah, GA, 2002, http://www.tes2training.com/handouts/search_line_survival_training.pdf (accessed 17.01.2011).
[7] National Fire Protection Association, NFPA 1500—Standard on Fire Department Occupational Safety and Health Program, National Fire Protection Association, Quincy, MA, 2002, Technical report.
[8] M.K. Donnelly, W.D. Davis, J.R. Lawson, M.J. Selepak, Thermal Environment for Electronic Equipment Used by First Responders, NIST, 2006, Technical report 1474, http://www.fire.nist.gov/bfrlpubs/fire06/art001.html (accessed 17.01.2011).
[9] International Association of Fire Chiefs, Fundamentals of Fire Fighter Skills, Jones and Bartlett Publishers, Inc., Sudbury, MA, 2004.
[10] W.E. Clark, Firefighting Principles & Practices, PennWell Corporation, Tulsa, OK, 1991.
[11] M. Worrell, A. MacFarlane, Phoenix Fire Department Radio System Safety Project, City of Phoenix Fire Department, Phoenix, AZ, 2004, Technical report.
[12] D. Steingart, J. Wilson, A. Redfern, P. Wright, R. Romero, L. Lim, Augmented cognition for fire emergency response: an iterative user study, in: Augmented Cognition, Proceedings of the 11th International Conference on Human-Computer Interaction (HCI), Las Vegas, NV, 2005.
[13] J. Wilson, V. Bhargava, A. Redfern, P. Wright, A Wireless Sensor Network and Incident Command Interface for Urban Firefighting, in: Proceedings of the Fourth Annual International Conference on Mobile and Ubiquitous Systems: Networking & Services, IEEE, 2007, pp. 1–7.
[14] G. Kantor, S. Singh, R. Peterson, D. Rus, A. Das, V. Kumar, et al., Distributed search and rescue with robot and sensor teams, in: Proceedings of the 4th International Conference on Field and Service Robotics, vol. 24, Springer, Berlin, 2003.
[15] J. Djugash, S. Singh, G. Kantor, W. Zhang, Range-only SLAM for Robots operating cooperatively with sensor networks, in: Proceedings of International Conference on Robotics and Automation (ICRA), IEEE, 2006, pp. 2078–2084.
[16] C. Fischer, K. Muthukrishnan, M. Hazas, H. Gellersen, Ultrasound-aided pedestrian dead reckoning for indoor navigation, in: Proceedings of International Workshop on Mobile Entity Localization and Tracking in GPS-less Environments (MELT), ACM, New York, NY, 2008, pp. 31–36.
[17] Dräger Safety AG & Co, Device and process for guiding a person along a path traveled—United States Patent 7209036, 2007, http://www.freepatentsonline.com/7209036.html (accessed 17.01.2011).
[18] V. Amendolare, D. Cyganski, R.J. Duckworth, S. Makarov, J. Coyne, H. Daempfling, et al., WPI precision personnel location system: inertial navigation supplementation, in: Position Location and Navigation Symposium, Monterey, CA, IEEE, 2008.
[19] D. Graham-Rowe, Indoor 'sat-nav' could save firefighters, New Sci. 196 (2634) 24. (2007).
[20] L.E. Miller, Indoor Navigation for First Responders: A Feasibility Study, National Institute of Standards and Technology, 2006, Technical report, http://www.antd.nist.gov/wctg/RFID/Report_indoornav_060210.pdf (accessed 17.01.2011).

[21] V. Renaudin, O. Yalak, P. Tomé, B. Merminod, Indoor Navigation of Emergency Agents, Eur. J. Navigation 5 (3) (2007) 36–45.
[22] W. Widyawan, M. Klepal, S. Beauregard, A backtracking particle filter for fusing building plans with PDR displacement estimates, in: Proceedings of the 5th Workshop on Positioning, Navigation and Communication (WPNC'08), Hannover, Germany, IEEE, 2008, pp. 207–212.
[23] O. Woodman, R. Harle, Pedestrian localisation for indoor environments, in: Proceedings of the 10th international conference on Ubiquitous computing (Ubicomp), vol. 344, ACM, New York, 2008, pp. 114–123.
[24] M.W.M. Gamini Dissanayake, P. Newman, S. Clark, H.F. Durrant-Whyte, M. Csorba, A solution to the simultaneous localization and map building (SLAM) problem, IEEE Trans. Rob. Autom. 17 (3) (2001) 229–241.
[25] J. Djugash, S. Singh, P. Corke, Further results with localization and mapping using range from radio, in: Proceedings of International Conference on Field & Service Robotics (FSR), 2005, pp. 231–242.
[26] J. Guivant, E. Nebot, S. Baiker, Autonomous navigation and map building using laser range sensors in outdoor applications, J. Rob. Syst. 17 (2000) 3817–3822.
[27] J. Kim, S. Sukkarieh, Real-time implementation of airborne inertial-SLAM, Rob. Auton. Syst. 55 (2007) 62–71.
[28] P. Newman, J. Leonard, Pure range-only sub-sea SLAM, in: Proceedings of the IEEE Conference on Robotics and Automation (ICRA), 2003, pp. 1921–1926.
[29] E. Olson, J. Leonard, S. Teller, Robust range-only beacon localization, IEEE J. Ocean. Eng. 31 (4) (2006) 949–958.
[30] S. Thrun, Robotic mapping: a survey, in: Exploring Artificial Intelligence in the New Millenium, Morgan Kaufmann, Burlington, MA, 2002.
[31] H. Durrant-Whyte, T. Bailey, Simultaneous localisation and mapping (SLAM): part I the essential algorithms, IEEE Rob. Autom. Mag. 13 (2006) 99–108.
[32] T. Bailey, H. Durrant-Whyte, Simultaneous localization and mapping (SLAM): part II, IEEE Rob. Autom. Mag. 13 (3) (2006) 108–117.
[33] J. Guivant, E. Nebot, Optimization of the simultaneous localization and map building algorithm for real time implementation, IEEE Trans. Rob. Autom. 2001.
[34] J.J. Leonard, H.J.S. Feder, A computationally efficient method for large-scale concurrent mapping and localization, in: Proceedings of the Ninth International Symposium on Robotics Research, Snowbird, UT, Springer-Verlag, 2000.
[35] S. Thrun, Y. Liu, D. Koller, A.Y. Ng, Z. Ghahramani, H. Durrant-Whyte, Simultaneous localization and mapping with sparse extended information filters, Int. J. Rob. Res. 23 (7/8) (2004) 693–716.
[36] M. Montemerlo, S. Thrun, D. Koller, B. Wegbreit, FastSLAM: a factored solution to the simultaneous localization and mapping problem, in: Proceedings of the AAAI National Conference on Artificial Intelligence, AAAI, Edmonton, Canada, 2002.
[37] M. Montemerlo, S. Thrun, Simultaneous localization and mapping with unknown data association using FastSLAM, in: Proceedings of IEEE International Conference on Robotics and Automation (ICRA), 2003, pp. 1985–1991.
[38] T. Bailey, J. Nieto, J. Guivant, M. Stevens, E. Nebot, Consistency of the EKF-SLAM algorithm, in: Proceedings of the IEEE/RSJ International Conference on Intelligent Robots and Systems (IROS), 2006.
[39] A. Brooks, T. Bailey, HybridSLAM: combining FastSLAM and EKF-SLAM for reliable mapping, in: Algorithmic Foundation of Robotics VIII, Springer Tracts in Advanced Robotics, vol. 57, Springer, Berlin, 2009, pp. 647–661.

[40] A. Kleiner, D. Sun, Decentralized SLAM for pedestrians without direct communication, in: Proceedings of the IEEE/RSJ International Conference on Intelligent Robots & Systems (IROS), 2007.
[41] S. Thrun, M. Montemerlo, The GraphSLAM algorithm with application to large-scale mapping of urban structures, Int. J. Rob. Res. 25 (5–6) (2006) 403–429.
[42] J.-P. Tardif, Y. Pavlidis, K. Daniilidis, Monocular visual odometry in urban environments using an omnidirectional camera, in: Proceedings of IEEE/RSJ International Conference on Intelligent Robots and Systems, 2008.
[43] J.-L. Blanco, J.-A. Fernandez-Madrigal, J. Gonzalez, Efficient probabilistic range-only SLAM, in: IEEE/RSJ International Conference on Intelligent Robots and Systems (IROS), 2008, pp. 1017–1022.
[44] M. Deans, M. Hebert, Experimental comparison of techniques for localization and mapping using a bearing-only sensor, in: Proceedings of the ISER '00 Seventh International Symposium on Experimental Robotics, Springer-Verlag, 2000.
[45] B. Cinaz, H. Kenn, HeadSLAM—Head-mounted simultaneous localization and mapping for wearable computing applications, in: Adjunct Proceedings of Pervasive 2008, Sydney, Australia, Austrian Computer Society, 2008, pp. 9–13.
[46] P. Robertson, M. Angermann, B. Krach, Simultaneous localization and mapping for pedestrians using only foot-mounted inertial sensors, in: Proceedings of UbiComp, Orlando, FL, ACM, 2009.
[47] E. Foxlin, Pedestrian tracking with shoe-mounted inertial sensors, Comp. Graph. Appl., IEEE 25 (6) (2005) 38–46.
[48] S. Beauregard, Omnidirectional pedestrian navigation for first responders, in: Proceedings of the 4th Workshop on Positioning, Navigation and Communication (WPNC'07), 2007, pp. 33–36.
[49] E. Foxlin, Motion tracking requirements and technologies, in: Kay Stanney (Ed.), Handbook of Virtual Environment: Design, Implementation, and Applications, Lawrence Erlbaum Associates, Hillsdale, NJ, 2002, pp. 163–210.
[50] S. Beauregard, Infrastructureless Pedestrian Positioning, 2009, Ph.D. thesis, University of Bremen.
[51] M. Hazas, C. Kray, H. Gellersen, H. Agbota, G. Kortuem, A. Krohn, A relative positioning system for co-located mobile devices, in: Proceedings of the 3rd international conference on Mobile systems, applications, and services (MobiSys), ACM Press, New York, NY, 2005, pp. 177–190.
[52] D. Fox, J. Hightower, L. Liao, D. Schulz, G. Borriello, Bayesian filtering for location estimation, IEEE Pervasive Comput. 2 (3) (2003) 24–33.
[53] D. Simon, Optimal State Estimation, first ed., John Wiley & Sons, Inc., Hoboken, NJ, 2006.
[54] K. Muthukrishnan, M. Hazas, Position estimation from UWB pseudorange and angle-of-arrival: a comparison of non-linear regression and Kalman filtering, in: Proceedings of Location and Context Awareness (LoCA), Tokyo, Japan, Springer-Verlag, 2009.
[55] T.L. Song, Observability of target tracking with range-only measurements, IEEE J. Ocean. Eng. 24 (1999) 383–387.
[56] P. Strömbäck, J. Rantakokko, S.-L. Wirkander, M. Alexandersson, K. Fors, I. Skog, et al., Foot-Mounted Inertial Navigation and Cooperative Sensor Fusion for Indoor Positioning, KTH, Stockholm, Sweden, 2009, Technical report.
[57] J.-S. Gutmann, K. Konolige, Incremental mapping of large cyclic environments, in: Proceedings of IEEE International Symposium on Computational Intelligence in Robotics and Automation, Daejeon, South Korea, 1999.
[58] S. Wan, E. Foxlin, Improved Pedestrian Navigation Based on Drift-Reduced MEMS IMU Chip, Proc. ION 2010, International Technical Meeting, San Diego, CA, 2010, pp. 220–229.

Feeling Bluetooth: From a Security Perspective

PAUL BRAECKEL

Identity Theft and Financial Fraud Research and Operations Center, University of Nevada, Las Vegas, Nevada, USA

Abstract

Wireless technologies are ubiquitous and an essential aspect in modern computing and mobile devices. In particular, Bluetooth (BT), if not only from its namesake, has captured end-user attention and acceptance; however, few possess familiarity with it in much depth aside from the obligatory accessory add-on when purchasing a new mobile phone. This article serves as both an advanced-level continuous-read self-study and a quick reference to BT technologies by presenting its history concisely, technical specifications, notable exploitations, audit utilities, and securing recommendations. One will notice similarities in comparison to other wireless network technologies, such as Wi-Fi (802.11). Briefly, BT is a device- independent open specification wireless networking technology, overseen by the BT Special Interest Group (SIG), which targets mobile computing and available to devices on a global scale. The specifications that define the BT protocols intend the technology for connectivity and communications within Personal Area Networks (PANs). BT is the amalgamation of a hardware description, an application framework, and interoperability requirements, with the main purposes of replacing cables, creating *ad hoc* networks, and establishing data/voice connectivity.

1. Bluetooth Details 162
   1.1. Wireless Basics 165
   1.2. Namesake Origin 169
   1.3. Specifications and Compliance 169
   1.4. Market Presence 170

1.5. Wireless Medium 172
  1.6. Network Topology 173
  1.7. Protocol Stack and Profiles 178
  1.8. Packet Format 183
  1.9. Transmission Range 188
2. BT Exploitation 189
  2.1. Exploiting the BT Protocol 190
  2.2. Exploiting a Wireless Device 191
  2.3. Limitations for Propagation 197
  2.4. Categories of BT Malware 200
  2.5. BT Malware Firsts 207
3. BT Auditing Tools 218
  3.1. BT Wardriving 219
  3.2. BlueScanner from Network Chemistry 221
  3.3. BlueSweep from AirMagnet 223
  3.4. Bluesniff from The Shmoo Group 223
  3.5. BTScanner from Pentest Ltd. 225
  3.6. BT_Audit from trifinite.org 226
4. Security Recommendations 228
  4.1. Device's Defaults 229
  4.2. BT Status 229
  4.3. Secure Pairing 230
  4.4. Secure Pass-Phrases 232
  4.5. Installing Applications 234
  4.6. Antivirus Application 234
5. Conclusion 235

1. Bluetooth Details

The technology niche for Bluetooth (BT) is its short-range wireless networking of computing devices and general ease of use. This niche has achieving acceptance on a global level, which is evident from the findings of two publications: a study by ABI Research, which is a New York-based technology market research firm, and a report from the BT Special Interest Group (SIG). These studies concluded that as of

November 2006, 1 billion BT devices have shipped worldwide.[1] The initial products to come to the consumer market hit two main applications: eliminated the physical medium for desktop connectivity known as a cabled environment and facilitated voice/data transmission between devices. An everyday BT device that has become omnipresent is the BT earpiece, which uses the BT hands-free profile to network wirelessly a headset and a mobile phone. The combination of this particular application's social acceptance as well as support from legislation passed or passing in numerous states to secure talking on the phone while driving will result in it making up 75% of the BT market.[2] These same reports also forecast that by the year 2013, 30% of new cars will support this hands-free feature. Therefore, based on these two forecasts for just one specific use of BT and the expectation for technology to evolve based on this moment, one can say that BT has indeed established a niche for itself.

In line with these forecasts, the BT market presence is expanding. The following list illustrates the range of the available BT applications. This list progresses from general to more specific and is by no means an exhaustive list of applications; following each application is a brief description as well as an example:

- *Mobile Computing*: Network specialized BT compliant devices that support complementary usage profiles. For example, the previously mentioned headset and phone scenario.
- *Personal Area Networking*: Connect peripheral devices to a networked computer without requiring physical attachment. For example, mouse and keyboard connect to a personal computer.
- *File Sharing*: Distribute data via dynamic *ad hoc* networking. For example, a device distributes marketing information directly to a perspective customer's device at a trade show.
- *Task Automating*: Mobile networks permit devices to travel in and out of connectivity range. Tasks synchronizing may be set to occur as devices enter or depart an established network. For example, PDA synchronizes to a PC.
- *Mobile Entertainment*: There is no need to have CDs, DVDs, other physical media, or even have access to broadcast radio as mobile phones act as the transport for one's favorites. For example, a car's stereo has access to the music synced to a passenger's mobile phone.
- *Residential Networking*: Remotely access and manage household appliances. For example, open the garage door for a UPS delivery. The eternal question concerning leaving the iron on needs no longer be posed.

---

[1] www.enterprisenetworksandservers.com/monthly/art.php?2838 (@ October 25, 2010).
[2] www.itfacts.biz/24-bln-bluetooth-enabled-devices-to-ship-worldwide-by-2013/11380 (@ October 25, 2010).

- *Gaming Applications*: Dynamically network handheld gaming devices and accessories resulting in an interactive multiplayer wireless gaming based on the proximity of gaming devices. For example, the iPad edition of Scrabble allows gamers with iPhones/iTouches using the Tile Rack app to flick tiles to the game board.
- *Academia Applications*: Administer students and information in an established classroom Personal Area Network (PAN) environment. For example, issue and collect examinations. *"Just one more minute"* from stalling students is eliminated.
- *Tracking Applications*: Inventory or locate objects and people through a network of sensors. For example, airports can confidently process luggage and passenger from passenger checking to baggage-claim redemption. This eliminates the need for airlines to make *last-call* for boarding because they are able to account for the passengers and their luggage.
- *Travel and Tourism Applications*: Eliminate paper handouts and create customized digital literature. For example, information kiosk at national parks issue background information, must-see points-of-interest, weather conditions, and suggested available parking when an automobile enters the park.
- *Surveillance Applications*: Monitor audio from microphones moving dynamically in/out range. For example, to obtain criminal evidence of a conversation in a hotel room, an operative conceals an audio hub in the room and records the conversation streamed wirelessly to the adjacent room.

This section presents details on the BT core specifications and profile usage models as specified by the BT SIG, which, as mentioned previously, is the trade association of device manufactures and governing body that manages and assures compliance to the BT specifications. The current version of this core specification is Version 4.0, was released June 30, 2010, defines BT v3.0+HR. The *core specifications* are the hardware and software implementation details that devices must adhere to classify as BT compliant. Each release of these core specifications builds on the previous versions and assures backward compatibility as the specifications evolve. The *profile usages models* are the services that utilize the BT technology, for example, dial-up network profile (DUN) would allow a laptop to establish Internet connectivity through a mobile phone's network connectivity. Since it's members are the manufactures that create BT devices, the BT SIG is very sensitive to the market demands for the technology and the profiles that permit its usage. As a result, the available profiles increase in number and variety as the BT specifications continue to evolve. It will also be shown, in this section, where BT falls within the wireless spectrum as it relates to other wireless technologies. Throughout this chapter, the terminology *device* refers to a BT-enabled computing device, such as a BT-enabled mobile phone.

1.1 Wireless Basics

The BT wireless technology, more commonly known as simply Bluetooth, is a member of the Wireless Networking Standards family of protocols. Approved by the Institute of Electrical and Electronic Engineers (IEEE) Standards Board in June 2002 based on v.1.1 of the BT Specifications, BT is the IEEE 802.15.1 standard.[3] The specifications that define the BT protocols intend the technology for connectivity and communications standards for PANs. These specifications apply to a large spectrum of devices that do not require high throughput, ranging from mobile phones to desktop computers. There are three design features[4] that differentiate BT from other members of wireless standards:

1. Low power consumption (10 μA–50 mA)
2. Inexpensive implementation (under $3 USD/radio transceiver)
3. Minimal physical footprint (chip size approximately 9 mm^2)

To understand how these design restrictions uniquely identity BT, one might compare the technology stands in comparison to other wireless technologies. Although these technologies tend to overlap in functionality, the main differentiating wireless factors are data throughput, power consumption, access range, and mobile application. To define each of these factors, data throughput is the maximum bandwidth achievable by the device. Power consumption is the amount of battery life required for contiguous usage. Each of these technologies achieves a different level of power consumption based on available power, for example, a wireless technology using a laptop device platform would have different low-power classification then a mobile phone platform. Access range is the maximum physical distance over which two wireless devices are able to communicate. Last, the mobile application is the end-user application where the technology satisfies a prominent need and has achieved acceptance.

Figure 1 and the list that follows break the most common wireless technologies into computing networks. This list provides the name of each technology and its defining wireless standard, starting with the widest wireless networks and progresses to the most narrow:

- *Wireless Wide Region Area Network—Wireless WRAN*
    a. IEEE 802.22 (2004)
        - Broadband (DSL and Cable) throughput, very large-range.

---

[3] http://ieee802.org/15/index.html (@ October 25, 2010).
[4] www.bluetooth.com/English/Technology/Works/Pages/Bluetooth_low_energy_technology.aspx (@ October 25, 2010).

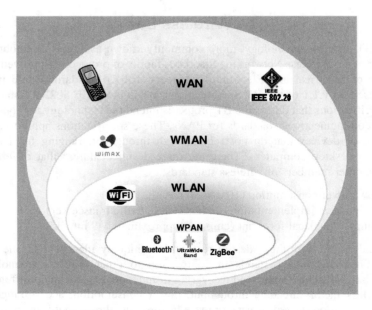

FIG. 1. Wireless standards networking and range relationships.

- Very new specifications that will achieve broadband throughput utilize the unused and unlicensed portion of the broadcast VHF and UHF TV channels.

- *Wireless Wide Area Network—Wireless WAN*
  b. Mobile/Cellular Telephone (0G, 1G, 2G, 3G, 4G...) (1946...)
     - Medium throughput, medium to high costs, medium to high power, long range.
     - A telephone standard for voice and nonvoice data.
     - Numerous standards exist depending on the generation of the cellular technology, but the 4th Generation promises a convergence to a more global standard.
  c. Mobil Broadband Wireless Access—WBWA (IEEE 802.20) (2002)
     - Broadband (DSL and Cable) throughput, medium cost, long range, and user moving at high speeds.
     - Design intention is to fill a niche for mobile WiMAX networking.

- *Wireless Metropolitan Area Network—Wireless MAN*
  d. Worldwide Interoperability for Microwave Access—WiMAX (IEEE 802.16) (2001)
     - Fixed broadband throughput, very long range (30 miles).
     - Application intentions are to connect Wi-Fi networks to achieve the wireless proverbial *last-mile* coverage.
  e. 3rd Generation Partnership Project Long Term Evolution - 3GPP LTE (UMTS/4GSM) (2004)
     - Broadband throughput, long range.
     - Application is mobile device Internet connectivity.
- *Wireless Local Area Network—WLAN*
  f. Wi-Fi (IEEE 802.11) (1997)
     - Medium to high throughput, medium cost,[5] medium power,[6] medium range.
     - Wi-Fi is a very well-accepted networking technology and is synonymous with WLAN. A common misnomer is that Wi-Fi is in direct competition with BT, but, although it may have similar applications, it in fact fills a separate wireless networking application.
- *Wireless Personal Area Network—WPAN*
  g. BT (IEEE 802.15.1) (2002)
     - Low throughput, very low cost, low power, medium to short range.
     - Details discussed within.
  h. Ultra-wideband—UWB (IEEE 802.15.3) (2003)
     - High throughput, low power, short range.
     - A category of networking that encompasses any wireless technology with a bandwidth exceeding 500 MHz.
  i. ZigBee (IEEE 802.15.4) (2003)
     - Low throughput, low cost, very low power, short range.
     - A very young wireless technology with products arriving at market in 2006 using its v.1.0 specification, released in 2005. Future applications planned specifically for embedded device using mesh networks.
  j. Wibree[7] (assimilated into BT standard) (2006)
     - Low throughput, low cost, ultra low power, short range.

---

[5] Bluetooth implementation costs are commonly estimated at one-third the price of Wi-Fi.
[6] Bluetooth technology power consumption is commonly estimated at one-fifth the amount of Wi-Fi.
[7] research.nokia.com/node/254 (@ November 20, 2009). Not depicted in Fig. 1 as it is part of the Bluetooth specification and called Bluetooth low energy as of June 2007.

- Initially developed by Nokia, optimized for its very low power consumption, smaller transceiver compared to Wi-Fi and BT, resulting in small form factor products.

- *Other Computing Networks*
    k. Radio Frequency Identification—RFID (ISO 18000, EPCglobal...)
        - Very low throughput, low cost, very low power wireless technology used primarily for wireless identification.
        - There is not global governing body for this technology and therefore numerous standards exist worldwide. ISO and EPC are the generally accepted standards.
    l. Infrared—IrDA (IrDA-SIR, IrDA-MIR, IrDA-FIR, IrDA-VFIR,...) (1993...)
        - Very low throughput, low cost, very short-range wireless networking technology that requires direct line of site between devices and prone to light interference.
        - Infrared Data Association (IrDA)[8] establishes these standards.
    m. Wireless Universal Serial Bus—Certified Wireless USB[9]
        - High throughput, low cost, low power consumption wireless networking technology built on the UWB platform.
        - This technology standard is currently still under development and is suggested that will incorporate UWB technologies as its foundation.

One quick note before continuing, as these technologies evolve they tend to borrow from the successes of each other, resulting in overlapping or converging technologies and the bounders between their usages becoming extremely mucky. For example, on April 21 2009, the BT Alliance announced the release of version 3.0 + HS, which will allow transfer speeds between supported devices of approximately 10 times that of BT 2.1, rivaling Wi-Fi throughput and at the same time retaining the lower power BT requirement. Subsequently, on October 14 2009 the Wi-Fi Alliance, the consortium that oversees 802.11, announced the 2010 release of a technology referred to as Wi-Fi Direct,[10] which will facilitate the creation of *ad hoc* networking of those supported devices rivaling the BT PAN. Technically, each of these technologies serves a specific target usage very well; however, time will determine which technology evolves past this challenge.

---

[8] www.irda.org (@ June 11, 2007).
[9] www.usb.org/developers/wusb/ (@ June 11, 2007).
[10] http://www.businessweek.com/technology/content/oct2009/tc20091013_683659.htm (@ November 20, 2009).

1.2 Namesake Origin

The Swedish telecommunications corporation Ericsson Mobile Communications developed the wireless communications standard BT in 1994 to allow laptops network access using a mobile phone's networking technologies[11] and hands-free, wireless mobile phone environments.[12] Through avid marketing for technology partnerships, by December 1999, the group of promoters included a strong corporate following including 3Com Corporation, Ericsson Corporation, IBM Corporation, Intel Corporation, Lucent Corporation, Microsoft Corporation, Motorola Inc., Nokia Corporation, and Toshiba Corporation.[13] The technology has experienced a slight lethargy while evolving into a strong market presence. However, the combination of more favorable implementation costs and a global explosion of mobile phone market have brought BT to general awareness and subsequent acceptance by the public.

Why did Ericksson select the name Bluetooth? Technology historians may dispute the finer details; however, the technology's namesake stems from a Viking/Danish king by the name Harald Blåtand II, in the tenth century AD. King Blåtand ruled a period from 940 to 981 AD and is credited with uniting the countries Denmark and Norway, converting these areas to Christianity, and was in the end dethroned by his own son. The king's surname literally translated to "Bluetooth" and was a reference to his dark hair and complexion. Aside from Ericsson Scandinavian heritage in the history, perhaps the name was in reference to King Blåtand's ability to control two adjacent countries with a distance of water between them. One could make the analogy to a BT-enabled device controlling another BT-enabled device over a short distance. Whatever the exact reasoning behind the name selection, BT was the name given to this Ericsson developed wireless technology.[14] The BT SIG owns the Bluetooth name, trademark, and logos.

1.3 Specifications and Compliance

The BT architectural combines a hardware and software solution with a logical division for simplification into three core components: a transceiver, a Baseband, and a protocol stack. The radio frequency (RF) transceiver is responsible for sending and receiving communications to and from the device. The BT Baseband is responsible for managing the BT networking between devices. The BT protocol stack is responsible for processing the communications and providing the service profiles for

---

[11] http://www.gsmfavorites.com/documents/bluetooth (@ November 25, 2010).
[12] http://www.techonline.com/article/192200674 (@ October 24, 2006).
[13] Brent Miller, Chatschik Bisdikian, *Bluetooth Revealed* (Prentice Hall, 2000), Chapter 16.
[14] www.ericsson.com/technology/tech_articles/Bluetooth.shtml (@ June 11, 2007).

BT functionality between networked devices. The BT SIG, which is the governing body over the BT technology, initially approved and released the document detailing the specifications and compliance for these components in 1999. Since this initial release, the BT technology and its specifications have evolved through several revisions. Table I lists the publicly released specification revisions, starting with this initial release in 1999. As per the BT SIG, the most recent official specification was released in April 2009 and they have slated several functionalities for future versions. This newest released and approved version of BT, v. 3.0+HS, which stands for *High Speed*, retains the simplicity and easy of connection already expected from BT but combines with the speed and efficiency of Wi-Fi.

1.4 Market Presence

Since its inception by the Ericsson Corporation in the early 1998, the BT SIG has grown to more than 12,000 member companies. A directory of these member companies is available on the BT SIG Web site (www.Bluetooth.com). The group has defined several membership levels for its member companies based on the amount of participation the member is able to lend throughout the life cycle of the BT technology. The higher membership levels are able to have more influence in its evolution. The BT specifications are openly available and provided for download on the "Official Bluetooth Wireless Info Site" (www.Bluetooth.com) Web site. The current specification version, Bluetooth 3.0+HR, is a detailed, voluminous document of 1712 pages. This open platform mentality has led to industry-wide support and dedication to maintenance, advancement, and encouragement to development of new BT applications. Due to this dedication and attention to detail, BT has achieved a significant level of global acceptance by both vendors and customers. Common applications that have incorporated BT hardware include mobile phones, GPS devices, streaming audio devices, automobiles, medical devices, desktop computing devices, and gaming and entertainment devices. This list of BT-enabled devices is growing rapidly, and, has a quasi brand name awareness and general market acceptance.

From the customer's perspective, BT products are usable out of the box. This is possible because, as long as a device is BT compliant and it supports the desired usage profiles, products are compatible without regard to their manufacturer. A subsequent section (1.7) details user profiles, but for now think of them as the software driver that supports using the device. This interoperability is functionality incorporated directly into the device. For those devices that do not possess built-in support, a BT adaptor facilitates achieving this interoperability. For example, mobile phones commonly will have a BT chip built-in, whereas a desktop PC that does not have built-in support may use a BT adaptor in the form of a USB dongle;

TABLE I
OFFICIAL BLUETOOTH SPECIFICATION REVISIONS[a]

Revision	Release date	Description
v1.0A	July 26, 1999	Short-lived, first public release of the Bluetooth Specifications and Compliance Requirements that incorporated limited profile functionality
v1.0B	December 1, 1999	First amendment to the Bluetooth Specification and Compliance, mainly addressed errata updates to v1.0A. The first products to reach market complied with this specification release
v1.1	February 22, 2001	Improved manufacturer interoperability of Bluetooth devices, improved general device usability, and increased the available functionality profiles
v1.2	November 5, 2003	Reduced interference with other devices on its host ISM band, increased throughput, and improved audio quality through implementation of FHSS. Referred to as Basic Rate: (transmission rate = 1 Mbit/s)
v2.0 + EDR	November 9, 2004	Improved compliance to basic design goals of higher throughput and lower power consumption. Referred to as Enhanced Date Rate (EDR): (transmission rate = 2–3 Mbit/s)
v2.1 + EDR	July 26, 2007	Increased available functionality profiles, increased network security (specifically pairing), lower power consumption, and modified specifications to adapt technology to incorporate high-speed throughput (transmission rate = 3 Mbit/s)
v3.0 + HS	April 21, 2009	Increased throughput and improved security (through Alternate MAC/PHY for transmitting large amounts of data), power management optimized (transmission rate = up to 24 Mbit/s)
v4.0	June 30, 2010	Bluetooth Low Energy (a.k.a. Wibree), Broadcast channel, improved QoS
Future	???	UWB to 60 GHz

[a] Bluetooth Special Interest Group, Specification of the Bluetooth System (11/4/2004): Part C, 55–57.

both of these devices are able to communication to other BT-enabled devices. The end-result is a technology that has achieved a public acceptance with limited resistance. The rapid expansion of the BT market exemplifies this acceptance. As for BT products presence, an ABI Research study performed in 2008, forecasted close to 2.4 billion BT-enabled units are expected to ship worldwide in 2013.[15] In addition to this forecast, as for the general awareness of the technology, an In-Stat, 2009 study found the following stats[16,17]:

---

[15] www.abiresearch.com/press/1146 (@ October 25, 2010).
[16] In-Stat is a research firm specializing in mobile internet and digital entertainment ecosystems.
[17] www.instat.com/newmk.asp?ID=2570 (@ October 25, 2010).

- BT semiconductor revenue will approach $4 billion by 2013. Bluetooth 3.0 will represent the largest share of semiconductor revenue by 2012.
- In 2009, the BT attach rate for mobile phones was estimated to be 63%. This attach rate refers to how many mobile phones have BT transceivers.
- 43% of those surveyed are very or extremely familiar with BT, and among those aged 34 and younger, the percentage is over 60.

1.5 Wireless Medium

BT devices communicate over the unlicensed and globally available Industrial, Scientific and Medical (ISM) radio band using an adaptive frequency-hopping spread spectrum (FHSS) algorithm. Other communications standards, such as cellular phone devices, require the usage of commercially owned bands. One impact for a device operating on a commercial owned band is some sort of monitory usage fee. The ISM band, however, is designated as a freely available radio band by the FCC and its equivalent worldwide agencies. This band is referred to as the 2.45 GHz band but is actually a bundled bandwidth of (79) 1 MHz[18] bands in the range of 2402–2480 MHz. The design intention for BT using this band was to allow universal availability since the ISM band is globally available without any usage licensing requirements from local authorities.

As a result, the ISM band's availability, many technologies operate at and communicate on it, for example, microwave ovens, garage doors, portable phones, and Wi-Fi devices. In highly populated areas, this may result in an overcrowding and saturation of the ISM radio band. To avoid potential collisions, bandwidth bottlenecks, and interference with other static ISM devices, BT employs a frequency-hopping algorithm called FHSS, which splits communications in a pseudorandom order among the (79) frequencies that make up the ISM band. The increment of time allotted on each frequency is a *slot* and for a connected device, this is 625 μs. This slot size allows the BT communications to change frequencies at a maximum rate of 1600 hops/s, with the actual hopping occurs at the beginning of each slot. BT refers to the summation of the frequency-hopping pattern as the BT single logical channel. The individual building blocks for BT communications are packets and these packets may span from one to five slots, during which time, the RF remains constant. A subsequent section (1.8) is devoted to the detailing of the BT packet format. The main benefits of the BT FHSS are its ability to operate in potentially noisy environments and limitation of signal degradation due to the signal strength fading.

---

[18] http://grouper.ieee.org/groups/802/11/Tutorial/90538S-WPAN-Bluetooth-Tutorial.pdf#search=%22IDC%20bluetooth%22 (@ October 25, 2010).

Since the ISM band is an entirely public band, BT must enforce a security policy. The SIG requires in the BT specifications that BT communications use password authentication and encryption. Pairing is the process when two devices initiating communications. During this process, each device authenticates with the other by entering a common pass-phrase. Once devices successfully pair with each other, they encrypt the payload of their packet transfers using 128-bit encryption. The frequency-hopping pattern discussed in this section introduces an additional level of security. The transmitted data is then ideally only accessible to devices, both transmitter and receiver, synchronized to the hopping pattern set by the FHSS.

In summary, what were the design intentions for employing the ISM band? In remaining true to its design goals, BT devices use the ISM band because it is economical and globally available. Since the ISM band is free to use, it is an obvious choice. However, to overcome traffic resulting from other technologies using it for the same economical reason, BT uses fast frequency-hopping and short-data packets. As a result, BT gains additional security because it appears as background noise to other technologies on the ISM band. The ISM band is available globally with a few exceptions in Australia, France, and Spain.[19,20]

1.6 Network Topology

The classification for a BT network is a PAN, which is a person- or device-centric network of computing devices that are within close proximity, usually a few meters, of the central person or device. These BT PAN are often mobile environments. Whereas Wired PAN are physically connected via a cabling medium such as USB or FireWire, Wireless PAN (WPAN) using BT wireless technology communication via a RF medium as discussed previously. The topologies[21] associated with a BT WPAN are Piconets and Scatternets. A Piconet[22] is an *ad hoc*, master/slave, wireless computing network that contains from two to eight devices, one of these devices being the master and the remaining the slaves. A Scatternet[23] collectively groups 2–10 piconets that are within PAN range of each other and allows communication between the groups of piconets. The individual devices that are members of a piconet fall into one of seven possible operational states: Standby, Inquiry, Page, Connected, Sniff, Park, and Hold.

---

[19] Due to different local range specifications, the ISM band is (23) 1 MHz channels in Spain and France.
[20] www.ce-mag.com/archive/01/05/didcott.html (@ October 25, 2010).
[21] Bluetooth Special Interest Group, *Specification of the Bluetooth System* (November 4, 2004): Part A, 51–53.
[22] www.bluetooth.com/English/Technology/Pages/Glossary.aspx (@ October 25, 2010).
[23] www.bluetooth.com/English/Technology/Pages/Glossary.aspx (@ October 25, 2010).

The piconet is the basic network building block associated with BT WPAN. The wired limitations associated with a fixed network infrastructure do not limit wireless devices. This mobility and the *ad hoc* nature of WPAN allow a network to be continuously changing. As a result, mobile devices are able to migrate in and out of each other's visible range. The two to eight devices that form a piconet are active devices. One may generalize the interaction between these devices into three possible configurations: point-to-point, point-to-multiple, or single-cell, multicell. Figure 2 depicts examples of these possible device configurations. Within each of these configurations, the piconet follows a master/slave communication protocol, and, since every BT chip is able to send and receive RF, every device is able to be either a master or a slave. The point-to-point configuration is between two devices: one master device and one slave device. The point-to-multiple or single-cell configuration is between two and eight devices: one master and two or more slaves. The multicell configuration is between two or more piconets, which would allow for up to 80 active devices.

The piconet's master, which is the device that initiates the connection, is responsible for establishing the links between itself and each of the slaves. It is also responsible for synchronizing communications while a slave is part of its piconet. All devices in the established piconet share this physical BT channel, which is the most basic layer in the BT architecture and serves as the means of device synchronization. This physical channel is a combination of a frequency-hopping pattern, a slot time increment, and an access code.[24] The frequency-hopping pattern is an index of RF channels that cumulatively add up to a BT channel. This means, as the

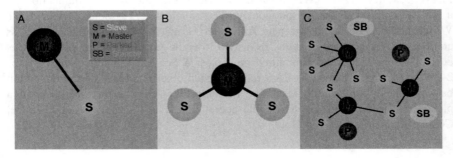

Fig. 2. Piconet topology.

---

[24] Bluetooth Special Interest Group, *Specification of the Bluetooth System* (November 4, 2004): Part B, 69-94.

name suggests, that a device hops in a set sequence over all the bands within the ISM band, spending a set amount of time on each band before hopping to the next band in the index. Section 1.5 discusses the slot time increment and Section 1.8 discusses the access code.

The process of synchronizing to a frequency-hopping pattern may be divided into two steps: creation of the RF channel index (Fig. 3) and synchronization of internal clocks (Fig. 4). To create the RF channel index, the master device uses as input its Bluetooth Device ID (BD_ADDR[25]) and internal clock. The BD_ADDR is a 48-bit IEEE standard address assigned to the device by its manufacturer, registered to the IEEE Registration Authority, and unique to all BT devices. The internal clock, also known as the device's native clock, is a 28-bit value based on the system to which it is a part. The output of this step is an index of RF channels, which is then mapped onto the ISM RF frequencies. There are slight variations in the derivation of this hopping sequences based, but Fig. 3 depicts the general case. Additional detail is beyond the scope of this chapter but may be referenced in the BT specification provided by the BT SIG. To synchronize internal clocks, each device calculates the offset of its own clock to that of the master device's clock. Each device then schedules its piconet activities, such as transmitting and receiving communications, to this synchronized clock. The design intention in using the master device's ID and

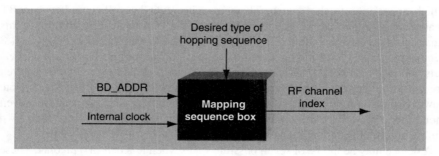

FIG. 3. Channel-hopping sequence selection. (Bluetooth Special Interest Group, *Specification of the Bluetooth System* (11/4/2004): Part B, 83.)

---

[25] IEEE assigns each BT chip manufacturer a 3-byte value, for example Sony Ericson is assigned 00:0A:D9; this is the first half of the BT device address. The manufacturer in turn assigns each chip created an additional 3-byte value. The concatenation of these two 3-byte values is the BD_ADDR of the device. Ideally, the result is a unique address for every BT device.

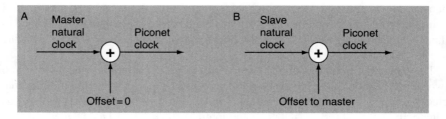

Fig. 4. Piconet clock synchronization. (Bluetooth Special Interest Group, *Specification of the Bluetooth System* (November 4, 2004): Part B, 72.)

clock is to create a pseudorandom BT channel since theoretically every device will have both a unique device ID and a unique native clock. When a master device connects with a slave device, the slave device synchronizes to the master's BT channel, and each device communicates at established time slots.

What happens when several piconets inhabit the same physical space? Numerous piconets may exist in the same physical space because they will possess different physical channels. If two devices were able to create the same physical channel, there would be constant communication conflicts. This is theoretically not possible, as the physical channel is based on the master device's ID and internal clock, both of which are unique to the device. There may however be collisions between BT channels hopping to the same frequency within the ISM band. The solution to this situation is the access code that precedes all piconet communications; Section 1.8 details this access code and the communications packet format. The use of scatternets allows BT devices to increase the number of networked devices within a given area, overcoming the maximum number of devices limitation of a piconet. Note however that multiple piconets do not have to interact to classify as a scatternet even though they may share the same coverage area. They each would have their own timing and frequency-hopping algorithm.

A device may be in one of seven possible states when it is part of a piconet. These are as operational states. These states facilitate the networking of BT devices and allow the devices to conserve power. Since power management is a concern for mobile devices, each of these states allows a device to minimize its power usage by only performing tasks relevant to that particular operational state. Figure 5 depicts theses states and the transitions that a device may take while interacting with a piconet. Four of the operational states correspond to a device creating a link to a piconet: standby, inquiry, page, and connected. A device is in a standby state if it is not associated with a piconet. A standby device consumes very low power, using only enough to maintain its native clock. This is the starting state for all devices; it is

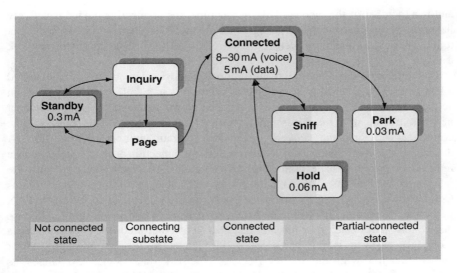

FIG. 5. Piconet operational states. The values shown with each state are the average power consumption for a device while in that particular state. For comparison purposes, the power consumption shown for a device in the Standby state is 3% of that used by a mobile phone.

a waiting state, and the most common operational state for a BT device to be. A device is in an inquire state when it is searching for other BT enables devices within PAN range. This inquiry process is an iterative process whereby a device's radio broadcasts an inquiry request and waits for responses from other devices within range. If there are other devices within range, they will respond to the device inquiry with specialized communications called a frequency-hopping synchronization (FHS) packet, which contains information necessary to synchronize to the same BT channel. This packet is a variation of the packet format discussed in Section 1.8. This broadcasting inquiry may result in communication collisions if more than one device is within range, but after an iterative query process, the master device will have the BT device address and clock of all devices within range. A device is in a page state when a master sends out a request for the device to join its piconet and it is transitioning to a connected state. The page state is a transitional state that follows the inquiry state, once a master device has determined that a device is present, it pages the device to join the piconet. A device is in a connected state if it is actively associated with a piconet. Once it establishes a link between itself and a slave, and the master device assigns an Active Member Address (AM_ADDR, 3 bits) to the slave device. As long as a device has an AM_ADDR, it is *active* to the master device and is a member of the piconet. During this time, the connective device is actively listening, receiving, and transmitting to its piconet's network traffic. In a piconet topology, devices are dependent on each other to relay messages, and network traffic.

The remaining three operation states correspond to a device after it has created a link to a piconet: sniff, hold, and park. A device is in a sniff state when it has an active connection to a piconet but in a reduced power state and listening to piconet network traffic at a master-assigned minimal rate of time. This rate of time correlates with ignoring network traffic for a certain number of time slots. A device is in a hold state when it has an active connection to a piconet but in a reduced power state. The hold-state device retains its synchronization with the master as well as its AM_ADDR, but is only able to listen to Synchronous Connection-Oriented (SCO) link packet transmission and able to return to PAN participate very quickly, within a 2-ms time frame. While in the hold state, the device may participate in or initiate the creation of other piconets. A device is in a park-state when it is not participating in piconet traffic, but it retains clock synchronization with the master device. The park-state device exchanges its AM_ADDR for a Parked Member Address (PM_ADDR, 8 bits) address, but it is able to return to PAN participate very quickly, within a 2-ms time frame. Placing a device in a park-state allows a piconet to overcome its eight active member limitation. The 8-bit PM_ADDR allows a master device to park up to 255 slave devices, all of which may return to active states when needed.

How do these states benefit the average BT user? The wireless networking infrastructure creation is dynamically and automatically performed when BT-enabled devices are within range of the BT PAN. The devices automatically also share their capability details when with range. The power management conserves a device's battery life to achieve the low-power design goal.

1.7 Protocol Stack and Profiles

The BT core specifications as defined by the BT SIG allow for both data and voice wireless transmission at actual data speeds up to 720 kbps and three simultaneous voice channels. These core specifications divide a BT implementation into a protocol stack and a series of usage profile specifications. This implementation may take a feature-complete form, allowing additions features through updates, and target desktop computers running an operating system (OS) such as Windows 7 or OSX 10.6 "Snow Leopard," or a specialized device form that using an embedded system, such as posting an entry in print queue in a photo printer from a camera. The BT protocol stack specifications are the layered functionality that defines the workings of the wireless technology. Layered architecture facilitates interoperability between new and existing BT devices as the BT SIG releases updates to the BT standard.

The BT usage profile specifications define the use-cases to apply the BT technology, specifically the dependencies of a profile on the stack protocols, potentially other profiles, and the necessary interfacings for this interaction. These are the end application usages for a device and define how the technology is used and how it is interoperable with other devices. Devices that support the same usage profile are able to interact. For example, the BT human interface device profile defines how a peripheral may interact with a computer. Supporting this BT usage profile, a computer, wireless keyboard, and mouse are able to communicate. These usage profiles allow for freedom during product development because vendors need only implement those profiles that benefit a particular product. Therefore, it is at vendor's discretion to implement these usage profiles, so this is the frontline when it comes to a secure BT network. The more access via open ports, which is introduced through usage profiles, the higher the potential for insecurities. Section 4 discusses end-user best practices. This current discussion presents the component protocols and profiles that compose the BT stack. The specifications provided by the BT SIG detailing each of these components is rather lengthy, so this discussion will serve as an introduction and a reference is provided to where more details may be found within the BT specifications document.

Designed in the layered protocol architecture, the BT protocol stack is a set of five core BT layers. The five core BT stack components, listed from lowest to highest on the stack, are the Physical Channel (Radio), the Link Controller (LC), the Link Manager Protocol (LMP), the Logical Link Control and Adaptation Protocol (L2CAP), and the Service Discovery Protocol (SDP). As is the case with stacked protocols, each level of the protocol stack communicates with its corresponding level of the stack on the target device. Table II summarizes these BT stack layers. At the physical link layer (LC), BT supports both synchronous and asynchronous communications based on the intent of the transmitted data. A synchronous link between a master and a slave on a piconet is known as a SCO link. This physical link uses circuit switching transmission and is primarily used for time-dependent data. The master device is able to support up to three simultaneous SCO links. An asynchronous link between a master and slave on a piconet is known as an Asynchronous Connectionless Link (ACL). This physical link uses packing switching transmission, error control through error detection, and retransmission to guarantee packet delivery. The BT protocols transmit voice communications synchronously using an SCO link to retain the coherency necessary in verbal conversations, and data communications asynchronously using an ACL. BT is capable of "supporting either one asynchronous data channel and up to three simultaneous synchronous speech channels, or one channel that transfers

TABLE II
CORE BLUETOOTH STACK LAYERS[a]

Item	Stack item description
Radio	*Manages the physical radio channel* The radio is responsible for managing the wireless interface details. This includes the RF band, the spread spectrum frequency-hopping pattern, and the power usage based on the device's power class. Section 1.5 details these concepts
LC	*Physical element that manages logical links* This is the physical link between a master device and a slave device for transmitting data. It acts as the transport for one or more logical links, which the LMP manages, and handles packet addressing, formatting, timing, and power control within the piconet. This is known as the Bluetooth Baseband
LMP	*Manages logical links between devices* This is the Logical Link responsible for managing the connections between devices in a piconet. This involves link creation and link management, which includes authentication between devices, encryption of the communications payload, and the sizing of packet transmission. Section 1.6 details these concepts
L2CAP	*Manages interface between Bluetooth stack and upper-level stack objects* This is the channel-based abstraction of the wireless communication technologies to the upper-level stack objects, which include profiles and applications. This includes managing packet assembly, packet segmentation, protocol multiplexing of channels, and protocol demultiplexing of channels across the logical links
SDP	*Manages a device details and its available service profiles* Each device has different service profiles based on its intended usage. When two devices create a connection, they will query each other for basic device information, supported service profiles, and the details associated with these profiles. The Device ID Profile (DID) is an extension of SDP that helps in identifying the device and provides branding details so its features may be better used. In a PAN, it facilitates the device pairing process

[a] Bluetooth Special Interest Group, Specification of the Bluetooth System (November 4, 2004): Part A, 21–26.

asynchronous data and synchronous speech simultaneously."[26] Table III summarizes these borrowed protocols, and Fig. 6 depicts a representation of the protocol stack showing the relationships of all listed protocols.

The upper-level stack contains standard protocols for performing various networking and communication tasks. As mentioned previously, these are referred to as adopted protocols. The adopted protocols are established protocols that BT borrows and adapts to incorporate their functionality into the BT technology. To allow these adopted protocols to function in concert with BT protocols, several protocols are

[26] www.niksula.hut.fi/~jiitv/bluesec.html (@ November 6, 2006).

TABLE III
ADOPTED BLUETOOTH STACK PROTOCOLS

Item	Stack item description
OBEX[a]	*Object exchange protocol* A transfer protocol commonly associated with IrDA. It allows the transfer of object data between devices. These objects are files commonly associated with Bluetooth devices and may include electronic business cards (vCard) and calendar events (vCalendar)
PPP[b]	*Point-to-point protocol* An Internet protocol used to establish a direct point-to-point link between two devices. It allows communications between the connected devices
TCP/UDP/ IP[c]	*Internet protocol suite* Communications protocols that implement the computing network protocol stack on which most networking is based. These protocols permit devices to establish, send, and receive communications
WAP/ WAE[d]	*Wireless application protocol/wireless application environment* WAP is a series of protocols that define a networking stack specific to mobile device thus bringing Internet content to mobile devices. WAE provides a vendor-independent platform for application development on mobile devices

[a] Bluetooth SIG, *Bluetooth Specification Version 1.1* (February 22, 2001): Part F:2.
[b] BNEP is replacing this protocol.
[c] Bluetooth SIG, Bluetooth Network Encapsulation Protocol (BNEP) Specification (February 14, 2003).
[d] Bluetooth SIG, *Bluetooth Specification Version 1.1* (February 22, 2001): Part F:4.

inserted that act as wrappers for interacting with these upper-level protocols. These interface, or wrapper, protocols include a virtual serial port emulator (RFCOMM) and a Telephony Control Specification (TCS BIN). The BT profiles use one or more of the adopted protocols or other profile specifications. Table IV lists these protocols, and Fig. 6 depicts these protocols within the BT stack.

The BT protocol stack is the industry standard as set by the BT SIG, and requires implementation of the entire core specifications and implementation of all or part of the adopted protocols and usage profiles. The intention is to ensure compatibility between devices that would otherwise be proprietary to each manufacturer and minimize any potential risk due to interoperability issues. There are several accepted and supported BT protocol stack implementations: for Microsoft Windows an XP stack,[27] a Vista stack, and a WIDCOMM stack[28] and for Linux the BlueZ[29] and

[27] http://msdn.microsoft.com/en-us/library/aa362932%28VS.85%29.aspx (@ October 25, 2010).
[28] www.widcomm.com (@ October 25, 2010).
[29] www.bluez.org (@ October 25, 2010).

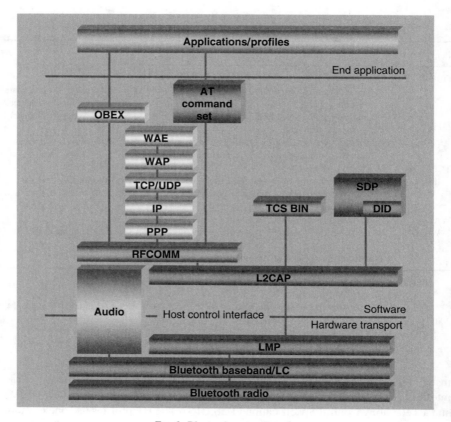

FIG. 6. Bluetooth protocol stack.

Affix stacks. The BT profiles then allow a device to establish and provide a service to another BT device. The profiles listed in Tables V and VI are not complete, because, as the BT technology evolves, the SIG introduces new profiles and deprecates other profiles. This does not result in changing the BT specifications; new profiles just describe new ways of using the existing BT specifications. Tables V and VI list the currently available profiles and an example of each profile. Figure 6 notates the location of these profiles within the BT protocol stack as *Profiles*.

TABLE IV
INTERFACE BLUETOOTH STACK PROTOCOLS

Item	Stack item description
RFCOMM[a]	*Radio frequency communications* A cable replacement protocol established over a virtual serial port connection (RS-232). This allows devices that commonly communicate over serial connections to connect via a BT connection. Serves as the basis for other protocols listed in this table to run over
TCS BIN[b]	*Binary telephony control protocol* A binary telephone control protocol that manages speech and data between BT devices. This protocol manages the specifics of a BT device usage in telephone applications. This protocol is also known as TCS

[a] Bluetooth Special Interest Group, *Bluetooth Specification Version 1.1* (June 5, 2003): Part F:1.
[b] Bluetooth Special Interest Group, *Bluetooth Specification Version 1.1* (February 22, 2001): Part F:3.

1.8 Packet Format

BT Network traffic,[30] as defined by the BT SIG within the core Baseband Specifications, is a packet-switching transport. There are two generalized packet formats based on a device's implemented version of the BT core specification: Basic Rate for BT v.1.2 and lower, Enhanced Data Rate (EDR) for BT v.2.0 and higher. A BT Basic Rate packet divides its contents into three bundles of data: the Access Code, the Packet Header, and Packet Payload.

Figure 7A shows this general Basic Rate packet format. Although there are some slight variations for the packet contents based on the intention for the communications, this is the general format for a BT packet. BT allows two possible types of physical link between BT-enabled devices, as mentioned in Section 1.7: SCO link, ACL. SCO links are connections that require time sensitive or full duplex communications and are typically used to route audio and voice data. ACL connections route data only when there is data to send; this is similar to the standard packet-switching network. These types of link between two devices dictate the encoded content within the Access Code and Header. It also determines the contents, type, and purpose of a packet within the BT operational states.

The Packet Access Code section is a 72-bit value that supports the radio communication signals between BT-enabled devices and acts as a wake-up signal to the receiving device. All packets on an established BT channel will have the same

---

[30] Bluetooth Special Interest Group, *Bluetooth Specification Version 1.1* (June 5, 2003): Vol. 3, Part B: 109–136.

TABLE V
BLUETOOTH PROFILES (PART I)[a]

Item	Profile description
A2DP	*Advanced audio distribution profile*
	Specifications that define the requirements for transmitting high-definition quality streaming audio. A common application is a Bluetooth audio source device streaming audio to a Bluetooth stereo headset
AVRCP	*Audio/video control transport protocol*
	Specifications that describe the transport mechanisms used in the exchange of messages in controlling audio and video devices
AVDTP	*Audio/video distribution transport protocol*
	Specifications that define the distribution of streaming audio and video
AVRCP	*Audio/video remote control profile*
	Specifications that define how a device acts as a standard interface to control the streaming audio/visual in media components. A common application is a Bluetooth device serving as a universal remote in a multimedia environment, for example, the Nintendo Wii and the Sony Playstation 3 wireless controllers
BIP	*Basic imaging profile*
	Specifications that define how a device may remotely manage an imaging device and the data stored on it. A common application is a Bluetooth device such as a mobile phone controlling a digital camera
BPP	*Basic printing profile*
	Specifications that define how a device may issue print jobs to a printing device without the need for printer-specific drivers. A common application is a PDA printing an e-mail, or a digital camera printing an image
CTP	*Cordless telephony profile*
	Specifications that define how a cordless phone connection may be implemented over a Bluetooth device connection. A common application is a wireless handheld phone in a residence
DUN	*Dial-up networking profile*
	Specifications that define how a device uses dial-up services over Bluetooth devices. A common application is allowing Internet access using dial-up connectivity by a PC through another Bluetooth device such as a mobile phone. This profile is also known as to as Internet Bridge Protocol
FAX	*Fax profile*
	Specifications that define how a device acts as a gateway and communicates similar to Facsimile. A common application may be a laptop using a fax application to use a mobile phone as a gateway to send a fax

[a] www.bluetooth.com/English/Technology/Building/Pages/Specification.aspx (@ October 25, 2010).

Access Code, and devices will disregard packets captured on their BT channel with an unexpected access code. The design intention is to identify a packet train since BT channels will inevitable overlap due to the limitation of frequencies within the ISM band. There are three possible access codes based on the purpose of the packet:

TABLE VI
BLUETOOTH PROFILES (PART II)[a]

Item	Profile description
FTP	*File transfer profile*
	Specifications that define the requirements to allow client-server file system management over a Bluetooth connection. A common application is browsing and transferring of objects on a server by a client device
HFP	*Hands-free profile*
	Specifications that define how a gateway device allows hands-free transmission and reception of audio data from a client device. Bluetooth headsets for mobile phones are a common application of this profile
HID	*Human interface device profile*
	Specifications that define how a Bluetooth device interfaces with another Bluetooth device to act as an input device. A common application is keyboards, mice, gaming, and other peripherals to communication with a desktop computer
HSP	*Headset profile*
	Specifications that define communications in the form of input and output audio between a Bluetooth headset device and another Bluetooth device. A common application is a headset linked to a mobile phone
ICP	*Intercom profile*
	Specifications that define how a device supports conference call functionality without use of a telephone carrier network. A common application is two mobile phones communicating as "walkie-talkies."
LAP/PAN	*Local area network access profile/personal area network profile*
	Specifications that define how a two or more Bluetooth devices create an *ad hoc* network as a piconet and translate connectivity over to a LAN. A common application is a PDA having Internet access via a networked desktop computer
SAP	*SIM access profile*
	Specifications that define how a device may connect to a SIM card in a mobile phone to share the information stored on it. A commonly application is a mobile phone sharing phonebook contact information to an automobiles phone services
SPP	*Serial port profile*
	Specifications that define how to emulate a virtual serial connection between two devices. A common application is using any device that requires an RS-232 connection to a computer
SYNC	*Synchronization profile*
	Specifications that define how to synchronize a set of data between two Bluetooth devices. A common application is a day-planner synchronizing its contents with a desktop computer
AT Command Set	The AT commands are a command-response executable commands used to manipulate and manage communications, in a similar means to a modem

[a] www.bluetooth.com/English/Technology/Building/Pages/Specification.aspx (@ October 25, 2010).

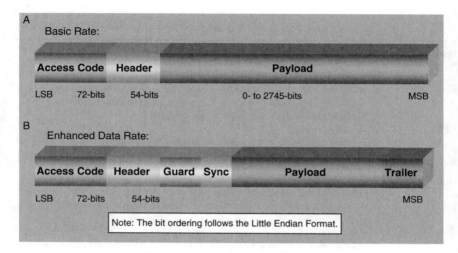

FIG. 7. Bluetooth packet format.

Channel Access Code (CAC), Device Access Code (DAC), and Inquiry Access Code (IAC). The CAC access code identifies an established piconet and is prepended to all packets from a master device to an active slave device. The DAC access code identifies a particular device and is prepended to a packet by a master device when paging a partially connected slave device. This page determines if this slave device is within range, activates it if it is within range, and then reconnects to it. An IAC determines if a nonconnected device is within range. When a device captures a packet, based on these access codes and its operational state, it will process the packet's payload, else it drops the packet.

The Packet Header is a 54-bit value that supports the BT physical links (SCO and ACL) and used by the LC. This header contains six fields: Logical Transport Address (LT_ADDR), Type, Flow, ARQN, SEQN, and Header Error Control (HEC). The LT_ADDR is a 3-bit value that is a temporary address assigned to an every active slave device in a piconet by its master. The master device then uses this LT_ADDR to communicate with a particular device but may also broadcast a packet to all slaves in the piconet by using an LT_ADDR of all zeros. The Type is a 4-bit value that indicates the type of data in the packet payload; there are 16 possible values based on the established physical link. This Type value helps a device determining the number of slots packets traverse, so it knows how much time it should listen to or ignore communications. The Flow, ARQN, and SEQN are all 1-bit values that, respectively, indicate ACL flow control, acknowledgment for successfully receiving a packet payload, and sequencing scheme for received packets. The HEC is an 8-bit value that verifies the integrity of the Packet Header. The careful reader will note that these values are 18 bits in length. All these values are appended in triplicate, resulting in the 54-bit length. This allows for redundancy to limit transmission errors.

The Packet Payload is a 0- to 2742-bit value that contains the data transmitted from one device to another. These contents are depended on the type of physical link between devices. For an SCO link, the payload is voice information, and for an ACL, the payload is data information. A cyclic redundancy check (CRC) is appended to the data in the payload for ACL packets. The receiving device uses this CRC as a tool to determine if the packet contains transmission errors. BT encrypts the entire payload with a 128-bit key resulting in an encrypted virtual private network.

The master/slave topology mentioned previously requires certain network restrictions for packet transfer between devices. In particular, the time slot, which is the smallest unit allotted to a device to communicate. The piconet master synchronizes time slots with its clock and is designates each slot as either even or odd. Then, the master device transmits packets on even number slots, and slave devices transmit packets on the odd number slots. Each packet must start at the beginning of a time slot, and may have a length of one to five slots, during which there is not frequency hopping. Slaves may also only transmit when polled by the master device.

The design of the packet is to take advantage of the frequency-hopping algorithm discussed earlier. The compact size of the packet allows rapid frequency hops within the BT channel. With other devices using a constant frequency within the ISM band, the small BT packet size also helps limit potential interference with other devices on the same radio frequencies. There is a side effect of this compact packet size: limited throughput. For the device to hop a frequency, the device must manage its radio. These radio tasks expend a certain amount of overhead, which take away from potential packet transfer time. ACL traffic will achieve higher transfer rates than SCO link traffic because it is able to use longer time slot allocation, since there is no concern in allowing full duplex communications. All in all, the reliability achieved through the frequency-hopping algorithm outweighs the potential limitation of the technology. To address this limitation and increase the throughput, the BT SIG introduced v.2.0.

BT v.2.0 offers an EDR, which improves on its predecessors by increasing the bandwidth and optimizing power management while retaining backwards compatibility. The bandwidth is increase from the BT v.1.2 specification of 1 Mbit/s three-fold to up to 3 Mbps. Better power management is achieved by transmitting the same amount of data more efficiently, using fewer packets or shorter payloads. The backwards compatibility is achieved because modifications are made only at the BT Baseband level, which is at the hardware level. These medications do not impact the higher levels of the BT protocol stack. Figure 7B shows the general EDR packet format. Note that the Access Code and Header are the same format as for Basic Rate packets, which keeps in line the backwards compatibility. However, EDR packets include three additional sections: Guard, Sync, and Trailer. The details associated with these sections are unique to the most recent version of the BT protocol, and rather complex. These details are referenced in the Bluetooth SIG v3.0+HR documentation on the Bluetooth.org Web site.

1.9 Transmission Range

The intention for BT is as a short range, RF communication protocol used to implement PANs. Devices networked within a PAN are stationary devices, such as a desktop PC, an iMac, a keyboard, a mouse, or a printer, as well as mobile devices, such as a PDA or camera. A stationary device does not face the power challenges associated with mobile device since it is able to take advantage of a constant power supply, plugging into the wall. A mobile device, however, has power limitations, which is a finite battery life. Anyone with a mobile phone may attest to cursing these limitations. The BT protocol takes into consideration this power limitation of mobile devices, allowing a mobile device to maximize its connectivity. There is however a direct relationship between power and range: lower power results in shorter range, more power results in greater range. Since the design target network is a PAN, a smaller amount of power usage by the device will achieve the desired range. The BT SIG designates three levels, or power classes, for BT devices; this allows for flexibility for the device implementation. Table VII summarizes these three levels.

As mentioned previously in Section 1.1, BT devices are design to operate consuming minimal power. The main difference in the BT classifications listed in Table VII is their power consumption. A BT device must fall within one of these SIG established Classes, and as one would expect, there is a direct correlation between power usage and the class's range. More power equates to longer range; however, this does limit device compatibility with other BT Classes. Each of the BT classes is compatible; for example, a Class-1 BT device may communicate with Class-2 or Class-3 devices. The principle of least privilege in regards to the maximum range may be used to the user's advantage when implementing an optimal BT environment. For example, one might use the range provided by Class-1 devices for their stationary devices since the power is not a concern, and then use a Class-2 on a mobile device to take advantage of the power savings. To put this power savings in perspective, during normal BT networking, a Class-2 device uses a small fraction of that used by cellular phones.

The short range is both a limitation and a security measure. The intended design is to implement PAN so the range is for devices within a smaller environment. Not only does this limit any potential interference with other devices on the same frequencies, but also, since a device must be within this range to communicate, the potential attacker must also be within a close proximity. By comparison, infrared

TABLE VII
BLUETOOTH POWER CLASSIFICATION

Power class	Maximum range	Application	Maximum power output[a]	v.2.1 Bandwidth[b,c]
1	100 m/330 ft	Industrial devices	0.1 W	v.1.2: 1 Mbit/s, v.2.0: 3 Mbit/s
2	10 m/33 ft	Mobile devices	0.0025 W	v.1.2: 1 Mbit/s, v.2.0: 3 Mbit/s
3	1 m/3 ft	N/A[d]	0.001 W	v.1.2: 1 Mbit/s, v.2.0: 3 Mbit/s

[a] This is the signal strength of the device while in an Active State, which is when a Bluetooth device is in constant communications within a PAN.
[b] This is raw data bandwidth. In practice, the bandwidth is slower due to Forward Error Correction (FEC), which limits the impact of random noise and the number of needed retransmissions.
[c] http://www.niksula.hut.fi/~jiitv/bluesec.html (@ October 18, 2006).
[d] Class-3 devices are relatively rare, and have no defined market application.

networking requires a line-of-site connection between devices, whereas BT uses radio frequencies, which are nondirectional. Nondirectional communication allows BT devices to communicate without the limitations of solid barriers such as walls and other physical structures. Throughout this chapter, references to BT devices refer to Class-2 devices because they are the most common power class. For example, mobile phones are generally BT Class-2 devices.

2. BT Exploitation

There is no magical shield to protect a computing device. As long as a device is in the ON state, an attack is possible. There are however general best practices when using any computing device; Section 4 addresses these best practices as they pertain to BT devices. Possible areas of concern for BT devices are the defining BT protocol, the wireless medium and the data transfer mechanisms over which devices communicates, and the implementation configuration settings that define BT usage. The following sections will present and introduce potential areas and methods of BT propagate methods, conditions that must be true for a device to be vulnerable, categories of known malware, and notable first attacks on the BT protocol. The combined understanding of these sections will provide the user a general awareness on how to protect their BT device(s).

Mobile devices are commonly restricted by size and power, and as a result, the device's OS and its management tools are not as extensive as with desktop computing devices. These limitations are changing as devices evolve, but attacks commonly exploit the naivety of the mobile devices. The potential impacts of a BT attack are numerous, and most of these attacks do not leave a trace on the host BT device. These attacks are also not partial to any particular device and affect both new and older devices equally. This may be viewed as a side effect of the SIG imposition of universal interoperability between BT devices. A short list of potential attack results is as follows:

- Eves dropping on telephone conversations
- Battery life depletion due to the malware propagation
- Usage of carrier offered services that may potentially incur usage fees
- Improper usage of the device resulting in frequent crashing
- Usage of the device as a gateway from which to make phone calls or send data

2.1 Exploiting the BT Protocol

One may generalize that there are three possible areas of exploiting the BT protocol: the core design by the BT SIG as described in Section 1, the implementation as performed by a device manufacturer, or both. The BT SIG provides the technology's specification standard and is responsible for incorporating updates and new functionality into this standard. There have been five main revisions in the lifetime of the technology; the current version is $3.0+HS$. The SIG has also established an aggressive plan for updating its current functionality, as dictated by market demands and competitive technologies, and current limitations, such as the WPAN size as dictated by a piconet. However, it is up to each device manufacture to implement the actual standard. In doing so, the device manufacturer must implement the core specifications but may be selective in offering BT functionality based on the intention of the device. BT offers functionality via what it refers to as service profiles, for example, a BT-enabled headset would offer hands-free service profile but not a file transfer service. Tables V and VI provide the details associated with both of these profiles.

Typically, an attack compromises a device due to vulnerabilities within the implementation of the core specifications or service profiles. The result of poor implementation is an attack or a series of attacks on a BT implementation until the manufacturer patches the faulty implementation. The end goal of an attack, aside from the brag-factor for its author, is to abuse the resources provided by the device. These resources may be either data that is stored on the device or a service that the device provides. Abusing stored data includes unauthorized access, stealing, deleting, or corrupting the integrity of the data on the device. Abusing a service on the device includes using a profile provided by the device such as eavesdropping on a service or denying a service on the phone. The vehicle for these attacks is malware commonly used on desktop computers that is adapted for the target device. This may include any of the commonly known types of malware: viruses, worms, spyware, Trojan horses. This strongly encourages device manufactures to use the principle of least privileges when designing their devices. This approach will limit potential vulnerability because only those services that are vital to the intention of the device are implementing. Up to this point, there have been limited attacks on devices using BT, but the BT SIG takes an active stance in addressing and patching the BT protocol.

An implementation may be in complete compliance with the specification yet be the target of an exploited through the technologies design requirements. For example, the specifications state the protocol stacks must contain the SDP (Section 1.7). The design intention is that BT devices are able to resolve supported profiles for those devices with their range. This is an automated convenience feature, which under normal usage standpoint is beneficial. However, from a malicious standpoint an easy avenue to attack because it serves as an open port that does not require user interaction to establish a connection.

2.2 Exploiting a Wireless Device

Wired networks limit communications to those devices that have physical connection to the network. BT networks, however, are a wireless technology, communicating over RF. Although the range of the wireless network serves as a limited factor, there is a greater potential for network vulnerable due to the lack of physical networking connection. This may result in surveillance, also known as sniffing, by any device that is within range of the wireless signal. The range of an attack is not limited to the range of the target BT device. An attacker may exploit the range associated with the three class tiers of BT devices. For example, a laptop device using a Class-1 device or a directional long distance antenna would be able to surmount the expected range limitations of a BT device and be able to communicate with devices from a greater distance. BT does provided security measures to protect against sniffing the wireless traffic: frequency-hopping (Section 1.5) and encrypted packet transmission. Section 3 discusses the software tools that one may use to sniff wireless traffic.

Numerous network data transfer protocols allow devices to communicate over a wireless medium. Malware deceitfully takes advantage of these services to propagate. This propagation may occur via such activities as text communications, file downloads, file transfers, or social engineering. Although these activities are by no means the only propagation options, especially since technologies come-and-go, they do include the common services provided by BT-enabled devices currently available. For a BT device to become infected, the malware must either install itself or convince the user to install itself on the host device. Once the malware is installed, the potential implications and abuse of the device is limited only by the services provided by the device. The more extensive the available services, the greater the potential for abuse. The following four subsections discuss these potential methods of malware propagation. Keep in mind that is common for malware to increase its propagation potential by combining more than one propagation methods. Once again, by no means are these the sole means of propagation, they are common methods as used by malware currently in the wild.

2.2.1 Propagation Method: SMS and MMS

Short Message Service (SMS), more commonly known in the United States as Text Messaging or texting, is a service that allows a computing device to send/receive a finite amount of text to/from another computing device. The maximum length of the SMS message is 140 octets,[31] which translates to 160 characters using a 7-bit ASCII encoding such as with the Latin alphabets or 70 characters using a 16-bit UCS-2 encoding such as with Asian and Slavic languages. The common application for SMS is manual entry of text communications between users of mobile devices, or automated messaging from a service to a mobile device such sports score updates, stock quotes, or device status. From a security perspective, an SMS message is limited to text data, which the receiving mobile device does not execute as an application. Therefore, an SMS message by itself is not dangerous. However, the contents of the SMS messages may be target of an exploit, as SMS messages are a common tool used to transfer privileged information. For example, system administrators commonly monitor the status of network devices through SMS issued to them by the device. An SMS message may be the tool used to exploit the trust associated with this privileged information. This abuse of the trusted content within an SMS message is a common target for malware.

Multimedia Messaging Service (MMS) messages overcome the text-only limitation of SMS by allowing attachment of multimedia objects, such as audio, images, and rich text.[32] The security concern with MMS is that it may contain multimedia format files; however, a message is able to contain any file format. Of particular concern to a device are the files necessary to install malware. Malware may propagate by

---

[31] 3GPP TS 03.40 V7.5.0 *Technical realization of the Short Message Service (SMS)* (12-2001): 11–12.

[32] 3GPP TS 23.140 V6.14.0 *Multimedia Messaging Services (MMS) Functional description Stage 2* (09-2006): 19–21.

exploiting the multimedia attachment content and attach itself to the MMS message. At this point, the malware is on the device and only requires installation.

There are challenges for devices to defend against malware that exploits messaging services. Malware may use an SMS message as a transport to send content from an infected device back to an attacker. This may include such information as personal user information, contact list information, calendar events, or any other data stored on the device. Aside from the obvious abuse of data integrity associated with the stolen information, the malware may also incur monitor impact for the device, as some of these services are chargeable. Devices often do not log messages sent using messaging services, as a call log does for a telephone service. The impact of this lack of message logging on the owner of the device is that he or she is not able to monitor the usage of the device's messaging service. Some devices do provide a visual or audio confirmation after sending a message; if a device allows this functionality, it is advisable to allow the device to confirm messaging with a confirmation.

It is commonplace for mobile devices to possess both BT and messaging (SMS and/or MMS) services. BT provides the networking functionality between devices within PAN range. Within these networks, services are commonly shared; this includes the messaging service. For example, pairing a desktop computer with a mobile phone creates the framework for the desktop computer to send and receive SMS messages over the computer via the mobile phone. This exemplifies how a malware author may combine both technologies to allow the propagation of the potential malware.

2.2.2 Propagation Method: Network Downloads/BT Transfers

There are different levels of connectivity possible for a device based on its available technologies and functionality. The level of vulnerability for a device is directly proportionate to the amount of networking technologies and services available to it. For a fixed device such as a desktop PC, there is vulnerability due to its physical network cable. A mobile device, however, is potentially vulnerable to not only a physical connection but also many wireless technologies. If the device has network connectivity, then it most likely is able to connect directly to the Internet and use some variation of an Internet browser. Wireless technologies, such as Wi-Fi and Cellular, or wired technologies, such as a physical connection to the device that possesses connectivity, will provide this connectivity. Using these services, the device is able to share data via downloads from the Internet. This is the potential attack vector for malware over the Internet. Once the malware has infected the device, it may then take advantage of the technologies on the device, which, among others, includes BT. For obvious reasons, if a device does not have network connectivity, then it will not be susceptible to downloading potential malware. This is the safest scenario but also not the most realistic.

Networking technologies associated with PAN also put a device in a vulnerable state. Common networking connectivity implemented on devices includes IrDA, BT, and via USB connections. Files may be transferred across these connections. Generally, when a device receives a request for a file transfer, it will prompt the user to accept or reject this transfer. This allows the user to take an active stance in managing the device, as they must consciously accept the file transfer. Blind acceptance unfortunately is commonly the norm for the inexperienced user, and malware plays on this inexperience to propagate. Malware may also use a creative name to deceive the user into thinking the file is something it really is not. For example, it is common for malware to use the process name of another application or to use a filename required by the device such as some sort of device system file. The user must remember that for a device to transfer data, it must be within range, so if the data being transferred is believed to be from the device manufacturer and the user is currently on a crowded subway train in downtown Manhattan, there is a good chance it is an exploit.

Unfortunately, BT malware may also propagate without the input or knowledge of the device user. Two possible areas of concern are the BT protocol stack and applications that network over a BT connection. If the malware exploits a bug in the BT protocol itself and the user is within range of the malicious device, the user will be at risk to the exploit. As mentioned previously, the IEEE has designated the BT SIG as the governing body for the BT technology. Their goal is universal acceptance and usage without hiccups, and their best intentions are with the users, as the group is composed of device manufacturers. As mentioned in Section 1.7, the BT protocol stack is a layered architecture with application sitting at the top. These applications use the BT stack as the networking tool to transfer data and BT serves as the vehicle for communications. In this situation, the communications are transparent to BT and it is the responsibility of the application author to secure their application. However, if an application is vulnerable and exploited, the user may be at risk. These two concerns are not limited to devices that propagate malware via downloads or transfers; they apply to any propagation method. Fortunately, the malware is not universally vulnerable to all devices; malware is platform dependent.

2.2.3 Propagation Method: Social Engineering

In reference to exploitation using BT, social engineering is a tool that uses human gullibility to gain access to a device. Since BT is a device feature that requires human interaction, it is possible to use social engineering techniques to exploit them. To do this, the malicious user will take advantage of the normal processes associated with using the BT protocol, such as authentication process for pairing. The pairing process is a process during which two devices establish a trusted relationship. It is a logical attack vector for malware because, once the two devices establish the trusted relationship, ongoing connectivity occurs automatically. The normal scenario for a device attempting to authenticate with another BT-enabled device starts with the target device displaying the name of the source device attempting to connect to it. The target device must first accept the pairing process to initiate it and then prompt the user for a pass-phrase to complete the authentication. To complete the pairing process, the user of the target device will enter a shared pass-phrase. If the user does not know who is initiating the pairing, they should not accept the authentication. If the user knows the source device, they may enter the shared pass-phrase, after which a trusted relationship is established. A trusted relationship allows each device access the services provided by the other device; herein lies the security concern. If a malicious device can deceive the user into creating this trusted relationship, they have access to the target device services, up until the target device removes this trusted relationship. Taking this scenario with a social engineer twist, it is a common tactic to exploit a novice user. Imagine the malicious source device has its name set to *Passphrase0000*, or *You have Won! Press 0000 to see what you have won*, and its pass-phrase to authenticate to *0000*. When the malicious user attempts to authenticate with a target, the target device prompts the novice user with the name of the malicious device, and either out of curiosity or uncertainty the user enters the value *0000* for the pass-phrase. This would result in the malicious source device successfully pairing with the target device, which puts them in the trusted list of target device and gives them access to the target device. This may sound like an unrealistic situation, but our experience with phishing and Nigerian 411 scams suggests that it is all too likely to lure a victim.

2.2.4 Propagation Method: Device Spoofing

Device spoofing is the act of impersonating a device and then convincing another device that they are indeed the impersonated device. To achieve this impersonation, a malicious device must first know and then mimic the information that uniquely identifies that particular device. Each manufacturer has proprietary gamut of device identification values, but a mobile device typically has a device name that the device uses to identify itself to other devices and a Hardware Identification that identifies the hardware itself. The device name is a software-set value, analogous to a name given to the desktop computing device, and is very easy to change. A manufacture provides a default value for the device name, which is usually descriptive of the device and identical for all devices of the same model. BT uses this device name when prompting a user to accept a connection to another device, and this default device name helps identify another device but it is not deterministic. Spoofing a device over a BT connection is as easy as referencing these default device names from an Internet resource and then changing the attacking device's device name within the device's software. Figure 8 shows the list of devices within PAN range for a device using Windows Mobile. How can the user be certain that the Pocket PC shown is the one with which they desire to connect? It is for this reason that Section 4 suggests always pairing a device in a secure location.

In the case of the Hardware Identification for mobile devices, every manufacturer issues each of its mobile devices a unique ID based on its service network[33]: International Mobile Equipment Identity (IMEI) or Electronic Serial Number (ESN). This ID includes information about the serial number, the model, and the manufacturer of the mobile device and is the equivalent of the MAC address associated with a Network adaptor. Assuming an attacker is able to compromise a device through either physical contact or a network connection, they will be able to harvest this unique ID. Not only would having this ID allow an attacker to duplicate a device, but also it would allow the attacker to then use the device and give the impression that it is the device from which the ID was stolen. This cloning as it is known may be the topic of its own investigation. To leave the reader with an

FIG. 8. Bluetooth device pairing.

---

[33] The GSM networked devices found globally but mainly in Europe use the IMEI and the CDMA networked devices found mainly in the United States use the ESN. The ESN is currently being deprecated due to its limitation of unique numbers to identify each device and being replaced by the Mobile Equipment Identifiers (MEID).

interesting thought on this topic, after cloning a device, imagine if the attacker sends an SMS message from the cloned device to the owner's employer saying something inappropriate.

2.3 Limitations for Propagation

Improper usages as well as vulnerabilities are both possibilities that may result in the occurrence of an exploit. Device complexity requires the user to be knowledgeable about the device's functionality, but all too often minimal user effort is devoted to familiarizing with the device. This results in improper use. Even with proper usage, not all devices are vulnerable to malware that exploits BT technologies; in fact, there is a short list of requirements that must exist for a device to be susceptible to and subsequently be infected by malware via BT services. Additionally, not all malware may propagate using BT, but they may use the service to some extent once it has infected a device. Last, malware may use some combination of propagation techniques. Malware is yet in the infantile stage for mobile devices, but as it evolves, time will allow its authors the chance to experiment and exploiting available technologies, of particular concern here, the BT protocol. Mobile devices and their potential vulnerabilities will inevitably advance to a stage of complexity similar to that of desktop computers.

For this discussion, there are three areas of limitations for propagation: hardware settings, user interaction, and device platform. The hardware setting follows a discussion of the configuration of the device. The user interaction sums up several ideas that have been touched on in reference to the user's knowledge of the device. The device platform is the actual hardware and each platform may be impacted differently by malware.

2.3.1 Device Configuration

The device configuration settings associated with the BT service are unique to each device; however, two settings that are always available for the service are the ON/OFF status and device's visibility status. How these settings are manipulated takes different forms based on the device. For example, a BT laptop may have a settings window listing these options, or a BT headset may require pressing a button for a set amount of time to perform a setting. The ON/OFF status is the power setting that determines if the BT service is currently available. Theoretically, if the BT service is set to OFF, an attacker cannot exploit it; however, to use BT services, this setting must be set to the ON position. For example, to use a wireless BT headset paired to a BT mobile phone to answer and talk on phone calls, both devices must have BT in the ON status. It is unavoidable to have the device in the ON status when using the BT service, but it is best to have the device in the OFF status when not being used.

The visibility status setting determines if another device is able *see* the device. A device's discoverable mode status indicates its visibility and is analogous to a Wi-Fi Access Point broadcasting its SSID. When a device is in discoverable mode, it is a beacon broadcasting that the device is present and any BT device within PAN range is able to see it. This is the single greatest possible risk for a BT device. It would seem obvious to keep this discoverable mode in the OFF position; however, when two BT devices are initially establishing a connection, one device must be in discoverable mode. When set to nondiscoverable mode, a device is theoretically not visible due to it not broadcasting its presence. However, if the device's BD_ADDR is known, perhaps from a previous pairing, then being in nondiscoverable mode does not eliminate another device within BT range of being able to locating it, in theory. Section 3.6 makes suggestions for both of these settings.

Devices commonly have an indication in the form of an icon on the device's home page to show the ON/OFF status of the BT service. If a device is infected, it will have the BT service in the ON status because the BT service must be ON for malware to spread over a BT connection. If the user did not place the device's BT in the ON status or if the icon display is not normal, this may serve as a clue to the user that a device is infected.

2.3.2 Hardware/Software

As with any computing technology, there is a hardware component and a software component. The hardware component that supports the BT technology is the BT radio chip transceiver. The form factor of this chip is approximately 9 mm^2 in size, adhering to the size design restrictions imposed on BT from the SIG (Section 1.1). There are numerous manufacturers of BT hardware based on products with different bus types, interfaces, power classes, and BT compliance levels; as of January 2011, there are 71 company suppliers of BT chips.[34]

The software component is the implementation of the BT protocol stack. This stack is the hierarchical layers of interconnected communication protocols that manage data flow for BT networking. There are numerous developers of the BT protocol stack based on different functionality profiles; as of January 2011, there are 53 company developers of BT protocol stack.[35] This stack implementation is commonly the area attacked by an exploit as a result of poor software implementation. This results in the devices that use that particular stack implementation to be vulnerable. A device will generally not be vulnerable to every exploit because the

---

[34] http://semiconductors.globalspec.com/SpecSearch/Suppliers/Semiconductors/Communications_RF_Wireless_Chips/Bluetooth_Chips (@ November 3, 2006).

[35] www.thewirelessdirectory.com/Bluetooth-Software/Bluetooth-Protocol-Stack.htm (@ October 25, 2010).

exploit plays on a specific stack implementation; this will help limit a device's vulnerability. It is common for one company to have a larger share of the devices; this would result in a larger number of devices being vulnerable.

2.3.3 User Interaction

There is a certain amount of expected interaction between the user and the device built into the BT protocol. This takes the form of a layering of prompts to establish an accurate level of user/device communications. The pairing process between two devices exemplifies this layered interaction. This process and its purpose within a PAN will be discussed in detail in Section 4.3. When a requesting source device initiates a pairing with a target device, this target device will prompt the user with the source device's name and a *yes or no* option to initiate the connection. This is the first prompt layer. The user has the option to accept or dismiss this request. If the user does not respond to this prompt within a device's prescribed time frame, the pairing is dismissed. Next, if the user accepts the pairing request, the target device prompts the user to authenticate with the requesting device. This is the second prompt layer. Both devices must enter a shared pass-phrase; if the pass-phrases do not match, then both devices will dismiss the pairing process. At this point, the user of the target device has had an opportunity to dismiss a potential attack by refusing the pairing. Continuing with this same example, let us assume a successful pairing. If the requesting source device then attempts to transfer a file to the target device, the target device will prompt the user to accept the file transfer. The user may *accept* or *dismiss* this transfer request. This is the third prompt layer. If the user dismisses the request, the device cancels the file transfer. If the user accepts the file transfer, the file transfer will occur to the target device. At this point, the transferred file is on the target device, but the user must select the file to open or execute it. This is not necessarily a prompt, but the user must consciously make the effort to select the file, so this will be called the fourth prompt. If the target device is able to associate the file with an installed application, it will open the file; else, it will give the user a message that it is an unknown file format. If the file requires installation, the target device will prompt the user if it is from an unknown publisher. These two prompts are the fifth prompt layer. If the user accepts to install the application or if the file was an executable file that did not require installation, and the file is a malware, the device becomes infected. During this entire scenario, the target device prompts the user five times. Different devices may have a slightly different approach for user interaction, but the approach presented is a very common scenario of interaction. To accept all five prompts requires considerable interaction from the user. However, an exploit may attempt to bypass the user interaction built into the BT process by using a disguised device name or file name to confuse the target device when prompting the user. In these situations, the users would be less likely to protect themselves from an exploit.

2.3.4 Device Platform

This last malware propagation limitation is common sense. Devices by themselves are not universally vulnerable to malware, but one may establish generalized factors that increase the device's potential vulnerability. These generalized factors will be based on the device's platform: OS, BT hardware and software implementation, and market share. There are several different OSs available for devices, and each requires different implementations of the BT software, for example, Mac OS X as supported since v10.2 released in 2002 and Windows since XP SP2 released in 2004. The limitations associated with each OS are similar to the ideas discussed previously for BT hardware and software (Section 2.3.2). Looking, in particular, at this last factor, market share, each platform possesses a different percentage of the total number of devices used in the market. This is often referred to as *the wild*. For obvious reasons, the number of devices in the market will typically be directly proportionate to the number of potential attacks. This falls along the same lines as why there is more malware in the wild for a personal computer than for the smaller market share OSs. Fortunately, fierce competition for market share has resulted in a large variety of device platforms with unique OSs. This then results in decreasing the potential to experience universal malware infestation. Additionally, the level of complexity for the malware has a direct proportionate relationship to the complexity of the device's platform. The more complex devices such as Smartphones combine all-in-one functionality. As of the second quarter of 2005, devices classified as Smartphones possessed only approximately 2% of the device market,[36] but this market share is changing rapidly. As this percentage increases, one would expect its associated malware to increase.

2.4 Categories of BT Malware

One may generalize BT malware into five categories, four of which the industry has officially accepted as names unique to BT. These five categories of names are as follows: *BlueBug*, *BlueJack*, *BlueSnarf*, *Car Whisperer*, and *Denial of Service (DoS)*. There is also malware whose classification may be a combination of these categories, such as *HeloMoto*, which combines BlueBug and BlueSnarf. The malware in these

---

[36] www.slate.com/id/2115118/, The Perfect Worm. (@ October 25, 2010).

categories often starts out as a proof-on-concept to demonstrate vulnerabilities in poor protocol design or implementation. It is then made available to the industry as a whole. However, sometimes intentionally or unintentionally, the malware escapes into the wild. Alternative to this protagonist perspective, for some users there is also certain amount of rewarding notoriety in pushing technologies to their limits and receiving the brag-factor that goes along with their exploitation. For example, a demonstration at DEFCON 2004 by Flexilis[37] captured 300+ mobile phone contact lists from unbeknownst pedestrians on Las Vegas Boulevard using a device called a BlueSniper, which is a riflescope and Yagi antenna attached to a BT-enabled laptop.[38] Eventually these categories will require recategorizing as new malware is born; until then, the following sections introduce these current categories. Once again, these are categories of attacks, not attacks themselves. The following section discusses each of these categories, mentioning the propagation methods discussed in Section 2.2.

2.4.1 Malware Category: BlueBug

A BlueBug attack allows a malicious device to use BT technology to access the core OS level commands of the target device without the consent or awareness of device's owner. The result of these attacks is full access to the core functionality of the compromised device by the malicious device. The target device uses a poor implementation of the BT protocol stack by the following device manufacturers: Nokia, Sony Ericsson, and Motorola.[39] The severity of this category of attacks is very serious due to its invasive nature.

The aspect of the BT implementation exploited is the pairing process. When two devices pair, BT requires that they authenticate to each other by entering a shared pass-phrase. BlueBug exploits a loophole in the faulty BT protocol implementation that allows the attacker to create a *hidden* serial communication channel that does not require device authentication. After bypassing the authentication, the malicious device then creates a virtual serial profile connection to the compromised device and has full access to the AT command set used to control this device. In Fig. 6, note that the AT Command Set utilizes a serial connection through the RFCOMM protocol. These AT commands are command–response executable transactions used to manipulate features on the mobile device. These commands have evolved from controlling modems and, although they differ slightly between devices, are a

---

[37] www.flexilis.com, Flexilis: Mobile Security Made Simple.
[38] www.tgdaily.com/2004/08/02/defcon_12/page6.html (@ November 1, 2006). www.wirelessve.org/entries/show/WVE-2005-0002 (@ November 1, 2006).
[39] www.wirelessve.org/entries/show/WVE-2005-0002 (@ November 1, 2006).

standard for controlling the functionality of mobile devices. For example, with a mobile phone, this functionality includes managing entries in the phone's contact list, initiating phone calls, accessing open calls, sending SMS messages, reading received SMS messages, using the network access such as the Internet, adjusting phone settings such as call forwarding. The simplest BlueBug attack may steal private information from the contact list, but an advanced attack may issue incriminating SMS messages to imposter the owner of the mobile phone. Imagine an attacker changing the ring tone volume to inaudible and then answer incoming calls without the user's awareness.

The main limitation for these sorts of attacks is that the malicious device must be within the BT range of the target device, and if a device having BT is in the OFF position, it is not vulnerable to this category of attacks. There is no legitimate usage of these attacks due to their invasive nature.

The namesake of this category derives from the ability of the attacker to turn the device into an electronic bug, from which the attacker is then able to listen from a distance. The device's owner is unaware that the device is being used as a bug.

2.4.2 Malware Category: BlueJack

Applications commonly use BT technologies to facilitate information sharing between devices. Based on the intent of the device, this information may take different forms. For PDAs, Smartphones, and mobile phones, it is common to share electronic business cards, which is the digital equivalent of the information that one would typically store on a business card. A normal transfer of an electronic business card occurs as follows: a source device attempts to send the electronic business card to a target device within BT range. The target device displays the contact name of the person on the electronic business card and then prompts the owner to either accept or dismiss this transfer. This is a simplified pairing handshake used to establish a connection between the two devices. If the target device accepts the business card, the target device creates a contact entry in its contact list with the electronic business card information. If the target device dismisses the business card prompt, there is no data exchange. This method of information sharing uses the *ad hoc* nature of BT networks to create a temporary connection between two devices to transfer data.

BlueJack attacks exploit the application performing the sharing of electronic business cards over a BT connection, in particular, the simple user interface to accept or dismiss the electronic business card. In a typical BlueJack attack, the source device will create an electronic business card with the contact name as something other than a first, middle, and last name. This may take the form of an interesting phrase, message, or joke and is possible because the BT protocol allows for a very large field size for the contact name (maximum 248 characters).

The source device then attempts to send this created electronic business card, and the target device will display this phrase, message, or joke. This is possible because BT uses an anonymous transfer when establishing this connection, which means that no authentication occurs between the two devices involved in the transfer. These attacks do not abuse an implementation bug because they are using the file sharing as it is intended; however, this might suggest that the application uses a better interface when transferring electronic business cards.

The main limitations for this category of attacks are that the target device must have BT in the ON state, in discoverable mode, and within PAN range. If the target device then refuses to accept the electronic business card, there are no adverse effects. If the target device accepts the electronic business card, the only effect is a cluttered contact list. The potential usage for this category of attacks is anonymous messaging in public areas, for example, mall retailers marketing when a customer is passing a store facade to inform the user of the current sale promotions. The typical user's response from BlueJacking is annoyance and is analogous to spam e-mails.

Although as of yet not widely used and potentially viewed more as an annoyance, there are applications available that are based on this type of attack that perform what is referred to as Proximity Advertising. Ever trying to reach a target audience, commercial entities want to use this form of advertising, which uses a BT adapter as a marketing tool, to communicate with customers with BT devices within range. Examples of these applications are Proximity Promoter 24×7[40] and Area Bluetooth Proximity Marketing Light 2.0.[41]

2.4.3 Malware Category: BlueSnarf

BlueSnarf attacks allow a malicious device to use BT technology to access data stored on a device without the consent or awareness of the device's owner. These attacks exploit poor implementations of the BT protocol and a vulnerability of devices with older implementations, in particular, certain models of Nokia and Sony Ericsson devices.[42] BlueSnarf attacks are similar to BlueBug attacks due to the amount of access the malicious device is able to achieve on a target device. Whereas BlueBug attacks achieve autonomous control of a device, BlueSnarf attacks abuse the File Transfer Profile,[43] limiting it to transferring files to/from the device. The severity of both categories of attacks is very serious due to its invasive nature.

---

[40] www.promotomobile.com (@ October 25, 2010).
[41] www.defconsolutions.com (@ November 23, 2009).
[42] www.wirelessve.org/entries/show/WVE-2005-0003 (@ November 1, 2006).
[43] Reference BT profiles in Section 1.7 and the FTP profile in Table 6, Bluetooth Profiles (Part II).

These attacks exploit a flaw in the implementation of the OBEX Push Profile (OPP) implementation, which bypasses the authentication process used in the pairing process. In Fig. 6, note that the OBEX utilizes a serial connection through the RFCOMM protocol. As mentioned previously, when two devices pair with each other, they share their available service profiles. OPP provides a method of information sharing that uses the *ad hoc* nature of BT networks to create a temporary connection between two devices to transfer data. The intended usage of the OPP is to facilitate the sharing of data between devices; it is a device-specific binary version of the File Transfer Protocol (FTP) where the target device prompts the user to receive a pushed file. BlueSnarf attacks exploit a loophole in this implementation, allowing the malicious device to avoid this prompting or authentication and to connect to a device via OPP and to transfer files to it, infecting the device. This may include disclosure or alteration of the information stored as a file on the device. However, the service that performs this file transfer does not allow file browsing so the attack is limited to the retrieval of files with known names. There is little security offered by the malicious device needing to know the name of the file to transfer because many of the files for storing system information on mobile devices use standard names, which are readily available over the Internet. For example, on most Symbian devices such as the Nokia 6600 series, the file system location and name of the device's phone book is *telecom/pb.vcf*.[44] The sort of files that may be retrieved are any known or guessable file name; this may include address-book entries, calendar events, digital media such as songs and images, the real-time clock, configuration settings, and the phone's IMEI or ESN.

The limitations for these sorts of attacks are that the target device must have the BT service ON, in discoverable mode, and within PAN range. Technically, it may still be possible to connect to a device running the BT service but not in discoverable mode, so the user should use caution when using the BT service. It is commonplace for mobile devices such as PDAs to offer telephony services, which may be potentially dangerous, as users store personal information on the device, such as passwords and financial information. The device owner is suggested not use obvious names for files that are stored on these devices; if the file name is guessable, a BlueSnarf attack may result in this information being hacked.

---

[44] Other common object names are available through open source listings on the internet. For example, OpenOBEX lists the objects as defined by the IrMC (http://openobex.triq.net/obexftp/services).

2.4.4 Malware Category: Car Whisperer

BT applications are standard features within new automobiles. These applications permit such functionality as rerouting phone calls and media stored on a mobile device directly to the audio system of the automobile. For example, when a mobile phone rings while the user is driving with their BT-enabled mobile phone in their briefcase, the user is able to accept the phone call using the audio system controls built into the steering wheel and a microphone built into the rearview mirror. Like any BT PAN, once the user establishes a pairing between devices, the car acts as a master device initiating the pairing process, and when the phone is within range of the car, ideally inside the car, the BT functionality is available.

The Car Whisperer category of attacks involves networking a malicious device to an automobile's BT kit. Once the malicious device establishes a connection, it is able to listen to or insert communications between itself and the automobile. This may include listening to audio through the microphone in the automobile BT kit or even playing audio over the sound system in the automobile. From its namesake, one would think this attack applies only to automobiles; however, any attack that establishes a two-way connection allowing listening and playing audio on a BT device falls into this category. The weakness exploited in these attacks is the pairing process for the devices involved; in fact, it is not a weakness in the BT protocol itself but a weakness in the pass-phrase used to pair devices. This pass-phrase is often very simple, to simplify BT usage and to overcome limitations of device interfaces. For example, where does one type a passphrase on an earpiece headset? For devices such as this that lack the ability to allow the user to input a pass-phrase, the manufacturers choose to hard-code a pass-phrase into the device. When the headset then pairs with another device, this device will enter the same pass-phrase that is hard-coded into the headset. The pass-phrase that manufacturers use quite often is 0000 or 1234. Knowing this pass-phrase and being within range of the automobile, the malicious device is able to connect to the automobile's PAN and exploit the BT profiles that it supports, specifically HFP and HSP.

Unfortunately, there is very little the end-user of the target device is able to do against Car Whisper attacks. The manufacturer is responsible for the hard-coded pass-phrase and tasked with implementing a more secure BT environment. One suggestion for the user is to prevent another mobile device from pairing with the automobile by establishing the initial pairing between a mobile device (the slave) and the automobile BT kit (the master) in a secure location. In doing so, if the car kit limits its network to one device, there is a reasonable assurance for the user that it is their device that is paired with the car kit.

2.4.5 Malware Category: DoS

A DoS attack is a common attack against network computers and Internet web services. The intention in mentioning this category of malware in reference to BT is to make the reader aware that BT services may be the target of a DoS malware; however, DoS attacks are a topic in themselves so this discussion will be limited in detail. A DoS on a BT device is a category of device malware where a malicious device abuses a target device's resource(s) resulting in the user's inability to use the target device for its intended purpose. The intentions behind this abuse of resources are varied, but quite often, the intention may be of no interest to the malicious device aside from causing the side effect that the target device has reduced usability. On a larger scale yet along the same lines, a Distributed Denial of Service (DDoS) is a category of malware where a service associated with the device is used in conjunction with a number of other infected devices to attack some other target. The intention behind this abuse of resources is typically to impair or shut down the resources on the target site. For example, if a large number of infected devices are simultaneously directed to the same Web site, the result may or may not be that the Web site is able to handle the unexpected amount of traffic. Twitter.com was the victim of such an attack in August of 2009.[45]

The resources commonly associated with a mobile device include services, battery reserves, and memory. The battery reserves and memory are both resources that a device depends to operate. Abusing their availability will result in a DoS attack. An example of DoS is a mobile device ping-of-death attack known as BlueSmack, which attacks a BT device's networking.[46] When a device is attempting to locate device within its PAN range, it will send out a request to which all discoverable devices will respond. BlueSmack abuses this request for response by making repeated requests for responses from all devices within PAN range. In doing so, it exploits the BT network's piconet topology due to their *ad hoc* topology (Section 1.6). There are two side effects from BlueSmack and other forms of DoS malware: loss of network connectivity due to overload of the BT link level (Fig. 6) and drainage of the device's battery. To address the abuse of battery resources, newer devices possess complex battery-conservation protocols to maximum available power while the device is in an active or in an inactive operational state. The BT SIG is addressing these forms of DoS attacks as a flaw in the BT stack design and future versions of the protocol will take precautions against them.

---

[45] www.searchenginejournal.com/was-the-twitter-ddos-attack-cyber-warfare/12362/ (@ October 25, 2010).
[46] www.infohacking.com/INFOHACKING_RESEARCH/Our_Advisories/bt/index.html (@ November 1, 2006).

Generally, DoS attacks are more annoyance and inconvenience rather than a physically damaging attack. The weaknesses that they commonly exploit are known bugs or limitations in the implantation of the BT protocol.

2.5 BT Malware Firsts

The mobile malware evolutions will most likely progress along a similar pattern to that experienced on desktop computers: initially very slow as malware authors learn the intricacies to exploiting devices, followed by a sudden explosion as other authors expound upon these first exploits. In an ideal world, device OS authors would learn from the desktop environment and create more secure products. In reality, device OSs are usually a subset of a complete OS due to a device's limited computing environment and resources and do not possess the security capabilities associated with the full desktop OS. As a result, devices are commonly vulnerable to malware, as they do not possess features such as authentication, encryption, and Access Control Lists. Newer versions of these OSs are including such security features, but this does not help the existing devices in the wild. Fortunately, there are *good-guys* (white-hat hackers) that help combat the *bad-guys* (black-hat hackers).

It is common for malware authors to write proof-of-concept malware that demonstrate weaknesses within the BT wireless technology and then to provide this malware to antivirus application developers. This malware is generally nonaggressive or destructive in terms of the payload and focuses more on BT propagation. These white-hats help the average user by researching different attack vectors so that measures may then be taken to prevent BT weakness from being exploited for malicious purposes. However, not all malware authors create exploits with such a benevolent approach. Some viruses are aggressive and contain a destructive payload. For example, in 2001, a virus in Japan blocked the ability of mobile phones to call emergency numbers.[47]

As the number of devices on the market grows, the authors developing their OSs are each striving for a larger portion of the user device market. Symbian OS currently holds the largest portion of the device market; however, iPhone OS, Blackberry OS, and Android are chipping away at market share and Window7 Mobile holds high expectations for some. A side effect of increased device population in the wild and these increased market shares by fewer OS authors is an ideal situation for would-be malware authors: more targets for the created exploits. With time on their side and more targets to attack, these black-hats are able to more easily

---

[47] www.bitfone.com/usa/presscov_20040415.shtml (@ November 14, 2006).

access and find insecurities and weakness in the devices. In addition to more targets, these devices are becoming more complex as customers demand more services. This complexity results in the potential for more avenues for attack as production time-lines do not allow for the required testing.

The malware to follow is for the OSs that possess a noteworthy malware in the wild: Symbian, Windows CE, and a few others. Table VIII lists a sampling of the malware for each of these platforms. To limit this discussion to a manageable portion, the discussion to follow is limited to the bolded malware. The intention in listing these pieces of malware is to introduce some existing attacks, to show that malware using BT technologies exists for each of these OSs, and, in doing so, demonstrate that no device is impervious to malware. The malware's detection date is displayed adjacent to each listing to provide a timeline perspective. More detail about each of these pieces of malware is available from the Web sites of any malware security firms, such as Symantec[48] and F-Secure.[49] Generally, each of them will require installation to infect the target device, and each OS uses different means to do this. Once installed, the malware will generally exploit a specific design or implementation error. This requires a certain number of factors to be present for the exploit to occur correctly. For example, most devices do not allow *autorun* from a memory card, but some devices view memory cards as a CD and will *autorun* when the card is inserted. Knowing this information tidbit, a malware author may use this *autorun* feature as an attacker vector to exploit a device reading a memory card.

2.5.1 Platform: Symbian OS

Symbian OS[50] is an OS based on open licensing standards, owned by the British Company Symbian Limited, and typically installed on mobile devices using the ARM CPU architecture. The key design features that make this OS favorable for mobile device vendors are extensive:

- Designed specifically for devices with limited resources yet offers features associated with a desktop OS
- Possesses memory management optimization
- Possesses power management features
- Multithreaded and multitasking
- Highly customizable

---

[48] www.symantec.com/index.jsp (@ October 25, 2010).
[49] www.f-secure.com (@ October 25, 2010).
[50] www.symbian.org (@ October 25, 2010).

TABLE VIII
SAMPLING OF DEVICE MALWARE AND THEIR TARGET OS, MALWARE IN ITALIC ARE DISCUSSED WITHIN

OS author	Example platform implementations	Malware in the wild
Symbian	• Nokia Series 60 (a.k.a. S60) • Nokia Series 80 • Nokia Series 90 • UIQ	• *Cabir* • Mosquit • Skuller • *Lasco* • Locknut • Dampig • *CommWarrior* • *CardTrap*
Windows	• Smartphone 2002 • Pocket PC 200–2002 • Mobile 5.0, 6.0 • Automotive 3.0–5.0	• *Duts* • *Brador* • Cxover • Letum
Palm	• Trēo Series of Smartphone • Palm Handhelds Z22, T|X, Tungsten E2	• *Liberty* • Phage • Vapor
Java (J2ME)	Platform independent	• *RedBrowser* • Smarm • Wesber
Mac (OSX)	• x86 • PowerPC	• *Inqtana* • Leap-A • Duh iPhone Worm • Ikee iPhone Worm
BlackBerry	• *Pearl (8130, 8120...)* • *Curve (8350, 8350i...)* • *Bold (9700, 9000...)* • *Storm (9550, 9530...)*	• *BBProxy*

Many mobile device platform implementations are Symbian OS based including the following interfaces: Nokia Series 60, Nokia Series 80, Nokia Series 90, and UIQ. Details on each of these interfaces and their implemented platforms are easily available to the interested reader. As of 2006, Symbian OS v.9.2 implements Bluetooth 2.0 support and in the most recent release, Symbian OS v.9.3 implements Wi-Fi 802.11 support. Per *canalys.com*, an independent marketing analysis expert for the technology industry, in October of 2006 Symbian's share of the mobile device market is 78.7%.[51]

2.5.1.1 Malware: Cabir (June 2004).

The Cabir worm is notorious for being the first known cell-phone virus. The research group by the name *A20* created the worm as a proof-of-concept virus and subsequently submitted it to antivirus companies for research. The worm was later detected in the wild. The intention behind its creation was to prove that BT connectivity might act as a vehicle to spread malware. The worm's target devices are several Nokia and Sony Ericsson phones based on the Symbian Series 60 Platform and propagate via a poor BT implementation. It is not dangerous; however, it does drain the infected phone's battery by continuously attempting to propagate. The malware's propagation limitation is ironically one of the design goals of BT; it is limited to PAN.

During a successful Cabir infection, target devices prompt the user to accept two actions: accepting the malware's installation and accepting the transfer of the malware onto the device. Cabir requires installation, and generally, devices will prompt the user to accept or dismiss installing an application. If the user accepts this installation, the worm infects the target device. Cabir is classified as a worm and replicates itself by sending itself using BT. Once installed, the worm will attempt to propagate by searching for a target device within PAN range and then trying to create and retain a connection with this device. If the target device moves out of range, the source device will attempt to locate another device, so it is continuously searching and attempting to connect to a target device. When source and target devices have established a connection, the source device will attempt to initiate a transfer with the target device to copy itself onto the target device. The target device will prompt the user to accept this transfer; the user must accept the transfer for the malware to transfer itself. If the user accepts this transfer, the worm propagates to the target device and the cycle continues.

---

[51] www.canalys.com/pr (@ October 24, 2006).

Future worms may take advantage of other device services, such as MMS, but this worm only utilizes BT to propagate. Symbian does not look up this as a flaw in the OS; however, they now offer a software-signing program whereby they verify and sign software as being authentic. This program builds a level of trust in the device OS because the user will be able to determine if the software is from a trusted source when they are prompted to install software.

2.5.1.2 Malware: Lasco (January 2005).

Lasco is a derivation of the same source code as the Cabir worm and therefore shares its same characteristics. Its main difference is that it is a file infector; in particular, it has the ability to infect installer archive files, which use a SIS file type. SIS files are standard ways to distribute Symbian applications. They are comparable to the CAB file type used by Windows CE. It gains notoriety for being the first piece of mobile malware for two reasons: being a file infector and the first to propagate via two methods. It is common for desktop computer malware to propagate via several methods but mobile malware is much less sophisticated.

When Lasco first infects a target device, it will search for all SIS files on the host device and insert itself into these SIS archives. When an infected SIS file is transferred to another device and installed, the Lasco worm will automatically attempt to install itself on the device. This installation will involve prompting the target device if they want to install the *Valsco* application. If the target device accepts the installation, the worm will continue propagating. This method of propagation is highly effective, as devices often share files. For example, devices commonly share games. Like the Cabir worm, Lasco also uses BT to transfer itself as an SIS file by the name *velasco.sis* to any device with range. The target device must accept this transfer of the file, select to install the malware, and then accept the warning that the application to be installed is not signed. When the device is infected, it will search for devices within BT range and attempt to send itself to them.

The Lasco worm does not have any destructive effects and infects the target device if the BT services are ON and it is in discoverable mode.

2.5.1.3 Malware: Commwarrior (January 2005).

Commwarrior is a worm malware classification. It originated from Russia and was the first mobile device worm to propagate via both MMS and BT. The worm infects the Symbian 60 Series Interface and has a unique twofold propagation method based on the internal clock of the device: from 8:00 AM to 11:59 PM, the worm spreads by establishing BT connections, and from 12:00 AM to 7:00 AM, the worm spreads using MMS messages. Using two methods of propagation allows the malware to propagate more aggressively.

When propagating via BT, the Commwarrior worm will attempt sending itself as a randomly named SIS file to other devices that are within range and will accept it over a BT connection. After a target device accepts or dismisses the malware transfer, the infected device will move onto another target device within range. This allows the worm to spread very rapidly, especially in highly populated device areas, such as in public transportation areas. The intention behind its renaming itself is to avoid the signatures used by virus scanners.

When propagating via MMS, the worm will send itself as an attached file to all entries in the device's contact list using the text message body from a preset list of possibilities. The name of the attached file is constant during MMS propagation (*commw.sis*). Using MMS allows the worm to overcome the limitations associated with BT. One may assume that the time frame selected to send messages from the device are times when the device will be less used, allowing the worm to spread without the knowledge of the target device's owner. By using the contact list in the device, the malware exploits trusted relationships between the device's owner and the entries in its contact list. Generally, if a user receives an attachment from a known user, it is apt to be opened.

The impact of the Commwarrior worm is not severe: it displays the message "OTMOP03KAM HET!" which translates from Russian to English as "No to brain-deads!" and uses the device's battery. There may also be monetary impact on the device's owner if there is usage fees associated with the device's MMS.

2.5.1.4 Malware: Cardtrap (September 2005).

Cardtrap is a malicious SIS-file Trojan malware classification. It is the first crossplatform pollination to propagate a piece of malware. In particular, it attempts to bridge the gap between devices using Symbian and desktop computers using Microsoft Windows via removable media. The Trojan infects target devices using Series 60 Symbian OS and then subsequently infects Microsoft Windows 32-bit devices. It uses either BT services or Internet download to propagate itself to new devices. Either methods require the user to accept the file transfer to the target device. Once on the target device, the user must install the malware. If the user accepts this installation, the device is then infected.

When the Cardtrap installation initiates, the target device will prompt the user to accept or dismiss installation of the file *Black_Symbian v0.10.sis*. If the user accepts the installation, Cardtrap performs several tasks once installed on the target device: disables the Applications Manager, installs Microsoft Windows malware on the phone's removable media, and maliciously attempts to disable the device's system and third party applications. First, it disables the target device's Application Manager as a protective measure to prevent the user from uninstalling the malware.

Second, it installs malware on the device's removable media. The malware installed on the removable media is a double-whammy punch of two Win32 worms: Padobot.Z and Rays. *Padobot.Z* is a buffer overflow worm that exploits the Microsoft Windows LSASS vulnerability. Cardtrap installs it on the removable media with an *autorun.inf* file that will attempt to launch the worm automatically when the user inserts the removable media into a Microsoft Windows PC. *Rays* is an e-mail worm and is installed on the removable memory disguised under the filename *System.exe*, which gives it a system file icon. Rays requires installation, so this system file masquerade is a trickery attempt to encourage the user to execute the malware when viewing the contents of the removable memory on a target Windows PC. Once the malware installs these worms on the device's removable memory, the exploit is established.

Third, it disables applications on the target device. The goal of Cardtrap is to force the user to insert the target device's removable memory into a Microsoft Windows PC and infect the personal computer with the two worms installed on this media. This goal is achieved by tricking the user into accessing this removable memory by encouraging them to insert the removable memory in the Windows PC. The approach taken is to impair usage of any applications installed on the target device, with the intention of rendering the device useless. To reinstall applications on the device, users could utilize the memory card to transfer them from their PC to the device. This is the ultimate goal of the Cardtrap malware.

Having introduced several pieces of Symbian OS malware, one can see the sophistication achieved in just over a year's time, as these malwares have evolved from the Cabir worm to this Cardtrap Trojan. Again, the intention in introducing these particular pieces of malware was to emphasize those that exploit BT technologies.

2.5.2 Platform: Windows CE

Windows CE,[52] also known as WinCE, is an OS owned by Microsoft, implemented on minimized device platforms and embedded systems, and supported on Intel x86 and compatible, MIPS, ARM, and Hitachi SuperH processors. WinCE is a

- Prefabricated modular components
- Thread based
- Optimized for devices with minimal resources

---

[52] msdn.microsoft.com/en-us/windowsembedded/default.aspx (accessed October 25, 2010).

WinCE serves as a foundation for numerous devices, for example, the Windows Mobile and Smartphone platforms are both subsets of WinCE. They each incorporate functionality of the OS's modular components and then implement functionality unique to their particular platform. Other platforms include AutoPC, Pocket PC 2000, Pocket PC 2002, Mobile 2003, Mobile 5.0, Mobile 6.0, Sega Dreamcast, and Portable Media Center among many other proprietary devices and embedded systems. Per *canalys.com*, an independent marketing analysis expert for the technology industry, in the second quarter of 2006, Microsoft's share of the smart mobile device market was 15%, which was second behind Symbian OS devices.[53]

Due to their design requirements, device OSs are often very minimalist. This has led to a general concern among security experts that WinCE is insecure. However, one can argue that it is just as insecure as any other mobile device OS. Due to Windows Mobile being one of the most popular platforms for devices such as PDAs and Smartphones, it is both a statistical and a logical target to exploit.

2.5.2.1 Malware: Duts (July 2004).

Duts is a parasitic file infecting virus for WinCE platforms that was developed by a member of the *29A* group as a proof-of-concept virus. This *29A* group is a collection of virus enthusiasts known for inventing, modifying, and testing technologies resulting in the creation of proof-of-concept malware, such as the Cabir worm mentioned previously. After it was developed, Duts was subsequently submitted to several antivirus security firms for analysis. The virus infects WinCE devices based on the ARM processor such as the PocketPC and Windows Mobile series and gains its notoriety for two reasons: it is the first virus for devices running WinCE and it demonstrated that Windows Mobile is indeed vulnerable to malware. Duts is not in the wild; in fact, the virus intention was not to propagate aside from a local propagation following a prompt to the user that it is about to do so.

Similar to all the malware presented thus far, the user of the target device must initiate the Duts virus's installation. When initiated, the virus prompts the user for a *YES* or *NO* response to the message, "Dear User, am I allowed to spread?" If the user selects *NO*, the virus terminates. If the user selects *YES*, the virus infects all WinCE executable files larger than 4 KB in the device's root directory by appending itself to each file and then adjusts the file's entry point to point of the virus. Executing any infected file will then result in a repeated virus propagating.

To propagate to another device, an infected executable file must transfer to another device. Potential transfer methods common for a device include through e-mail attachment, Internet download, removable media transfer, device

[53] www.canalys.com/pr (@ November 07, 2006).

synchronization, or BT file exchange. Executable files commonly face resistance to transfer over e-mail and Internet download due to their reputation as potentially harmful. Removable media and file transfer are both typically transfer methods that involve one-user transfer between trusted devices. This one-user transfer is in a sense similar to a sneaker-net where the user transfers something to another device by first copying the file onto removable media. BT is potentially the most invasive transfer method because a second person could push a file to another device. If the user accepts it, the infected file transfer is complete and available for execution by the unsuspecting user.

2.5.2.2 Malware: BRADOR (August 2004).

Brador is a malicious Backdoor Trojan Horse malware classification that infects WinCE platforms. Its notoriety is twofold: it is the first known Backdoor Trojan Horse for WinCE devices and is the first malicious piece of malware for the WinCE platform. It is a classic Backdoor Trojan Horse, which means it is a tool that allows a remote malicious device (client-role) to control the infected target device (server-role). Whereas previously created malware were proof-of-concept malware used for testing and research such as with Duts, Brador is the first of WinCE devices with a potential malicious impact. The malware infects devices based on the ARM processors using WinCE 4.3 or newer. It originated in Russia and some reports indicate that its author is aspiring for capitalistic gain by selling the client portion of the Trojan. Thus far, it is not in the wild.

Once again, for the device to become infected, the user must initiate the Brador infection process by clicking on its executable. When launched, Brador performs the following tasks: it establishes itself as a Trojan server service by copying itself to *Windows\Startup\svchost.exe*, e-mails the compromised IP of the target device to the client software, opens TCP port 2989, and then idly listens for instructions. Copying itself to the Startup directory is a survivalist measure to assure automatic start-up every time the host device starts. This allows the Trojan to retain controls of the device. E-mailing the address of the device to a client informs the malicious device that the target device is active and connected to the Internet. If the device does not have network connectivity, it repeatedly attempts to e-mail the infected device's IP until it is successful. It is common to use TCP port 2989 for Remote Administration of devices; in this situation, the port allows the malicious device to communicate with the Trojan device. Opening the TCP port will only be detectable by the user of the target device if they are monitoring the port status on the device. If the target device is using an Internet-security software solution, its firewall should detect this action and prompt the user to create a firewall rule. If the target device is not using an Internet-security software solution, the user will not be aware that this external connection has been established. Ironically, the port number 2989 is *BAD* in hexadecimal notation. When listening, the Trojan will respond to a series of commands that the client is able to issue to the target device via this backdoor: list directory contents, upload and download files, execute command, display message. The details of each of these types of commands are easily understood.

Brador will not propagate on its own due to its backdoor nature but requires user interaction to transfer itself to the target device. Potential transfer methods common for a device include through e-mail, Internet download, removable media transfer, device synchronization, or BT file exchange. As discussed with the Duts malware, BT is potentially the best way for it to propagate.

2.5.3 Platform: Others

Generally, malware tends to hone in on OSs or environments that possess a larger portion of the device market. In particular, its authors prefer more sophisticated devices because they offer more potentially exploitable technologies, but it is not always limited to these devices. Considering the massive number of devices on the market, even malware that targets a small portion of this market-share pie will affect a large population of devices. As a result, malware exists for devices based on these smaller market-share OS (Palm OS, Java J2ME, and Mac OS X) and it has been in the wild for quite some time. The Palm OS, for example, in August 2000 experienced a Trojan horse called *Liberty* that tried to delete all applications on the device and then reboot. Palm offered a certain amount of security through obscurity, but this has changed with J2ME environment whose details will be discussed briefly. BT enables devices, even if they do not have the same OS, to be vulnerable to malware transfer. If the user chooses to accept a BT file transfer, the file will transfer.

2.5.3.1 Malware: RedBrowser (February 2006).

RedBrowser is a malicious Trojan horse malware classification, and it has gained notoriety as the first malicious malware for the Java 2 Micro Edition (J2ME) environment. The benefit gained for developers by utilizing the Java Runtime Environment (JRE) is that it allows a Java application to execute on any platform as long as there is a runtime available and installed on that particular platform. Unfortunately, this benefit is also available for the malware authors. The JRE available for various mobile devices are the J2ME environments. Therefore, RedBrowser is able to infect any device, regardless of the platform OS, that has a J2ME environment installed. This ranges from the more simplistic mobile phone devices to the more advanced PDA and Smartphones devices. Since Java is a widely implemented coding language due to its ability to function on numerous platforms, many devices have the J2ME preinstalled. Its potential to allow widespread malware propagation from devices to desktop computing is a grave concern.

To propagate to another device, RedBrowser must be transferred to another device. It cannot propagate by itself. Potential transfer methods common for devices include through e-mail attachment, Internet download, removable media transfer, device synchronization, or BT file exchange. As mentioned previously for malware, BT is potentially the most invasive transfer method because a second person could push a file to another device. If the user of the target device accepts the transfer, the device is then infected.

RedBrowser arrives on the target device as the archive file *REDBROWSER.JAR*. Similar to all the malware presented thus far, the user of the target device must initiate the infection by extracting and installing it. When launched, RedBrowser displays a splash screen of a red moon rising while it installs, followed by a message in Russian telling the user that it will provide free Internet browsing without access to a WAP using SMS messaging. This is an attractive proposal for the target device that does not have Internet access yet does possess SMS messaging functionality. This is, however, a trick to allow the Trojan to use Java-based SMS functionality on the device. If the user accepts the invitation to get Internet access, the exploit will randomly send an SMS message to a preselected list of premium-rate numbers, accruing a service charge for the device of about $5.00(USD). The user will incur charges per SMS message sent to a premium-rate number, and this request to send and sending process repeats until the user does not accept the request. As the user is lead to believe that Internet access is achieved through repeated SMS messages, the average user will accept the repeated requests to send SMS messages resulting in hefty fees.

The target device for RedBrowser is an unsuspecting Russian; the malware will not affect devices outside of Russia because it does not use country codes when sending an SMS message to the premium-rate number.

2.5.3.2 Malware: Inqtana (February 2006).

Inqtana is a worm malware classification for Mac devices, in particular, those running OS X 10.3 and 10.4. It is a proof-of-concept piece of Java-based malware that attempts to propagate using the BT OBEX Push protocol and exploits the Apple Bluetooth Directory Traversal Vulnerability. For it to propagate, the malicious device must be within PAN range and BT must be active. Like all file transfer using this protocol, the user must accept the file transfer. Once accepted by the user, the worm then creates itself an user account, modifies the device's boot process, and continues propagation. The user account it creates for itself is under the user name *Bluetooth*, does not require a password, and grants this created account *root* system access to the device. This new account is then accessible remotely using a backdoor that is installed at the same time. It modifies the boot process so that the next time the device reboots, the exploit will start automatically. It is able to install itself and alter the boot process due to the Directory Traversal Vulnerability. Once the device is infected, it attempts to establish a BT connection with another device to continue propagation. This exploit was never found in the wild, and will no longer propagate in its current strain due to an internal limitation for it to stop in February 24, 2006. Users may protect themselves through security updates released by Apple.

3. BT Auditing Tools

The potential attackers will always take the path of least resistance available to them. Why break in the house when the door is wide open? Specific to BT devices, an attacker will commonly start with the physical aspects of the device: the device manufacturer and the device's heads up display BT indicator. Manufacturer and device-specific vulnerabilities are readily available on the Internet. It is commonplace for manufacturers to have certain consistencies in their products, for example, a default device name given to every device of a certain model. It is also commonplace for certain products and versions of software to have known vulnerabilities; Section 2.5 mentions the platforms that each malware exploits. The path of least resistance to locate this common information is through personal and community vulnerability archives, which can be extensive due to community support.

Every device has some variation of a heads up display that communicates the device's status to the user; some of these displays are more complicated, whereas others are very simple. On this display, an indicator shows the status of the device's BT service as either active or inactive. This may take the form of an icon if the device has a user-interface display or a flashing light or an LED in a conspicuous location on the device if the device is less complicated. The general color of choice is blue for obvious reasons, so if an attacker sees the device indicating that BT is active, such as a blinking blue LED, they may attempt to identify the device by scanning for those in discoverable mode. If the device is discoverable, the attacker may then turn to exploiting the device. How is a user to protect himself or herself against this potential attacks? They may use software-auditing tool.

Similar to white-hat malware, software-auditing tools typically start out as proof-of-concept utilities before they become commercially available. The goals of these tools are numerous, but they generally attempt to detect any potentially exploitable loophole in a device's line of defense. These utilities play on the two BT characteristic: it is easy to talk to BT devices because they are eager to talk when in discoverable mode, and BT is commonly on all the time, unlike a Wi-Fi on a laptop, which turns itself off to save power. These goals commonly address, but are not limited to, the following functionality:

- Assessing if the device is functioning correctly
- Examining a BT network infrastructure
- Determine the types of devices in the WPAN
- Locating and tracking a device within a WPAN
- Detecting rogue devices on the WPAN
- Maintaining network connectivity records of all discovered devices
- Protecting data stored on the device, such as BlueSnarf exploits (BlueSnarf Category)
- Protecting against DoS, such as BlueSmack exploits (BlueSnarf Category)
- Creating awareness for users of the million BT radios shipped each week

Software auditing tools are platform dependent to achieve different goals. As these tools evolve, their functionality boundaries may merge and expand to evaluate other wireless networking technologies. Although there are numerous tools available, Table IX lists a few of the more accepted and widely used wireless network-auditing utilities.

This list includes the more commonly used and noteworthy BT auditing tools. Others are available, for example, Mac OS X offers an open source wireless discovery tool similar to networking tool NetStumbler called iStumbler. NetStumbler[54] is a free, Windows-based, 802.11 WLAN diagnostics utility that has established itself as an industry standard, cant-live without tool. Other than auditing tools, there are also tools available for other purposes such as penetration testing; for example, Bluediving available on *soureforge.net* is a tool that implements several categories of attacks on a BT device such as BlueBug, BlueSnarf, and BlueSmack. This discussion narrows itself to introducing the above-mentioned five auditing tools, starting with the method by which a user would go about performing a BT audit: BT War Driving.

3.1 BT Wardriving

Wardriving is the act of attempting to discover Wi-Fi networks while using some sort of mobile transportation such as automobile and software utilities such as NetStumbler. The term was coined when Wi-Fi networks first started becoming omnipresent.

---

[54] www.netstumbler.com (@ October 25, 2010).

TABLE IX
BLUETOOTH AUDITING TOOL SUMMARY

	Bluetooth auditing tool			Discussion section			Platform implementation		
	Audits discoverable devices	Audits nondiscoverable devices	Performs live scanning	Logs discovered devices	Enumerates offered services	Enumerates device properties	Audit location assignment	Bluetooth protocol stack used (MS/Linux)	Device port scan
BlueScanner				3.2		✓	Microsoft Windows		
BlueSweep	✓	✓	✓	✓ 3.3	✓	✓	✓ Microsoft Windows	(✓, −)	
BlueSniff	✓		✓	✓ 3.4	✓	✓	✓ Linux	(✓, −)	
BTScanner	✓	✓		0		✓	Java (Windows, Linux, Mac)	(−, ✓)	
BT_audit	✓		✓	✓ 3.6	✓	✓	✓ Linux	(✓, ✓)	✓

From a technical standpoint, some may argue the exact definition of this term is based on the extent to which the user discovers and uses a Wi-Fi network.

The BT version of *Wardriving* is based on this same concept but instead the user is attempting to identify and map BT devices. This practice may be referred to by numerous names including Bluedriving, Bluewalking, Bluecasing, War Nibbling, among others. As with the definition of *Wardriving*, the extent that one takes this discovery depends on whom you ask. In fact, that legalities associated with it are changing rapidly, for example, a Windows Mobile BT tool by the name *btCrawler* is no longer available on its developer's site due to 2007 updates in German cybercrime laws. The goal may be to merely discover a device or to discover a device and steal some information from the device. This is the origin of the term War Nibbling in that small amounts of data are grabbed from the device; for example, this data stolen may be an entry in the device's contact list. No matter what the name of this practice, it involves using one or some of the software utilities introduced below. The hardware tools required to BT Wardrive are a device that has a BT adaptor and, depending on the range that one is hoping to search, a longer-range, directional antenna, or directional adaptor, as the BT radio by nature is nondirectional.

As BT falls more under mobile networking and the devices that contain their transceivers are mobile devices, the audit tools themselves are also available on mobile devices. To enumerate a few of these mobile audit tools, the above-mentioned *btCrawler* and CIHwBT[55] for Windows Mobile, BlueRadar and Super Bluetooth Hack 2009 for Symbian, and Bluetooth Browser in J2ME. All too often, there is a very fine line between well-intentioned auditing and ill-intentioned hacking; with this in mind, these mobile tools no matter what their initial intentions serve as good BT audit tools. Their main limitation, as with many mobile devices, is the displaying of information, which in this case is the audit information and as a result the preference for not using them for auditing.

3.2 BlueScanner,[56,57] from Network Chemistry

Originally developed by Network Chemistry, a California-based company that specializes in and creates a series of wireless vulnerability assessment products, and subsequently acquired by Aruba Networks, Inc., BlueScanner is both a free Windows-based utility and modular part of their RFprotect Mobile Suite, which is a set of BT

---

[55] Can I Hack With Bluetooth (CIHwBT) is a Security Framework for Bluetooth on Windows Mobile 2005 available at Soureforge.net.

[56] Reviewed BlueScanner Version 1.1.1. The Linux counterpart to this Windows utility is a utility called BTScanner, and based on the BlueZ Bluetooth Stack.

[57] labs.arubanetworks.com (@ October 25, 2010).

Vulnerability testing application. In particular, BlueScanner operates with Windows XP SP2, the BT Protocol Stack built into SP2, and any BT adaptor that supports this stack.

BlueScanner is an active tool that sends out polling requests for any discoverable BT device within range. The intention is to identify these devices without actually authenticating with them. As presented previously, discoverable mode is a BT device setting that allows other devices to see the device and initiate communications with it. When launched, the utility will scan for devices every 10 s within the class range of the BT adaptor. It then retains a log of the identified devices and certain information specific to each device. This includes the device's given name, its unique address (BD_ADDR), the type of device (computer, phone, keyboard, ...), its advertised service profiles, the discovery time, and the time last seen. Figure 9 shows a scan of a location that detected six devices and then the service profiles available for a device that it was not able to determine. If the device discovery occurred while the device was in discoverable mode

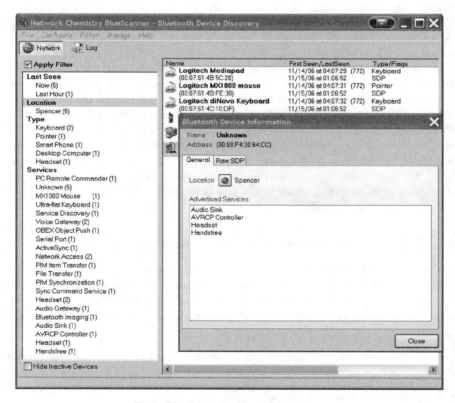

FIG. 9. BlueScanner by Network Chemistry.

and it is later set to nondiscoverable, it will still be identified when within range. The utility also categorizes the captured device data according to an assigned location, so, for example, when installed on a laptop device, BlueScanner could scan different locations within a corporation to log all BT devices based on physical locations within the corporation. This is the true intention for using the utility: to establish one's exposure or vulnerability to potential BT threats or to determine one's compliance to an established security policy for BT devices. Continuing with this example, if the corporate security policy states that a device may not be ON or in discoverable mode, when installed on a laptop and allowed to scan at strategic locations, this utility will facilitate determining the level of compliance by its employees. Considering the strict compliance guidelines imposed by such standards as HIPAA, SOX, and GLB, there should be a tool to assess a company's vulnerabilities to all wireless products.

3.3 BlueSweep from AirMagnet

AirMagnet is a Sunnyvale, California-based company that researches and produces mobile workplace assessment products. They previously released a free Windows utility called BlueSweep; it is no longer available on AirMagnet's Web site. This utility produces comparable results to Network Chemistry's BlueScanner. BlueScanner and BlueSweep were released at the same time and, aside from personal preference and user interface, achieve the similar output. Figure 10 shows the utility and a summary of a scanned environment. Both products allow for live scanning of an environment for BT devices, saving of a live scan for later review, and multiple views of scanned devices and their services. While BlueScanner creates more logging during scanning and allows scanning location specification, BlueSweep provides more device details about each scanned device as well as an overall summary of the environment. This summary includes the device's manufacturer, the device's pairing state, and the device's authentication usage during communications. This utility is also bundled with AirMagnet's Spectrum Analyzer suite of tools, which will detect all RF in the ISM band and identify all radios devices. The result is a marriage of utilities as a product suite that will detect if any BT devices are within range regardless of encryption status of the communications.

3.4 Bluesniff[58] from The Shmoo Group

Bluesniff is a Linux-based tool used for BT Wardriving. In particular, it is a front-end interface for a proof-of-concept utility called Redfang. Ollie Whitehouse with @Stake developed Redfang from the idea that a BT-enabled device that is not in

---

[58] http://bluesniff.shmoo.com/ (@ October 25, 2010).

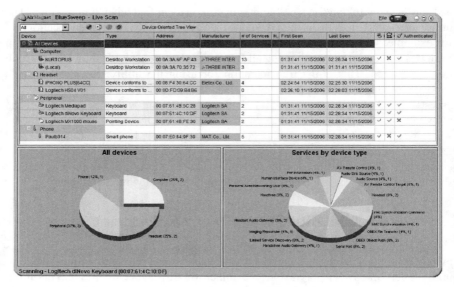

Fig. 10. BlueSweep by AirMagnet.

discoverable mode will still respond to a request stimulus. This is to say, if a second device knows the address (BD_ADDR) of a device, this second device may initiate communications with this device and the device will respond. The utility uses a brute-force method to guess the BD_ADDR and identify all devices within range.

This BD_ADDR is a 48-bit value that is unique to every device similar to a network adaptor's MAC address. Typically, these values displayed are in hexadecimal form as shown below in Eq. (1). The first 3 bytes of this address are specific to each manufacturer; the three *MM* represents this value. This 3 byte value may easily be determined from published resources if the manufacture of the device is known. As these 3 bytes are easy to determine, this requires only to guess the last 3 bytes of the BD_ADDR. The pattern of these last three bytes is somewhat predictable based on knowing the manufacturer pattern of naming devices. Redfang tries to guess a device's address by rotating through these last three bytes of BD_ADDR. Once it identifies a device, this process is equivalent to the device being in discoverable mode and Bluesniff will be able to scan that device whenever it is within range. In addition to this brute-force method to identify devices, Bluesniff adds the functionality to find devices in discoverable mode. The result is a comprehensive auditing tool. Figure 11 depicts the Bluesniff results from a scan identifying two devices.

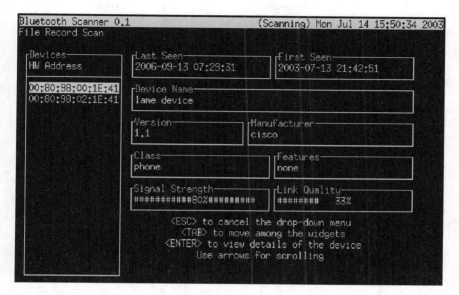

FIG. 11. Bluesniff by The Shmoo Group (www.rootsecure.net/content/downloads/pdf/atstake_war_nibbling.pdf; accessed October 25, 2010).

$$MM : MM : MM : XX : XX : XX \tag{1}$$

$$16^6 = 16,777,216 \text{ possibilities} \tag{2}$$

As one would expect, this brute-force method is a slow process. There are three reasons for this time-intensive process. First, the utility must discover the frequency-hopping pattern used by the devices (Section 1.5). Second, there are a large number of potential address permutations, 3-byte Hex values and 16 possible Hexadecimal characters for each result in a large number of possible permutations to guess (Eq. (2)). Third, probing for an individual address requires a specific amount of time, approximately 2–10 s. There are, however, methods to decrease this required time: such as using multiple BT adapters or a multithreading computing environment, so this is a viable audit tool.

3.5 BTScanner[59] from Pentest Ltd.

BTScanner is a Java-based BT auditing utility designed by Pentest Ltd. It is OS platform independent as long as the appropriate JRE is installed and available for both the Microsoft and BlueZ BT protocol stacks. The utility is able to scan and

---

[59] www.pentest.co.uk (@ October 25, 2010).

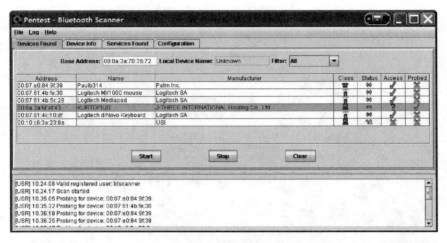

Fig. 12. BTScanner by Pentest Limited.

extract the maximum details from BT devices within a PAN range without actually authenticating with the devices. This includes device-specific information as well as the profiles the device offers. It displays this device information in a tabbed interface and updates this information by monitoring each device as well as its radio signal through an open connection. In addition to live scanning, it allows saving scan sessions to compare results over time (Fig. 12).

3.6 BT_Audit[60] from trifinite.org

trifinite.org is a not-for-profit group of hobbyists that perform research on wireless communications. Their BT_audit utility is a *NIX-based BT auditing suite of tools that provides the ability to scan a device's BT protocol stack and determine the services currently available on that particular BT device. This utility is a console-based tool that attempts to gather as much information as possible about BT device within PAN range without actually pairing with these devices. Although the SDP is responsible for reporting what services, also referred to as ports, are available on a device, it does not always tell a complete story. For example, the BlueSnarf attacks discussed previously exploit a port that is open yet not reported by a query to the BT protocol stack's SDP.

---

[60] trifinite.org/trifinite_stuff_btaudit.html (@ October 25, 2010).

A BT device communicates to other devices via open ports, similar to the TCP/UDP/IP architecture. BT_audit allows the user to perform a port scan on a device to determine which communication ports are open. The goal is to identify these open ports because they could potentially translate to vulnerabilities that may be exploited. This is analogous to scanning a computer, such as with the freeware Network Mapper (Nmap) tool by *insecure.org*, during a security audit of a computer. The two main protocols in the BT protocol stack are the L2CAP and RFCOMM protocols. The L2CAP protocol multiplexes services into odd-numbered ports ranging from 1 to 65525, and the RFCOMM provides serial emulation on up to 30. The casual reader can easily see the potential for danger, as there are so many available ports over these two protocols. There are two main tools within the suite: PSM_scan (Fig. 13) and RFCOMM_scan (Fig. 14). PSM_scan scans the L2CAP protocol and returns the open port numbers and RCOMM_scan scans the RFCOMM protocol and returns the open channels. Figures 13 and 14 show the output of each of these tools from sample scanning a device.

This utility advances on the previous mentioned utilities in that it looks at the raw BT protocol stack communications. Newer utilities are continuously available and improved upon but trifinite.org has set the standard for auditing BT technology.

```
scanning, this will take some time...
psm: 0x0001 (00001) status: L2CAP_CS_NO_INFO    result: L2CAP_CR_SUCCESS
psm: 0x0003 (00003) status: L2CAP_CS_NO_INFO    result: L2CAP_CR_SUCCESS
psm: 0xfffd (65533) status: L2CAP_CS_NO_INFO    result: L2CAP_CR_BAD_PSM
psm: 0xffff (65535) status: L2CAP_CS_NO_INFO    result: L2CAP_CR_BAD_PSM
```

FIG. 13. *PSM_scan* tool in BT_audit.

```
rfcomm: 01 closed
rfcomm: 09 closed
rfcomm: 10 open
rfcomm: 11 closed
rfcomm: 12 closed
rfcomm: 13 closed
rfcomm: 14 closed
rfcomm: 15 open
rfcomm: 16 closed
```

FIG. 14. *RFCOMM_scan* tool in BT_audit.

4. Security Recommendations

Security is a large consideration in the design and ongoing development of the BT protocol and the SIG has established a large community effort to help avoid potential BT exploits. BT's main design specification, which some view as its limiting factor, is also its most securing feature: range. The short-range nature of BT limits a device's potential exposure compared to other wireless technologies. The convenience factor of BT's automated networking is also a security concern. However, a user should always practice some basic precautions and not overlook them as they appear as common sense:

- Establish a general awareness of the device. The user should educate themselves on the device by reading the literature made available from the device's manufacturer.
- Establish a general understanding of BT due to its ubiquity. The user should educate themselves on the basics of the BT technology through the numerous sources available on the Internet.
- Be aware of and pay attention to device activity that may be suspicious. There is a certain normal activity for a device, and the user should know when the device deviates from this normal.
- Stay updated. Some devices with Internet connectivity allow the device to update by the user; otherwise, the manufacturer must service the device.

A certain comfort level of security is achieved with a wired environment, so in migrating to a wireless environment, certain measures must be taken to convince the everyday device user that a similar level of security is achieved. The security level of the device also dictates the security level of the wireless network. If one device enforces strong security measures, yet another enforces more relaxed security measures, security may be limited. In particular reference to BT, there are six recommendations that will be discussed in detail: Device's Defaults, BT Status, Secure Pairing, Secure Pass-Phrases, Installing Applications, Antivirus Application. Based on the finer details discussed up to this point, each of these recommendations will provide a user-friendly reasoning for its usage. An attacker will generally take the easiest route, or the most vulnerable target, so by heeding these precautions as well as implementing the following recommendations, the user has established a high level of security. In the following recommendations, the term *user* is a reference to the person to whom the recommendations are being made.

4.1 Device's Defaults

Out of the box, BT-enabled devices have certain default settings; of particular concern is the name of the device. The manufacturer creates this settings default values, which is typically the same for all the devices of the same model. Among other reasons, the manufacturer's intention is to facilitate usage of the device by the average user in that they do not have to alter settings but are quickly able to make use of the device. As one would expect, these values are easily available to a potential attacker. It is advisable to change the device's name to avoid an attacker from spoofing the device by using a default device name. Additionally, it is advisable to keep one's device name private.

Why is this important? The manufacture default name is commonly a descriptive name associated with the model of phone. When pairing two devices, a device scans for visible devices and lists these visible devices based on the device names. Although this descriptive name may facilitate selecting the desired device, a device may be mistakenly trusted as a known device when in fact it is a malicious device spoofing the desired device's name. It is advisable to keep one's device name private because that value is so easy to spoof, as it is so easily changed. The device resolves this name from the BD_ADDR unique to each device. Some devices will display this BD_ADDR when searching for visible devices; this is advantageous when pairing because it allows the user to see the device address in addition to the device name. As the device address is more static on the device, it lends more assurance that the device is pairing with the desired device. For several reasons, including that this device address is vital in the pairing process to calculate algorithmically the frequency-hopping pattern, one may argue that it is better for this value to remain unpublished.

4.2 BT Status

The main settings associated with BT usage were discussed in Section 2.3.1. Using the premise of learning by repetition, it is suggested to turn the BT discoverable mode OFF when not actively communicating with other devices and turn BT OFF when not using the BT service. Turn the BT service ON only when needed; this would include only when the device is communicating within a PAN.

Why is this important for discoverable mode? Discoverable mode allows a device to broadcast one's presence to all devices within PAN range. It is during this discoverable mode that a device is most vulnerable, yet it is necessary to pair two devices; the following recommendation (Section 4.3) will detail this pairing process. Whether it is a misunderstanding of how to use the BT technology or just forgetfulness to turn off BT discoverable mode after using it, devices are often left in discoverable mode. As a result, many newer BT devices will either automatically turn this mode OFF after a short time or allow the user to specify a window to be in discoverable mode when it is turned ON. The user should be still cautious when the discoverable mode is OFF but the BT service is ON. Although the device is not as easy to detect in this state, if another device knows the device's ID, it is still potentially visible to these devices. One possible method to locate a device in this state is through a brute-force method, such as with the Bluesniff tool discussed above (Section 3.4). A second possibility is for previously paired devices; during the pairing process, devices share their device IDs, so they would be privy to having and retaining this information.

Why is this important for BT services? Turning the BT service OFF when not in use will not affect the functionality of the device; it will save battery life and allow the user to actively manage the device because a conscience effort is then required to use the BT service. However, if the BT service is not ON, the device cannot participate in the PAN, so the user must balance the need for its usage with need for a secure computing environment. By default, all BT communications do not require authentication, so to avoid a device automatically connecting, this is another reason to turn the BT OFF when not in use.

4.3 Secure Pairing

Device *pairing* is the process by which two devices establish a trusted relationship, a.k.a. secure communications. This process ensures a secure connection between two devices because, when successfully paired, each device views the other as trusted. Using the master/slave relationship introduced previously (Section 1.6), one device plays the role of the master and the other a slave. During the pairing process, the master device creates a link key from an algorithm that uses a pass-phrase, the BD_ADDR of each device, the internal clock of the mater device, and a random number. This link key is a 128-bit secret key used to authenticate and then encrypt all future communications between the two devices. It will always be unique between two devices. This link key and the address of the device to which it allows communications is stored securely in a database within each device and used to create subsequent connections automatically. Section 4.4 discusses details concerning this pass-phrase. This pairing process may vary slightly from device to device but will generally occur as described in Table X. This process allows the two devices to skip future pairing process because they have previously established a trusted relationship.

There are three recommendations associated with securing the pairing process: only pair with known devices, only pair devices in a secure location, and review the list of trusted devices regularly.

TABLE X
PAIRING PROCESS

Step	Device	Action	Description
1.	SLAVE	Change visibility to discoverable	Discoverable mode allows the master device to identify the slave device when it performs a search
2.	MASTER	Search for all discoverable devices	This search will return all devices with PAN range that are in discoverable mode
3.	MASTER	If the slave device is found, select that device	Select desired slave device from the enumerated list of devices located within PAN range. Devices are listed based on their given device name
4.	MASTER	Create secure pass-phrase and enter when prompted	These details are associated with the pass-phrase discussed in Section 4.4
5.	SLAVE	Enter pass-phrase shared with master device	Slave device will prompt the user to enter the pass-phrase created by the master device

Why is this important to pair with known devices? Any device may initiate the pairing process, but the receiver of the invitation has the ability to dismiss this request to pair. When prompted for a pair request, the user sees the name of the device making the request. If the name of the device is unknown, dismiss this invitation. The range of the BT device will also limit its potential for pairing. If a device is requesting a pairing, it is most likely within eyesight. For example, it is commonplace to receive a request in busy public areas such as subway train. As mentioned previously, successful pairing creates a list of trusted users on the device. A trusted device has access to data and BT services on the target device, so for obvious reasons, it is the goal of the malicious device to be on this list. For these reasons, the user should not accept invitations to pair from unknown devices.

Why is this important to pair in a secure location? This pairing process puts the devices in a potentially vulnerable state because it is during this process that the security is established. The communications involved at this point are over an unencrypted link. If a third party were able to sniff the pairing communications, they could potentially extract the created link key and the BD_ADDR of each device. Not only is privileged information abused in this situation, but it also compromises the trusted relationship between the two devices, for example, knowing the link key would allow for man-in-the-middle attacks and knowing the link key as well as each device's BD_ADDR would allow for device spoofing. Similarly, if communications between two devices are interrupted, the user should wait until in a secure location to re-pair a device. Theoretically, a malicious device could force two devices to re-pair by removing a device from a trusted list to sniff the pairing traffic. To avoid this potential malicious action, the recommendation is for the user to pair devices in secure locations. The BT range allows the user the peace of mind that two devices will pair safely, for example, in one's place of residence.

Why is this important to review the device's list of trusted devices? A device will provide the user a list of devices to which it has paired. The user should review this list on a periodic basis to check for unknown devices. If a malicious device is able to sneak itself onto the trusted list of a device, regular checks of this list by the user will eliminate potential future damage. Additionally, if a device is lost, the user should remove this lost device from the trusted list on any other device because the malicious user will be able to impersonate a trusted device to any device on the lost device's trusted list. With the same thought in mind, it is suggested to also update the pass-phrases of the trusted devices on regular basis, which is synonymous to updating one's personal passwords to avoid the possibility of them potentially being compromised.

4.4 Secure Pass-Phrases

BT imposes three security mechanisms all of which take place at the Link Layer: authentication, encryption, and key management. Authentication is the E1 algorithm and challenge/response system used during the pairing process. Encryption is the 128-bit E0 stream cipher used on the payload during communication between authenticated devices. Key management is the database of link key for known devices to allow quick association when devices are within range. This section discusses the first security mechanism, Authentication, which is where the user's actions are able to influence secure usage.

As mentioned in Section 4.3, a pass-phrase is the tool to establishing trusted relationships between devices. In particular, the pass-phrase is the phrase transmitted between the devices during the initial handshake, and it is the primary mechanism used to authenticate devices during the pairing process. This authentication is a challenge/response process built into the BT protocol. This pass-phrase, also commonly referred to as a password or key, is a user-selected phrase that is 4–16 alphanumeric characters. For those devices that do not allow for alphanumeric user input, such as hands-free headsets or mice, the pass-phrase is predetermined and hard-coded into the device. The device manufacturer documents the device's pass-phrase and the master device enters it during the pairing process. The pass-phrase for these devices is commonly four zeros (0000) or a sequential series of numbers, such as 1234.

Why is this important? The reason for pairing in a secure location becomes more apparent when considering that the pass-phrase is the main variable part of the link key. The BD_ADDR of each device and the internal clock of the master device do not change and are easily available, as a device shares them when attempting to pair.

When in a nonsecure environment, an attacker may sniff the pairing traffic between the two devices during this handshake process and obtain the pass-phrase using brute-force methods. If an attacker were able to guess the link key using the pass-phrase, they would be able to spoof the device. Any pass-phrase is guessable, given enough time to try every possibility, which is the goal of the brute-force methods. For example, if a pass-phrase is only four digits long, there are only 10,000 possibilities. This is an invitation for a brute-force attack, as it is so easily cracked. Table XI shows the number of guesses required for several pass-phrase lengths and contents. These calculations make it readily obvious how a more complex pass-phrase will reduce the likelihood of compromise from a brute-force attack. The user should use a long pass-phrase and, as with standard password policies, should not use birthdays, anniversaries, addresses, or pass-phrases that may be easily guessed by a malicious device. Heading these two warnings will increase the time required to crack a pass-phrase using the brute-force methods and increase the target device's security level.

The manufacturer-issued pass-phrases are the same for every unit of a particular model and readily made available by that particular manufacturer. Once a malicious device knows this pass-phrase, they then have a greater ability to pair with any of the exact model. In these situations, the user does not have the option to create a more secure pass-phrase, so performing the pairing process in a secure location is even more important.

Many view the authentication as being the downside of the BT protocol. Therefore, some applications will also choose to enforce their own authentication in addition to the pass-phrase discussed as part of the BT protocol. However, some of the services provided by BT do not require authentication. For example, the transfer of an electronic business card only requires the user to accept the business card, and even worse, some older devices accept these electronic business card transfers without prompting the user. Imposing a long and complex

TABLE XI
PASS-PHRASE PERZMUTATIONS BASED ON LENGTH

Pass-phrase contents	Pass-phrase possibilities
4 D	10,000
4 C	456,976
10 D or C	3,656,158,440,062,976
16 D	10,000,000,000,000,000
16 D or C	7,958,661,109,946,400,884,391,936

Key: D, a single digit; C, a single character.

pass-phrase as well as changing it on a regular basis in a secure environment will reduce the potential for attack.

4.5 Installing Applications

Malware may propagate by numerous methods; several of these methods were discussed previously (Section 2.2). Once the malware is able to deceive its way into the device, the user must then install or execute it. The recommendation is that the user should install or execute applications only from trusted sources and not blindly accept and install data transmitted wirelessly from unknown devices. These sources may include known devices or reputable software vendor Web sites if the device has direct Internet access.

Why is this important? The goal of malware is generally to infect a device and then propagate to another device to thrive. The repercussions along the way may be adverse to the devices that they infect. The BT protocol itself has several built-in security features to help the user along the way. During file transfer to a device over a BT connection, the file must be either accepted or denied by the user of the device. This allows the user to participate actively in the data sharing procedure. If the user dismisses the file transfer, no harm has come to the target device. If the user accepts the file transfer and the file is from a malicious source, the malware is then on the target device. Some device OSs will warn the user when an application installation is from an unknown source. The user should consider if they view the source as a trusted source. If the user dismisses this installation, no harm has come to the target device and the user may then delete the conspicuous file from the device. If the user accepts the installation and the file is from a malicious source, the malware has infected the target device. The user should also ensure they update the software they choose to install on the device. To assure the device has the latest version of the software, visit the device's manufacturer to update the device's software often. As malware is released into the wild, the device manufacturers will issue updates to help protect devices from potential vulnerabilities.

4.6 Antivirus Application

The recommendation is for device users to install an antivirus application and update its malware signatures often. Possible antivirus applications are the F-Secure Mobile Security, the McAfee Mobile Security, and Trend Micro's Mobile Security.

Why is this important? Similar to the antivirus applications available for desktop computers but on a smaller scale, mobile antivirus applications are available to help protect devices from known viruses. The trend is for device malware to increase due to, among other reasons, the sophistication level of the devices that allows for malware development, the fact that mobile device is not physically limited as desktops are, and the potential for damage pinging the radar of the malware authors. This suggestion would appear to be common sense, but these products are relatively new and the average user is not even aware the antivirus solutions exist for mobile platforms.

5. Conclusion

Among the wireless technologies, BT's acceptance has exhibited tremendous growth and has the appearance of gaining momentum. In November 2005, 9.5 million BT radios were being shipping each week,[61] by 2012, the number of BT enabled devices shipped to market is expected to reach over 1.8 billion. This is considerable greater market presence than potentially competitive technologies such as Wi-Fi. This presence will also increase with such events as states passing legislation to not allow handheld cell phones while driving and the medical field planning BT-enabled devices with the release of the Health Device Profile (HDP). This momentum also becomes obvious if one were to look at the numerous BT products currently available to the end-user. Established end-user applications are cable-replace solutions or audio headset connectivity applications, yet numerous emerging applications are and will soon be available. BT is evolving to fit usage profiles requested by end-user, and BT applications are becoming available throughout the house and work place. The BT SIG provides a comprehensive list of available devices:
www.bluetooth.com/English/Products/Pages/default.aspx

The BT SIG is also continually updating the technology to make it more streamline, more power friendly, more functional, and more competitive with comparable technologies. The most abundant power class of BT devices is Class-2, which allows for a range of 30 ft. However, Class-1 BT radios are able to achieve ranges comperable to Wi-Fi radios, and future releases plan to incorporate UWB functionality to overcome throughput concerns. This allows the technology to extend past its original short-range limitation. Additionally, these radios adhere to the low power consumption associated with BT devices. The combination of this range and power usage makes BT very attractive for devices to incorporate the technology into their devices.

BT-enabled devices also often manage a tremendous amount of sensitive information. Used in such applications as smart phones, laptops, and tablets these devices often manage account login information, financial information, contact information, and scheduling information, all of which are expected to be secure. All of this sensitive information is the proverbial *low-hanging fruit* that is the target for

---

[61] http://ipv6.com/articles/applications/Bluetooth.htm (@ February 9, 2011).

potential exploits. Combined with BT's growing presence and increased functionality, this attracts the attention of the potential malware author. As a result of this concern, the end-user is recommended to practice the discussed computing best practices.

All in all, this rapid growth of BT applications, its strong presence in its own niche market, its comparable functionality to other wireless technologies, and its managing of sensitive contents, all demand that security be a serious concern. As the total number of devices continues to increase and the networking and wireless abilities of the devices improve, these devices become increasingly viable targets for the potential malware author. Nokia and Sony Ericsson are able to attest to the fact that the more abundant the device, the more vulnerable they will be. Although BT malware attacks are very much in the infantile stage of development, if the end-user does not exhibit caution as malware sophistication increases, forthcoming BT exploits will have grave impact on BT-enabled device computing.

Glossary

29A group	A collection of virus enthusiasts known for inventing, modifying, and testing technologies resulting in the creation of proof-of-concept malware
DoS	Denial of Service
FHSS	Adaptive Frequency-hopping spread spectrum
IEEE	Institute of Electrical and Electronic Engineers
IrDA	Infrared
PDA	Personal Digital Assistant
QoS	Quality of Service
RF	Radio Frequency
RFID	Radio Frequency Identification
UWB	Ultra-Wideband
Wi-Fi	Wireless Fidelity
WiMAX	Worldwide Interoperability Microwave Access
Social Engineering	Exploiting someone's general sense of trust through such activities as deceitful coaxing or impersonation

Digital Feudalism: Enclosures and Erasures from Digital Rights Management to the Digital Divide

SASCHA D. MEINRATH

New America Foundation, Washington, District of Columbia, USA

JAMES W. LOSEY

New America Foundation, Washington, District of Columbia, USA

VICTOR W. PICKARD

New York University, New York, USA

Abstract

The Internet is argued to be the most participatory telecommunications medium ever developed—it is, in essence, a vast commons—originally developed with taxpayer funding and built to ensure that no one entity owned (or could control) the entire network. Developed using an end-to-end architecture, the Internet places the intelligence at the network's edges, allowing users to define what applications and services to run and which hardware and devices to use. However, the democratic potential of the Internet is being threatened by structural changes that, if left unchecked, will limit future innovation and participation. If these trends continue, the Internet will devolve into a feudalized space—one that limits democratic freedoms while enriching an oligopoly of powerful gatekeepers. While incumbent phone, cable, and software companies stand to gain financially from these enclosures, as this chapter documents, the negative outcomes stemming from this digital feudalization will have profound detrimental impacts on democratic and affordable communications. This chapter highlights specific policy debates, both within United States and internationally, that undergird these vulnerabilities, and illuminates normative understandings about the role of the Internet in supporting a democratic, civil society. Using the seven-layer OSI model as a framework, our analysis catalogs current threats to this telecommunications commons and examines the policy provisions that should be implemented to prevent the feudalization of the Internet.

1. Introduction 238
    1.1. The Political Economy of the Internet 240
    1.2. Feudalizing Public Space and the Art of Enclosure 242
    1.3. Critiquing the Internet 243
    1.4. The OSI Model 245
2. Enclosures Along OSI Dimensions 245
    2.1. Physical Layer Problems 245
    2.2. Data Link and Network Layer Problems 250
    2.3. Transport Layer Problems 259
    2.4. Session Layer Problems 261
    2.5. Presentation Layer Problems 262
    2.6. Application Layer Problems 263
3. The Need for Open Technology 266
    3.1. Limitations on Today's Closed Networks 268
4. Policy Recommendations for the New Critical Juncture in
   Telecommunications 270
    4.1. Physical Layer Solutions 271
    4.2. Data Link and Network Layer Solutions 274
    4.3. Transport Layer Solutions 278
    4.4. Session, Presentation, and Application Layer Solutions 279
5. The Need for a New Paradigm 280
    5.1. Looking Forward 282
    References 283

1. Introduction

As we enter the second decade of the twenty-first century, we find ourselves at a rare historical moment—a time of great opportunity fraught with substantial pitfalls. Numerous potential trajectories for the Internet may unfold before us. Decentralized information and participatory platforms have birthed a revived movement for democratized media production. These phenomena depend on the common resource of the Internet, common not in ownership of the integrated networks, but because access and use of the network does not discriminate toward a particular use or user group [1]. However, as markets have evolved, uncertainty is growing that policy decisions will be made that will benefit the general public. Even as developments associated with the much celebrated "Web 2.0" empower users with new social networking and media production capabilities—transforming politics as well as everyday life—less visible structural changes threaten to foreclose many of the Internet's democratic possibilities. While some scholars continue to herald the brave new world of digital networks [2], others suggest more cautionary tales of opportunities lost, market failure, and corporate mismanagement [3]. Focusing on this core tension, this chapter examines a number of recent and ongoing Internet policy battles, ranging from net neutrality to intellectual property rights, which will help determine the future of the Internet's fundamental structures. These fights come at a critical juncture in Internet's development. If history can be taken as a reliable predictor, it is not hyperbolic to expect that the outcomes of these crucial debates will help shape the contours of the Internet for decades, if not generations, to come.

Despite the general excitement around the popularity and political power of online services like YouTube, Facebook, Twitter, and other innovations, we are also facing unprecedented attacks on basic Internet freedoms. Recent digital rights management (DRM) rules have criminalized lawful behavior on the Web, threats to net neutrality and user privacy are being exposed on an almost monthly basis, and worsening digital divides are undermining optimistic, yet uncritical, assessments. The oft-lamented fact that the United States has plummeted in its international rankings on broadband penetration rates in recent years indicates that something is undermining the participatory ideal of universal broadband connectivity and suggests that not all is well with the state of the Internet. Indeed, less than a decade ago the United States was respected as a top country in terms of Internet adoption.[1] An October 2000 report by the Danish National IT and Telecom Agency, *The Status of Broadband Access Services for Consumers and SMEs*, estimated that the United States was "12–24 months ahead any European Country" in terms of broadband penetration and access. By 2009, the United States had slipped rank to 15th, based on OECD ranking, for overall nationwide broadband penetration, and Denmark was first.[2] These are just a

---

[1] See Free Press *International Broadband Data*. http://www.freepress.net/files/international-broadband-data.pdf (accessed 10.05.2010).
[2] See OECD *Broadband Subscribers per 100, by technology (Dec. 2008)*. http://www.oecd.org/dataoecd/21/35/39574709.xls (accessed 26.04.2010).

few of the indicators, part of a longer list that we catalog below. To understand how this contemporary broadband morass arose, we must first examine the political economy of broadband markets and the policies that structure them.

1.1 The Political Economy of the Internet

The Internet includes a diverse array of stakeholders and participants. From the physical infrastructure to the users of applications, the ecosystem builds on the participation of different players and contains myriad dependencies. The data transmission depends on access to the physical network and application functionality depends on the transport of data. Thus, the influence of different stakeholders, from network operators to protocol developers, to regulate access to or the functionality of different network layers can define the end-user experience. And seemingly innocuous interventions can have profound implications for the flow of information, what applications or hardware will run, and the specific content or messages allowed over a network.

Challenges can be seen in the pricing and availability of access to broadband Internet. In most markets across the United States, people must rely upon a duopoly for their broadband connection options, having to choose between one cable provider and one telephone company for their Internet services.[3] This lack of choice and competition is one of the key reasons why broadband services in the United States have lagged behind a growing number of other industrialized countries and why customer service has been so remarkably substandard where broadband connectivity is available [4]. This significant market failure largely accounts for the fact that Americans typically pay several times more a month for a fraction of the broadband speeds available in other countries. In 2010, a typical 50 Mbps connection in the United States cost as much as $145 a month, compared with $60 a month in Japan, $29 a month in South Korea, and $38 a month in Hong Kong [5]. In Sweden, open access networks have created vibrant competitive markets, dropping the pricing for 100 Mbps symmetric lines to $46 a month in Stockholm [8]. Given such stark contrasts between the United States and other countries, it is reasonable to assume that if the former implemented similar functional markets, Americans could have access to broadband connectivity an order of magnitude faster for a fraction of the price, as evidenced by the pattern borne out in a growing list of other

---

[3] The National Broadband Plan released by the Federal Communications Commission notes that 96% of Americans have a choice of two or fewer wireline broadband providers. Federal Communications Commission, *Connecting America: The National Broadband Plan*, p. 37 (2010). http://download.broadband.gov/plan/national-broadband-plan.pdf (*National Broadband Plan*).

countries.[4] Wherever competition is fostered, prices drop and broadband speeds increase dramatically.

The cost and availability of broadband is only an introduction to the dilemma of digital feudalism. Network operators have exhibited a growing resistance to implement network management techniques that increase capacity, choosing instead to ration "existing capacity among competing network users or uses" [6]. In the United States, users might share a local node with over 200 other connections[5] and experience average speeds that are half the advertised price [7]. In this chapter, we document examples where network traffic has been monitored, communication for specific applications has been terminated or blocked from accessing a network connection at all. Additionally, we document where Internet standards and protocols can define what types of languages can be used or the terms of distributing information.

If these negative trends continue, the Internet might become, in effect, a feudalized space—one that limits democratic potential while enriching a handful of corporate interests, for example, incumbent phone, cable, and software companies that stand to gain from policy-enabled enclosures. This analysis illuminates the specific policy debates connected to these vulnerabilities, while uncovering normative understandings about the role of the Internet in a democratic society. Our earlier work examined fundamental Internet tensions between structure and agency, and between encroaching commercialization and democratic possibilities [8]. Our more recent work has focused specifically on the net neutrality debate, linking it to a larger set of normative criteria for democratizing the Internet [9], and to the democratic potentials for opening up new commons and "defeudalizing" government spectrum [10]. In our view, what is still lacking from current scholarship is a comprehensive analysis of the key policy debates, both U.S. and global, around multiple layers of the Internet. By cataloging current threats to a democratic Internet and closely examining the linkages between intersecting policy battles, this chapter illuminates both what is at stake and what policy provisions should be implemented to prevent the feudalization of the Internet.

---

[4] As Benkler concludes, "Our most surprising and significant finding is that 'open access' policies—unbundling, bitstream access, collocation requirements, wholesaling, and/or functional separation—are almost universally understood as having played a core role in the first generation transition to broadband in most of the high performing countries; that they now play a core role in planning for the next generation transition; and that the positive impact of such policies is strongly supported by the evidence of the first generation broadband transition. The importance of these policies in other countries is particularly surprising in the context of U.S. policy debates throughout most of this decade. While Congress adopted various open access provisions in the almost unanimously approved Telecommunications Act of 1996, the FCC decided to abandon this mode of regulation for broadband in a series of decisions beginning in 2001 and 2002. Open access has been largely treated as a closed issue in U.S. policy debates ever since." (Ref. [4], pg. 3).

[5] See Ref. [11]; footnote 3.

1.2 Feudalizing Public Space and the Art of Enclosure

The popular metaphor of the Internet as a public sphere often overlooks the darker side of this formulation. In discussing the structural transformations of the public sphere, Habermas clarified that while the market helped create the initial space for civic engagement, it also constantly threatened to colonize public spheres through privatization [11]. He referred to this phenomenon as the "refeudalization of the public sphere"—a process in which the newly created public space would succumb to commercial pressures, becoming reorganized along familiar power hierarchies. In recent years, Habermas has increasingly underscored the risk of market colonization, decrying the tendency toward treating the public sphere as merely another location for commercial relations to take hold [12].

A related metaphor for a similar phenomenon is referred to as "enclosure," a process by which common or public lands are taken over by private interests to be exploited in ways that exclude others. The common example used to define enclosure is looking at the English poor laws. In the fifteenth and sixteenth centuries, the English countryside witnessed a transformation in which land that had since time immemorial been treated as a commons was suddenly privatized. This change in the legal standing of the land in turn criminalized a range of behavior that had previously been accepted as cultural norms, such as maintaining livestock and harvesting food or gleaning on common lands [13]. With the advent of poor laws, these behaviors were suddenly recategorized as poaching. In similar fashion, enclosure systematically removes resources out of the public sphere and replaces a general notion of maximize the public good with a logic of profit maximization, thus excluding the majority of people and furthering the profits of a minority.

Debates over digital commons often assume a false dichotomy predicated on thinking of digital goods as traditional commodities. Most commodities are rivalrous—their use by one entity excludes their use by another. Thus, if one buys and eats a fish, then that fish is not available to anyone else. There is a natural rivalry among consumers for access to these goods. Likewise, most rivalrous goods are excludable—one can prevent other consumers from eating the fish by charging a price for it (thus laying the foundation for the traditional "Economics 101" assumption that pricing will seek equilibrium between supply and demand). Another foundational pillar of this traditional thinking is that nonrivalrous goods are also nonexcludable—that it is impossible to stop an individual from utilizing this resource (e.g., daylight, air, learning). In addition, rivalrous, yet nonexcludable goods have given rise both to the "commons" and to the dystopian "tragedy of the commons," exemplified by problems like overfishing, overgrazing, and pollution of the environment.

	Excludable	Nonexcludable
Rivalrous	Private goods	Common goods
Nonrivalrous		Public goods

What is at stake in this increasingly feudalized space is a shifting concept of ownership. What happens when public goods and common goods are reenvisioned as private goods? And how do regulatory process and technological innovations spur these shifts? The "digital commons" metaphor may serve as a poignant reminder that the Internet's unique power has rested largely on its openness, on the fact that it is our most public media, and that it was created as a result of public support through DARPA and other tax-supported entities, but it oversimplifies the multilayered nature of the technology and the potential for enclosures to manifest themselves in seemingly innocuous ways. Built using a "stupid" infrastructure,[6] the Internet has been defined by protocols that transfer data packets on a "first-in first-out," best-effort basis. The success of the Internet has been defined by the range of uses and application freedoms facilitated by its openness. Under this framework, application usage by two users with access to bandwidth (such as neighbors both subscribing to 5 Mbps connections), or two viewers watching the same online video, is *de facto* nonrivalrous. However, as the fundamental structures of the Internet undergo transformation, nonrivalrousness is quickly being supplanted by the same forces that drove wedges and created power pyramids in the English countryside in the fourteenth century. The freedom to define the Internet to the needs of the user, to share video, and to choose the appropriate method of communication over TCP/IP is quickly becoming enclosed.

1.3 Critiquing the Internet

An expanding corpus of research has documented various points where corporate encroachment has already been occurring. One of the best known examples is Lessig's distinction between read-only culture and rewritable culture, where he notes that creativity is being sacrificed for private profits as an intellectual policy regime runs amok [14]. Perelman makes a similar argument that the public domain's digital commons are undergoing a kind of enclosure and becoming increasingly impoverished by a proprietary mentality [15]. Other scholars focus on how the Internet is being transformed at the network operation and content layers [16].

---

[6] See D. Isenberg's, Rise of the Stupid Network. http://www.hyperorg.com/misc/stupidnet.html (accessed 17.09.2010).

Lessig listed threats imposed by gatekeeping Internet service providers (ISPs) in *The Future of Ideas* [17]. One recent article categorized these criticisms as falling along structural or individual levels [18]. Our analysis focuses on enclosure tactics that are more structural in nature (as opposed to the ways in which content providers have become increasingly commercial in recent years). Though some attempts have been directed toward these areas, relatively few efforts have tried to systematically model the various critiques or try to connect them to larger systemic analyses. While many of these critiques deal with the oft-mentioned "digital divide" and focus on issues related to access, other critiques emphasize deeper systemic issues.

Dan Schiller leveled one of the first critiques aimed at delineating the neoliberal shift in market expansion and political economic transition encompassing the Internet, which he called "Digital Capitalism." Schiller noted that Internet networks increasingly serve the aims of transnational corporations via strict privatization of content and unregulated transborder data flow. Likewise, he advocates for the creation of a "communications commons," and efforts toward the "... financing of a multiplicity of decentralized but collectively or cooperatively operated media outlets, licensed on the basis of commitment to encouraging participatory involvement in all levels of their activity" to "... more fully [release] the democratic and participatory potential of digital technologies."[7] These critical trends presaged a growing body of work that addresses normative concerns like open architecture, open access, and online ethics. For example, Benkler's *Wealth of Networks* advocates for a commons-based policy orientation. Along with Lessig and others, this approach is aligned with the notion of Cooper's "open architecture." Frequently referred to as a commons-based approach to the management of communications systems, this model emphasizes cooperation and innovation as opposed to privatization and enclosure.

More recently, Jonathan Zittrain has intervened in these debates to argue that the United States is allowing the Internet to be treated like an appliance instead of something that produces generative technology [23]. Zittrain sees the development of the Internet as a history of lost opportunity. To underscore this shift, he uses a three-part-layered model that distinguishes between physical, protocol, and application layers. He also allows for content and social layers above these three.[8] Although this model is useful for highlighting areas of enclosure—and provides yet another typology for understanding different theories of technology—we prefer an adaptation of an older schematic that has been utilized as the foundation of the Internet and has the advantage of highlighting certain aspects of the Internet that other models fail to capture: the OSI model.

---

[7] See Ref. [13], pp. 204–205.
[8] See Ref. [23], pp. 67–68.

1.4 The OSI Model

The seven-layered OSI model serves as a convenient framework through which we can see how the Internet is under threat at different points from different sources. Neither is digital feudalism limited to the threats at the OSI levels, nor is the OSI model a strict cross section of the Internet, but the model is a useful heuristic for documenting encroachments at individual or across multiple of these layers directly influencing surrounding layers and end-user control of the end-to-end Internet.

OPEN SYSTEMS INTERCONNECTION BASIC REFERENCE MODEL

		OSI model	
	Data unit	Layer	Function
Host layers	Data	7. Application	Network process to application
		6. Presentation	Data representation and encryption
		5. Session	Interhost communication
	Segment	4. Transport	End-to-end connections and reliability
Media layers	Packet	3. Network	Path determination and logical addressing
	Frame	2. Data link	Physical addressing (MAC and LLC)
	Bit	1. Physical	Media, signal, and binary transmission

Source: Wikipedia.

2. Enclosures Along OSI Dimensions

2.1 Physical Layer Problems

The foundational elements of networks, the physical layer, can define the network itself. These transport mediums—infrastructure built of copper and fiber, switches, routers, and slices of radio spectrum—can be open or closed, but any restrictions will define the foundation of communications over the network. Just as AT&T's

blocking of the Hush-a-Phone[9] and Carterfone devices on their handsets was overcome through regulatory and legal intervention, policies can and do help define limitations to network operations, with far-reaching implications. When the Supreme Court ruled in 1968 that the Carterfone device must be allowed, it created a legal precedent that facilitated the development of numerous, so-called foreign attachments. These devices included everything from office telephone systems to answering machines, but the single most important newly legalized device was the computer modem. Without knowing it, the Supreme Court had created the conditions under which the Internet as we know it could develop, and it was only after a prolonged court battle and unequivocal mandate that AT&T open up its network that the foundation for the Internet was possible. However, next-generation networking systems may not be so lucky. As we shall see, AT&T and other telecommunications incumbents have recreated the same pre-Carterfone conditions that existed over 40 years ago on today's wireless networks.

2.1.1 Open Access and Common Carriage

In addition to the mandate to allow "foreign attachments" on the telephone network, two key elements fueled the establishment and growth of the Internet: open access and common carriage. Open access policies required "existing carriers to lease access to their networks to their competitors, mostly at regulated rates."[10] Open access meant that anyone could create an ISP and AT&T had to provide access to facilities and interconnection for these new rivals. In addition, because of Common Carriage, AT&T had to provide connectivity over its own network between end users' computer modems and these new competitive ISPs. In many regards, these key factors are what helped the Internet develop into "a network of networks" as opposed to a singular system controlled by one entity.

However, these provisions have been systematically eroded through a concerted effort by the telecom incumbents. The Telecommunications Act of 1996 codified a regulatory binary of telecommunications services (e.g., telephone) and information services (e.g., Internet), with only the former being subject to common carriage and unbundled access to a telecommunications network at reasonable rates.[11] However, as the new technologies came into use that provided different services and the convergence of cable and copper (and later fiber) picked up steam, regulatory

---

[9] In the case of the Hush-a-Phone device, a small plastic cup that clipped onto the telephone receiver in order to facilitate private ("hushed") conversations, AT&T argued that this device infringed upon their right to control their network and must be made illegal.

[10] See Ref. [4], pg. 11.

[11] Telecommunications Act of 1996 47 U.S.C. 251 (c)(3) (2000).

challenges began to emerge. Due to the myopia of the 1996 Act, questions over whether to treat IP-based voice services (e.g., VoIP) as a telecommunications or information service began to cause more and more headaches for legislators and regulators.

On June 27, 2005, the Supreme Court's "Brand X" decision upheld the authority of the FCC in reclassifying cable broadband service as an "information service."[12] As a result of these decisions, the FCC had effectively "deregulated" carriage for cable ISPs. This meant that cable ISPs could offer telephone services on their networks without providing access to AT&T (because they were not covered by common carriage provisions), but that AT&T had to allow cable ISPs access to its own infrastructure (since they were considered a telecommunications service). In turn, AT&T argued that this provided an unfair competitive advantage to the cable ISPs and the FCC quickly ruled that digital subscriber line (DSL) ISPs and others were no longer required either to unbundle their services or sell network access to potential competitors.[13] Since these deregulatory decisions, broadband competition in the United States has collapsed.

According to the U.S. census, nearly 75% of independent ISPs have gone out of business over the past decade, and the United States' international broadband rankings have plummeted over the same period. In the National Broadband Plan released in March 2010, according to the FCC's most recent data, only 4% of Americans have more than two wireline ISPs to choose from.[3] The noncompetitiveness of this market is contributing to both the slow speeds and high costs of wireline connectivity in the United States. Yet when it comes to wireless communications, the prognosis may be even worse.

2.1.2 Spectrum Resources

Spectrum capacity is essential for wireless networks, but current licensure and distribution of spectrum is both archaic and hyper-inefficient. Almost invariably, current spectrum allocation models assume that single entities need to have absolute control over their spectrum band at all times. In essence, our license regime assumes the same technological capabilities as existed in the 1920s and 1930s when it was first developed. Decades later, spectrum allocation and frequency assignments largely ignore today's technical realities and gross inefficiencies of spectrum

---

[12] *Inquiry Concerning High-Speed Access to the Internet over Cable and Other Facilities, Internet Over Cable Declatory Ruling*, 17 FCC Rcd. 4798, 4841–4842 (2002).
[13] *Appropriate Framework for Broadband Access Over Wireline Facilities, Report & Order & Notice of Proposed Rulemaking*, 20 FCC Rcd. 14866, 14904 (2005).

usage currently exist. Between January 2004 and August 2005, the National Science Foundation commissioned a report from Shared Spectrum looking at actual spectrum use in six cities across the United States—their results showed that the amount of spectrum actually being used as pitifully low:

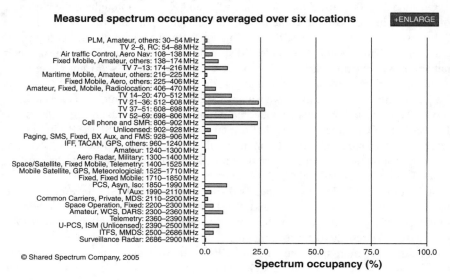

Source: Shared Spectrum Company.

The overall results of this analysis were that, of the sites surveyed, spectrum occupancy was highest in New York City, with 13.1% of the frequencies in use, and that the average utilization rate was 5.2%.[14] In their most recent study of spectral efficiency in 2007, they conducted measurements in Limestone, Maine, and found a usage rating of 1.7%.[15] Given these real-world measurements, why do we have so few opportunities to access the public airwaves? Especially since preventing access to clearly underutilized spectrum resources prevents device developers from creating innovative new technologies and the public from expanding communication opportunities.

---

[14] See http://www.sharedspectrum.com/measurements for the full analysis, methodology and reports for each survey site.
[15] Report available at http://www.sharedspectrum.com/measurements/download/Loring_Spectrum_Occupancy_Measurements_v2_3.pdf (last visited 09/17/10).

Wi-Fi serves as a striking example of the transformative potential and remarkable innovation of unlicensed spectrum. Wi-Fi utilizes just two small unlicensed bands and has already had profound impacts in how we communicate and use twenty-first century technologies. In addition to easy connectivity in homes, coffee shops, airports, and now airplanes, Wi-Fi has expanded the reach of the Internet through community and municipal initiatives to provide free or affordable Internet access. However, Wi-Fi operates on at 2.4 and 5.8 GHz—frequencies that are not ideal for penetrating obstacles like trees and buildings—and unlicensed is very much the exception to the rule. Today, over 95% of the public airwaves ($<$ 30 GHz) are either reserved for governmental use or licensed to private parties.[16]

The spectrum allocation regime ignores the technological reality that unlicensed radios can communicate concomitantly with licensed users, sharing a block of frequencies by listening and adapting to utilize what space is available in real time, as our earlier research in this area underscored:

> The FCC and NTIA have continued to privilege a model for licensure that allows only a single entity to broadcast on a given frequency, often at a specific power level and geographic location. This "command and control" mentality of spectrum management—by which the FCC and NTIA allocate spectrum into bands, assign and prescribe how these bands will be used, and oversee the method of giving exclusive rights to specific licensees—is woefully outdated given current technologies and spectrum needs. While digital technologies have radically transformed almost every facet of current society, the U.S. licensure regime is predicated on use of the public airwaves as if we were still utilizing precomputer (analog) technologies dating back to the World War I era. Tim Wu wrote an op-ed in the *New York Times* likening U.S. spectrum policies to "Soviet Style Rules ... [governing] ... a command and control system dating from the 1920s." Wu estimates that "At any given moment, more than 90 percent of the nation's airwaves are empty" (2008). Other analysts referred to current spectrum management policy as a "paradigm for economic inefficiency" (Weiser and Hatfield, 2005). Whether one looks at the debate over low-power FM radio licensure, interference temperature, or unlicensed devices in unused television broadcast bands, the story is invariably the same: incumbent interests already invested in licensed frequencies seek to prevent competition by maintaining the antiquated regulatory status

---

[16] In cases like the citizens' band (CB), spectrum is set aside for amateur use, or according to "Part 15" rules which allow some public wireless devices such as garage door openers and microwave ovens to operate in unlicensed spectrum. See Bennett Z. Kobb, *Wireless Spectrum Finder: Telecommunications, Government and Scientific Radio Frequency Allocations in the US 30 MHz–300 GHz* (McGraw-Hill, 2001), and National Telecommunications and Information Administration, *Manual of Regulations and Procedures for Federal Radio Frequency Management (Redbook)* (Washington: US Government Printing Office, 2008).

quo. In this way, incumbents dramatically slow down change or stop it altogether. "Among neutral observers," Nuechterlein & Weiser note, "there is little dispute that ... the current spectrum regime requires a comprehensive overhaul" [10].

From a digital feudalism perspective, the physical layer of the OSI model is rife with examples where control over this facet of the Internet's architecture is today being utilized to impede competitors, lock customers into expensive service tiers, and control innovation. However, as we move up the OSI stack, numerous additional exemplars of this phenomenon emerge.

2.2 Data Link and Network Layer Problems

The data link layer creates the foundation for TCP/IP transmission. Transferring data between different components of the network, the second OSI layer creates the framework for additional protocols like UDP to communicate. Enclosures to the data link layer can make communication inoperable or, by "leapfrogging" into the functionality of other layers, differentiate between types of communication and create virtual circuits that can control or even break end-to-end functionality. The network layer marks the beginning of full network communication. Bridging node-to-node communication of the data link layer and often helping maintain quality of service requests of the transport layer, the third OSI layer provides the foundation for most end-to-end communications.

2.2.1 Internet Protocol Addresses

Internet protocol is central to connecting devices over their physical networks. Much like a telephone number rings a specific device or address, Internet protocol addresses help route data to specific destinations around the network. Introduced in 1981, IPv4 uses a 32-bit address space and can support a maximum of 4.3 billion addresses (2^{32}). This increase in address space was a large leap from the early days of the ARPANET (which only allowed for 256 unique address), but just 10 years after its introduction, fears are mounting about "address space exhaustion." While not completely exhausted, the IPv4 address space is running dangerously low and may, for all intents and purposes, run out by 2012. While work-arounds like NAT (network address translation) has helped slow the rate of exhaustion, as the number of IPv4 addresses decreases, it will become increasingly difficult to add publicly addressable devices, Web sites, and destinations [19].

To help address this problem, a new IP address standard, IPv6, has been developed. IPv6 contains 2^{128} addresses (about 3.4×1038). That's enough IP addresses to assign one to every atom on the surface of the Earth (and then do the same with 100 more Earths).[17] It's enough IP addresses to give each of the 6.5 billion people alive today more IPs than existed in the entirety of IPv4 and still have hundreds of billions of IPs left over. Globally, IP address allocation has been Americentric. North America currently has 32% of IPv4 addresses, and at one point, Stanford University had more address allocations than China.[18,19] The IPv6 distribution may also be very Western-centric, with 73% of recent address allocations going to Europe and North America, compared to 63% in IPv4.[18] While Western nations may have a greater need for addresses initially, concentrating the addresses in the Western world can exclude developing nations when faced with another shortage of address space. In addition, with a *de facto* unlimited supply, one would expect that IPv6 addresses should cost next to nothing, yet this is not what has happened. Instead, almost the exact same pricing regime that has existed for IPv4 address space has been carried over into IPv6 address space[20]—creating, for all intents and purposes, a cost barrier for a commodity that is far more plentiful than atoms on the face of the Earth. As a quick view of the American Registry for Internet Numbers documents, the fee schedule for IPv4 and IPv6 begins at a cost of $1250 per year for an "X-small" allocation to $18,000 per year for an "X-large" allocation[21]:

IPv4 ISP Annual Fees

Size category	Fee (U.S. dollars)	Block size
X-small	1250	Smaller than /20
Small	2250	/20 to /19
Medium	4500	Larger than /19, up to and including /16
Large	9000	Larger than /16, up to and including /14
X-large	18,000	Larger than /14

---

[17] See http://itknowledgeexchange.techtarget.com/whatis/ipv6-addresses-how-many-is-that-in-numbers/ (accessed 17.09.2010).
[18] Protocol Politics, p. 173.
[19] Protocol Politics, p. 155.
[20] Though it should be noted that one does receive substantially more IPv6 IP addresses for the same price vis-a-vis IPv4 IP addresses, the main barrier for many new entrants is often the cost, not the number of IP addresses.
[21] See American Registry for Internet Numbers, Fee Schedule at https://www.arin.net/fees/fee_schedule.html (accessed 17.09.2010).

IPv6 Annual Fees[22]

Size category	Fee (U.S. dollars)	Block size
X-small	1250	Smaller than /40
Small	2250	/40 to /32
Medium	4500	/31 to /30
Large	9000	/29 to /27
X-large	18,000	/26 to /22
XX-large	36,000	/22 and larger

In the United States, the lack of national policy for transitioning to IPv6 leaves ambiguity about how the distribution of remaining IPv4 addresses will take place. In fact, IPv4 exhaustion may be creating a grey market for addresses, which will inevitably lead to these increasingly valuable addresses creating a digital divide between those with the ability to pay for them and those who will be unable to [20]. Because IPv6 and IPv4 cannot communicate directly with one another, special arrangements will need to be made for networks that are on legacy IPv4 networks that connect through IPv6 enabled providers (and likewise, early adopters of IPv6 may find themselves having problems if their upstream provider is still using IPv4). Finally, users of IPv4-only networks may find themselves unable to reach IPv6 destinations, further exacerbating the digital divide.[23] In essence, IP addresses become another way that dominant market players (e.g., those with control over key physical layer assets) can leverage control over higher layers of the OSI stack, including into the very content of the Internet.

Communication can also be disrupted by blocking IP addresses. IP addresses can be blocked individually or as blocks by a variety of entities—network administrators at schools or businesses can block access to certain Web sites or Web sites can block user access to content or purchases of goods and services, for example. IP addresses can also be used to block both legitimate and illegitimate communication—for example, spam, the bane of e-mail communication since April 12, 1994 when a pair of lawyers sent out an unsolicited message to every Usenet group,[24] but valid

---

[22] To incent adoption of IPv6, the American Registry for Internet Numbers instituted fee waivers that pay for a diminishing amount of these fees year to year and phase out entirely in 2012. See footnote 25.

[23] See http://www.ripe.net/info/faq/IPv6-deployment.html (last visited 17.09.2010).

[24] *How to Make a Fortune on the Information Superhigheway: Everyone's Guerilla Guide to marketing on the Internet and other On-Line Services* (New York: HarperCollins, 1995) qtd in Alexander r. Galloway *Protocol.*

communication can also be blocked by the same means. The greatest challenge, however, is when IP addresses are blocked by an ISP.

On December 22, 2004, Verizon started blocking e-mail sourced from IP addresses from European ISPs [21]. Although IP addresses can be used to identify and block spam, many of the addresses blocked were not included lists of IP addresses from known sources of spam.[25] The embargo was later identified to be a result of overvigilant spam filters. However, the damage to communication was evident to users resulting in a class action suit [22]. Verizon is by no means alone in challenges with blocking legitimate e-mail communication. Two years earlier in October 2002, a similarly overly sensitive spam filter blocked a week's worth of incoming e-mail for EarthLink subscribers [23]. IP addresses can also be used to block access to certain Web sites by an ISP, or by a Web site to block access to certain users. In July 2009, AT&T blocked the imageboard Web site 4Chan [24]. Wikipedia sometimes blocks users from editing the Web site,[26] such as in the case of scientology [25]. Ticketmaster uses IP addresses to identify and block bulk purchasing of tickets [26]. Left unchecked, IP blocking can have profound impacts on people's ability to freely communicate and can be utilized to censor content, specific users, and entire regions of the Internet.

2.2.2 IP Multimedia Subsystem

IP Multimedia Subsystem (IMS) is a still-evolving feature set for architecting wireline and wireless networks that has great potential to enclose portions of the next-generation communications systems. Telecommunications firms have been particularly focused on deploying IMS in their wireless networks. The Internet is a packet-switched network—information is broken into packets on one end and can travel independently over multiple pathways to be reassembled on the receiving end. However, IMS acts by making communication resemble more of a circuit-switched network, where data flows are each given an earmarked end-to-end channel for communicating [27]. Like the days of copper-wire telephone, IMS allows a carrier to earmark specific channels for specific communications, creating the ability for bill differentiation among data types that are actually traversing the same network architecture. Thus, while the Internet may allow a user to send e-mail, use Skype, surf the Web, and IM with friends for a single fee, IMS is being used on wireless networks to differentiate these services—thus, a user may have to pay once for your

---

[25] See reports at http://www.flyertalk.com/forum/travel-technology/383299-verizon-blocking-all-incoming-e-mail-outside-usa-4.html (accessed 22.04.2010).
[26] See Wikipedia *Blocking Policy* at http://en.wikipedia.org/wiki/Blocking_policy.

voice plan, a second time to surf the Web or send e-mail, and a third time to send text messages to friends. In essence, users pay three different times for use of the exact same network.

The low barriers to entry of the open Internet allow developers to innovate and create new ways to use bandwidth resources, but quality of service implementations like IMS predefine the value of certain uses of network bandwidth and freeze prioritization for specific current services and applications [28]. When one looks behind the different charges on cellular phone networks, the nefariousness of the IMS-style pricing regime becomes immediately apparent. While residential broadband connectivity costs $0.01 per megabyte regardless of use, wireless voice costs $1.00 per megabyte of bandwidth and text messages, however, are extraordinarily priced at over $1000 per megabyte [29].

Although 3G cellular networks have separate channels for different types of network access, to allow prioritization of voice and differentiated billing, 4G networks are based on IP, allowing for end-to-end communication, and voice will be operated by using VoIP, creating the flattened network dynamics seen in wireline connectivity. IMS adds complexity to wireless networks in the interest of managing scarcity of bandwidth—which has become the focus of network operators, who seem more interested in centralized control over an end-to-end network than upgrading their networks with the additional capacity that would both make these prioritizations irrelevant and would greatly benefit consumers. As John Waclowski writes: "With IMS, you will never know if you are getting the advertised broadband capacity you think you are paying for. The actual bit rate will be a function of what IMS thinks you are doing" [27].

2.2.3 Media Access Controller

Every network interface controller, such as wireless cards and Ethernet cards, has a unique identifier built into the device. This identifier, the Media Access Controller address (MAC address), is often used as a method of identifying a device on the network and is often more static than IP addresses and other identifiers. For example, many common security systems utilize MAC addresses as the identifier for users who have paid to use an access point and those who have not.

If connectivity is tied to a MAC address, a user can be restricted from transferring connectivity to another device. Thus, if a user purchases Internet connectivity in their hotel room under a user name, the individual could work on a laptop during the evening and then connect a smart phone or tablet to watch a film in bed. In contrast, if the connectivity is defined by the MAC address, a user cannot transfer connectivity but must repurchase it for each device. As connectivity is included in an increasing array of portable devices, restrictive pricing schemes serve to create more opportunities to charge consumers for connectivity based upon each and every device they want to connect.

This was attempted in the wireline residential connectivity world as well. A number of ISPs attempted to control the number of devices their users could connect to their network (in much the same way AT&T did this prior to the Carterfone Supreme Court decision) by locking in a specific MAC address as the only authorized device allowed on the home connection. This, in turn, leads to the widespread use of something called MAC spoofing—a process whereby one can manually change the MAC address of one device (and can even make one device's MAC address the same as another device). As with the hotel example, often, users would change the MAC address of their second device to mirror the MAC address of the device they originally paid for Wi-Fi on, thus being able to get online using the same account. For home users, many users took a home router and spoofed the MAC address of the computer that was registered with their ISP, thus enabling them to connect multiple devices through the router to their Internet uplink. Of course, one can combine these efforts to route around the barriers created by service providers—adding a wireless access point with a spoofed MAC address will often allow you to freely rebroadcast connectivity to everyone in your room.

2.2.4 IMEI

Much like MAC addresses, cell phones also have unique identifiers embedded in their firmware. The International Mobile Equipment Identity (IMEI) is unique for each phone or Subscriber Identity Module (SIM) card. The IMEI authenticates a cellular device with a network and allows the device to communicate. These identifiers can be used to track phone use, or restrict connection to a network. Just recently, Mexico passed legislation requiring all phones to be registered in order to operate at all [30]. However, in addition to being useful as unique identifiers of devices, because cell phones tend to be carried around by their users, IMEIs can also function as a tracking identifier—a major concern for privacy. Since IMEIs are broadcast regularly by these devices during normal operations (even when the cell phone is not being actively used by its owner), even if you are neither a law enforcement official nor a telephone operator, it is possible to track users. In fact, some store have begun to use cell signals to track a shopper's movement and patterns in a store, such as which aisles they visited and how much time is spent in front of a display [31]. These types of surveillance are automatic, and it is difficult to know when one's shopping activity is being monitored, or to know how to opt out. Over time, these actions can allow companies to build up remarkably detailed profiles of cell phone users and, by partnering with any carrier that has your home address, target advertisements directly to your front door.

2.2.5 Copyright Enforcement Versus Fair Use

Attempts to stem transfer of copyrighted material over Internet connection have begun utilizing technologies that take advantage of certain facets of the OSI model. Countries like France have proposed (or passed) laws to terminate the Internet connection to a household if a user of that household's Internet connection transfers copyrighted material three times. France's legislation has been on the forefront of blocking Internet access over copyright violations. Originally proposed in March of 2009, the law "Création et Internet" creates a new group, HADOPI, to maintain a blacklist of accused households. After a third accusation of infringing on copyright, a household is blacklisted and cut off from Internet access from any ISP for 3 months to 1 year [32]. After failing to gain support, a revised version of the bill was passed in September 2009, making users responsible for any use on their network [33]. Ireland has followed with similar legislation [34]. Only after vociferous objection was language recommending that all countries follow suit removed from the draft release of the Anti-Counterfeiting Trade Agreement released in April 2010.[27]

In addition to equating accusations of copyright violations of any kind, known or unknown, to the same value as Internet access, this type of legislation places the burden of proof on the customer. Violations of copyright can easily be unintentional, either by accident or not fully understanding the convoluted laws that regulate intellectual property. Perhaps the most relevant example would be the misappropriation of a copyrighted font by HADOPI, the French agency charged with overseeing intellectual property. In January 2010, it was discovered that the font used in HADOPI's logo had been used without permission [35]. The copyright infringement was later attributed to an unwitting mistake by an employee. As with many laws passed by legislators who do not understand the technologies involved, the actual detection mechanisms that underlie enforcement of these laws are left undefined. However, due to the ways in which the OSI stack works, would almost certainly require some form of deep packet inspection, which is the technological equivalent within a packet-switched network to a wire tap on a circuit-based telephone system. In essence, implementing the required law enforcement mechanisms requires a widespread privacy-invasive surveillance regime that would look at which devices are accessing specific materials and what the actual payload of individual packets contains. As we explain below, even were we to set aside the privacy issues involved, what would at first appear to be a straightforward enforcement mechanism can become quite problematic.

---

[27] See Consolidated Text Prepared for Public Release Anti-Counterfeiting Trade Agreement. http://www.ustr.gov/webfm_send/1883.

2.2.6 Tampering and Forging of Packets

The Internet, and most other packet-switched networks, utilizes data packets containing two distinct pieces of information: the header and the content or payload. One can think of the header as the digital equivalent of the address on an envelope you send through the mail and the payload as the content of that letter. Packets interact with multiple OSI layers and those seeking who control the network layer of a network can tamper with and forge packet headers in much the same way that a postal employee could do likewise with a piece of mail. However, unlike the strict laws against tampering with mail, no such rules or regulations hold ISPs accountable for nefarious behavior.

In 2007, an engineer named Robb Topolski noticed that file transfers using the peer-to-peer software BitTorrent were not transferring properly. In fact, none of the early-twentieth-century barbershop quartet songs he was trying to share would upload. Topolski discovered that his ISP, Comcast Communications, was intercepting packets sent from his computer and inserting reset packets that cause connections to abort and start over. In essence, Comcast was pretending to be one party involved in the file transfer and purposeful "hung up" on the communication that had been established. These forged reset packets would interrupt normal TCP/IP communication and terminate Topolski's BitTorrent transfers. The Associated Press and the Electronic Frontier Foundation were able to duplicate Topolski's results transferring content including the Bible. At the time, Comcast had not disclosed that it was engaging in this practice and, when caught, actually denied inferring with BitTorrent transfers at all. Eventually, Comcast settled a class action lawsuit for $16 million [36].

A user's ability to define how to use Internet services depends on the freedom to use the application of their choice. An ISP does not need to have direct access to a user's computer to control which applications they can run if those applications interact with the network. As Topolski and thousands of other Comcast subscribers found, their ISP was able to terminate the communication of a selected application by forging those application data transmissions—all from layer three of the OSI model.

2.2.7 Blocking Video

The increased use of the Internet to distribute video content has revealed new methods of blocking this medium. Despite the near-universal uptake of visual media, video content available online is often not universally accessible. Some types of content is blocked based where the computer trying to access the content is located or where the content is being downloaded from. In essence, many ISPs treat their users as digital vassals whose content access should be defined by the network owners instead of the end users of these systems.

Blocked video content is commonplace on Major League Baseball's (MLB) Web site, MLB.tv. As part of the MLB's content Web site, MLB.tv offers subscription service to allow fans to watch "every out of market game" and advertises a total availability of 2430 games.[28] However, the services stipulate that games are subject to local black outs "in each applicable Club's home television territory."[29] This is significant not just because games are blacked out based on where you live, but the use of technology used to do so. These blackouts are determined based upon users' IP addresses, which MLB.tv uses to identify the location of each Web user.[30] In turn, knowing your IP address only works when your IPs are controlled by your local ISPs (in essence, if IPs were distributed to end users, it would become near-on impossible to know your actual geographic location), which is yet another reason why control over IPs can directly impact the content you are allowed to view.

Access to content is also being determined by agreements between content providers and ISPs. ESPN recently launched ESPN360.com, which offers streaming coverage of a wide variety of live sporting events. These services offered "free of charge," but only if a user receives service from a "participating high speed Internet service provider."[31] DSLreports.com reports that these participating ISPs have paid ESPN for access to the content for their uses[32]; thus, the customers of these ISPs are being forced to pay for a service whether they use it, want it, or even know of its existence. The ESPN360 model has been used by other broadcasters such as NBC. When airing the 2010 Winter Olympics, NBC limited access to online content to users accessing the Internet through ISPs that also have videos that include NBC content.[33] NBC also asked Canadian ISPs to block access from U.S. users so that Americans could not watch key events live, but had to watch them during prime time hours (when NBC could charge a premium for advertising on their network).

---

[28] Quoted from demo video available at http://mlb.mlb.com/mlb/subscriptions/index.jsp?product=mlbtv&affiliateId=MLBTVREDIRECT (accessed 22.04.2010).

[29] See http://mlb.mlb.com/mlb/subscriptions/index.jsp?product=mlbtv&affiliateId=MLBTVREDIRECT (accessed 22.04.2010).

[30] Note MLB.com live game blackouts are determined in part by IP address. MLB.com At Bat live game blackouts are determined using one or more reference points, such as GPS and software within your mobile device. The Zip Code search is offered for general reference only. http://mlb.mlb.com/mlb/subscriptions/index.jsp?product=mlbtv&affiliateId=MLBTVREDIRECT#blackout (accessed 22.04.2010).

[31] See FAQ at http://espn.go.com/broadband/espn360/faq#2 (accessed 22.04.2010).

[32] Small ISPs Revolt Against ESPN360 Model, DSL Reports, February 12, 2009. http://www.dslreports.com/shownews/100843, ESPN 360 ISP Model Spreads To HBO, Olympics, DSL Reports, Febrary 17, 2010. http://www.dslreports.com/shownews/106949. 25.03.2010

[33] See http://www.nbcolympics.com/entitlement/select-provider.html (accessed March 25.03.2010).

Finally, it is important to note that content blocking and declarations that ISPs should have the right to do so is not just being propagated by ESPN, NBC, and MLB—the single biggest purveyor of this technology is China, which uses the same technologies to control access to content by its citizenry. While the rationale may be different, the technical underpinnings of this form of digital feudalism, irregardless of who perpetuates it, are the same.

2.3 Transport Layer Problems

The transport layer is responsible for quality control and reliability—helping ensure that the data transmitted and data received are the same. The transport layer is the transmit control protocol (TCP) component of a TCP/IP network and has been under attack by telecommunications companies striving to integrate "quality of service" techniques that would oversee how transport is controlled.

Several telcos have begun a campaign to create a false dichotomy between speed and openness (e.g., that capacity limitations or the exoflood, quality of service, or network management requirements necessitate a more closed approach to the transport layer). Whereas for the incumbents, there are obvious benefits if that dichotomy gains traction, for the general public, it could have disastrous consequences. The fact remains that openness, by eliminating barriers to innovation, facilitates packet flow, and upgrades paths, which in turn fosters higher speed networking [1].

Cable network operators are facing challenges over the medium-term as they attempt to deal with severe architectural limitations and upgrade to DOCSIS 3.0. Cell phone providers are struggling to implement 3G and 4G networks but are falling far behind other industrial nations. In 2004, the proportion of 3G subscribers to total mobile subscriptions stood at 0% in the United States compared to 13% in Japan; by 2007, the United States had 16% of subscribers on 3G services compared to 83% in Japan [37]. On the other hand, Verizon and other fiber-heavy ISPs are in a good position to leverage their speed into *de facto* monopolies, yet have dramatically slowed their planned rollouts over the past 2 years. Speed alone does not lend itself to monopoly, but once speed becomes a salient differentiator among networks (and without structural separation to ensure that telcos cannot leverage Layer 1 control to lock down everything else), this is an area that will necessitate close observation in coming years.

2.3.1 Port Blocking

A port is a software construct utilized at the Transport layer and often used by applications to streamline communication over protocols like TCP. Much like boats or planes tend to use the same birth or gate when loading and offloading customers, applications often use specific ports for sending and receiving data packets. Thus, by blocking a specific port, an ISP will block any application specifying that port for communication. Though technically savvy users can route around this problem using port forwarding (where you manually tell an application to use a different port than its default), for the average user, port blocking is often the equivalent of a denial of service for specific applications.

In early 2005, VoIP provider Vonage reported that a local ISP was blocking the use of the application [38]. Vonage requested investigation by the FCC. On February 11, 2005, the FCC began an investigation into why Vonage service was nonfunctional on Madison Internet service.[34] The FCC eventually found that Madison River Telephone Company LCC "was cutting off access to Vonage and other VoIP services by blocking certain IP ports."[35] Madison River Telephone Company was fined $15,000 and ordered to not "block ports used for VoIP applications or otherwise prevent customers from using VoIP applications."[36]

Port blocking can negatively impact numerous other applications as well. Port 25 is used for SMTP (Simple Mail Transfer Protocol), the protocol used to send outgoing e-mail from an e-mail client such as Mozilla Thunderbird or Microsoft Outlook. In 2004, Comcast began blocking port 25 to stop computers that were sending spam (so-called zombie computers) [39]. Comcast reported a 35% decrease in spam [40]. However, by blocking port 25, Comcast also made it difficult for individual users to send legitimate e-mail using their own e-mail servers. Faced with port 25 blocks when using SBC DSL, users found they were unable to send mail through their own servers as well and it remains unclear how much of the 35% in "spam" was actually legitimate e-mail traffic.[37]

Presentation layer protocols can also be utilized for malfeasance—invading user privacy and mining personal information. Important data, such as passwords and account numbers, can be intercepted if transferred in plain text. Some individuals also prefer to ensure communication is private, much like sending a letter in a sealed envelope rather than on a postcard. Telnet, an early protocol used to create virtual terminals, transmitted data, including passwords, in plain text. Secure Shell (SSH) has replaced Telnet in most instances and creates secure communication between

---

[34] Federal Communications Commission, *Madison River Communications, LLCand affiliated companies, Consent Decree*, 2005. http://hraunfoss.fcc.gov/edocs_public/attachmatch/DA-05-543A2.pdf.

[35] Madison River to Pay FCC $15,000 for Port Blocking, Von, March 7, 2005 http://www.von.com/news/53h785615.html.

[36] See Federal Communications Commission, *supra* note 76 and *Madison River Communications, LLCand affiliated companies, Order*. http://hraunfoss.fcc.gov/edocs_public/attachmatch/DA-05-543A1.pdf.

[37] See FAQ at DSL Reports: http://www.dslreports.com/faq/12321.

two devices. By creating a shell to encrypt data bits, communication can be resistant to deep packet inspection or snooping of malicious hackers or those seeking to look at what content you are transmitting or receiving. Default operation of SSH requires port 22. When this port is blocked, individuals lose the ability to direct traffic to different places, conduct point-to-point tunneling, or maintaining their security over the Internet. Thus, port blocking had directly infringed on users' right to privacy. Hypertext transfer protocol (HTTP), a vital protocol for displaying Web pages, can be blocked by blocking port 80, thus limiting the ability for an individual to run a webhosting server in their own home.

Ports are vital channels of communication for applications and servers. While interfering with ports can at times be used for good purposes (e.g., preventing a denial of service attack or stemming spam), these restrictions not only hinder normal operation of the Transport layer, but they also lead to the development of more advanced malicious code. Since any decent programmer knows how to do port forwarding, the assumption that an individual capable of creating a worm that can create a botnet of zombie computers is unable to change the port used for communication is rather inane. At best, you gain a momentary reprieve; but in the end, network operators end up punishing users, preventing them from using legitimate services and applications, while those who are doing illegitimate actions route around the problem. In the end, users are left wondering why applications are not functioning as prescribed, or certain aspects of the Internet are no longer working. Most ISPS often block various ports without disclosure or advanced notice, leaving Web savvy users scrambling for work-arounds and the rest of consumers unable to trust an application to work when needed.

2.4 Session Layer Problems

The session layer determines connections between computers and/or devices. A session is a single connection, or transfer of packets, between connected NICs. As an example, a user operating a Web browser may only need one session to download a Web page; however, the increasing complexity of today's Web pages often necessitates multiple connections, or sessions. By running multiple sessions in parallel, one can greatly increase the page load speed. Factoring in multitasking, such as loading a streaming video, sending an e-mail, or even loading multiple pages, the benefit of multiple sessions is easily recognizable. Instant messaging clients, for example, often use a session per simultaneous conversation. With multitasking becoming normative among so-called netizens, limitations to the number of sessions can dramatically impact how applications can engage with the Internet, the functionality of those applications, and an end user's experience of those services and applications.

2.4.1 Session Limits

Beginning in 2005, and widely deployed in 2007, Comcast began monitoring the number of open sessions of specific applications in a region.[38] Using switching equipment from Sandvine, the PTS 8210, Comcast was able to "identify unidirectional P2P uploads" of predefined protocols, such as Ares, BitTorrent, eDonkey, FastTrack, and Gnutella.[38] The Sandvine PTS 8210 is capable of inspecting packet header information through stateful packet inspection (SPI)[39] and, as described in a filing to the FCC regarding the practice, "Comcast established thresholds for the number of simultaneous unidirectional uploads that can be initiated for each of the managed protocols in any given geographic area."[44] When the thresholds were reached, Comcast began terminating communication of the applications such as BitTorrent.

Limiting sessions negatively impacts a user's ability to control their communication over the network. In Comcast's case, they confounded the problem by creating thresholds for blocks of users in specific geographic areas—thus, so-called overuse of a specific application by one user can detrimentally impact legitimate use of that same application by another user in the neighborhood. For example, when Comcast found BitTorrent sessions that exceeded the Uni Threshold of 8 among a block of users, their network management systems blocked additional functionality.[44] Of course, SPI can be circumvented by directly connecting to a device by secure tunneling, a technique Topolksi used to discover that BitTorrent was being blocked by his network provider [41]. However, implementing thresholds that are beyond any users ability to control (i.e., a single user in a network cannot prevent another user from running the application of their choice, yet Comcast engaged in a collective reprisal against an entire geographic area when overuse was identified). In Comcast's case, when these practices were discovered, they then provided false information to consumers and the media, stating that traffic was not being blocked, only "delayed" (the equivalent logic of stating that hanging up the phone on someone does not terminate the call, only delays it) [42].

2.5 Presentation Layer Problems

The presentation layer creates the framework for how information is displayed in the application layer. Examples can include protocols for the display of text, such as ASCII, or controlling streaming in media content. Presentation layer components can also include encryption and compression.

---

[38] See Attachment A: Comcast Corporatoin Description of Current Network Management Practices, 5. http://www.wired.com/images_blogs/threatlevel/files/comcastic.pdf.

[39] See Sandvine Policy Traffic Switch 8210. http://www.sandvine.com/downloads/documents/PTS8210_Datasheet.pdf.

2.5.1 ASCII and MIME—Websurfing and E-mail

ASCII (American Standard Code for Information Interchange) is a character-encoding scheme, the foundation for turning bits to text. Many character sets are based on ASCII, but the code is intrinsically Americentric. In applications where ASCII is used as the character-encoding scheme, non-Latin languages cannot be used. This became particularly problematic with domain name addresses. Thus, every URL had to be displayed in ASCII, leaving entire language groups unable to build URLs in their native tongues.

Historically, domain names were ASCII dependent—only Western characters were included in addresses. On November 16, 2009, the International Corporation for Assigned Names and Numbers (ICANN), the body that coordinates naming schemes for the Internet, launched the IDN ccTLD Fast Track Process, the first step to including internationalized domain names (IDNs).[40] IDNs are domain names and extensions using non-Latin characters, meaning that domain names could include local languages for the first time [43]. Previously, even if a Web site was entirely in Arabic or Chinese text, the domain name required the use of Latin characters. Although this a dramatic step forward, not all languages are included. Arabic, Chinese, Greek, and Japanese are among the 10 additional languages added thus far[41]; but additional languages must be individually added through a request process.[42]

Multipurpose Internet Media Extension (MIME) is a standard for formatting e-mails and providing support for body and headers in different character sets and attachments. Data included in a MIME header define what type of content the e-mail contains, whether the message includes multiple parts or whether the data are encrypted. Like ASCII, the application function of MIME must be included in the MIME protocol. Defining the acceptable language of e-mails, if a character or character set is not included, it functionally does not exist, affecting both addresses and e-mail content. Thus, MIME created an electronic communications medium where certain languages can not be simply used to send e-mail to recipients.

2.6 Application Layer Problems

The application layer bridges the presentation of data with the end user. Providing the foundation for software, application layer elements include HTTP and file transfer protocol (FTP). Enclosure to the application layer can cripple applications and communication.

---

[40] See *IDN ccTLD Fast Track Process* at: http://www.icann.org/en/topics/idn/.
[41] See Internet Corporation for Assigned Names and Numbers, *IDNs: Internationalized Domain Names*, available at: http://www.icann.org/en/topics/idn/factsheet-idn-program-05jun09.pdf.
[42] See ICANN Fast Track Process at: http://www.icann.org/en/topics/idn/fast-track/.

2.6.1 DNS Hijacking

The Domain Name System Protocol (DNS protocol) is the translation of IP addresses to text. Creating a system for more easily recognizable Web addresses, users submit a DNS query, such as www.newamerica.net, rather than needing to remember 69.174.51.225. In this example, the Web address is matched to an IP address run by Mzima Networks and the New America Foundation's Web server. If the query is resolved by an NXDomain response, the user is directed to the Web site. If a domain name does not exist, is not associated with an IP address, such as misspellings like www.newamerca.net, a HTTP gives a 404 error indicating that the Web page or server does not exist. Applications layer responses can interpret this error to the user through messages such as Firefox's "Check the address for typing errors such as ww.example.com instead of www.example.com."

Some ISPs have begun implemented services where a server synthesizes an NXDomain response if a DNS query is not resolved, redirecting traffic rather than transmitting the error protocol to the application. In 2007, Cox Communications began experimenting with DNS redirection. In June 2007, Verizon began trials under their Web Search Service,[43] and Comcast began trials in July 2009 for their Domain Name Helper service, redirecting mistyped URLs to an advertisement heavy Web site with a search function [44]. This service was rolled out nationally in August 2009 [44]. Cox, EarthLink, and Charter have all used redirection as well.[44] In a preliminary report on DNS modification, the ICANN Security and Stability Committee notes: "any party involved in the resolution process can perform NXDOMAIN redirection for *every* name which it determines or is notified does not exist, regardless of whether an authoritative server gives an NXDOMAIN."[45]

DNS redirection can be a profitable endeavor where enclosure of traditional application layer functionality (and neutral error messages) can generate significant revenue. Often, Web users would need to opt out of these redirection "services" if they want to actually know what errors they are actually receiving when a Web page will not load—and sometimes opting out is not possible at all. According to

---

[43] Cox Tests DNS Redirection, DSL Reports, May 19, 2007, http://www.justbroadband.org/shownews/83929; Verizon DNS Redirection—The latest ISP to profit off your butterfingers, DSL Reports, June 20, 2007. http://www.dslreports.com/shownews/Verizon-DNS-Redirection-85063 (accessed 06.05.2010).

[44] Verizon DNS Redirection 'Service' Spreads, DSL Reports, November 5, 2007, http://www.dslreports.com/shownews/Verizon-DNS-Redirection-Service-Spreads-89137 (accessed 06.05.2010).

[45] ICANN Security and Stability Advisory Committee, *SAC 032Preliminary Report on DNS Response Modification*, June 2008. http://www.icann.org/en/committees/security/sac032.pdf (accessed 06.05.2010).

DSLreports.com, DNS redirection can boost revenue for ISPs by $5 per month for every user.[46] Senior Security Technologist for ICANN Dave Piscitello has expanded the possibility for redirection, suggesting that it could be used for other IP-based applications such as redirecting e-mail or a VoIP phone call to a wrong number.[47] One example of more invasive redirects occurred in April 2010 when Windstream users found that searches using the Google search bar were also redirected to a competitor's search service[48]; though Windstream called this redirection accidental and changed it the next day, it demonstrates how easily one could hijack traffic to send it to another site (as examples, an ISP could redirect all traffic from Ford.com to Chevy.com or all traffic to McDonalds.com to BurgerKing.com).[49] The ICANN board has banned the practice of redirection for new top level domains; however, as of this writing, ISPs such as Verizon and Comcast continue redirecting mistyped domains to their own services.

2.6.2 H.264 and the Future of Online Video

The rising prevalence of online video has drawn attention to its dependency on third-party support such as Adobe Flash or Microsoft Silverlight. The latest standards for hypertext markup language, HTML, include new tags for making video support native. This new tag allows HTML 5 browsers to natively play video content; however, one included video standard, the H.264 codec, is owned by the MPEG LA group. With licenses held by Microsoft, Apple, and others,[50] this codec has become a privately owned Internet standard.

Licenses for the H.264 are needed for encoders and decoders, such as video software, browsers, or video capable recording devices like digital cameras, or for content providers such as YouTube or over the air television.[51] The availability of no-cost H.264 licenses was set to expire at the end of 2010, but MPEG LA in February 2010 announced an extension for no-costs licenses for "Internet Video that

---

[46] ICANN Slams DNS Redirection, DSL Reports, November 25, 2009, http://www.dslreports.com/shownews/ICANN-Slams-DNS-Redirection-105651 (accessed 06.05.2010).
[47] *ICANN Start*, Episode 1: Redirection and Wildcarding, http://www.icann.org/en/learning/transcript-icann-start-01-22mar10-en.pdf (accessed 06.05.2010).
[48] Windstream Hijacking Firefox Google Toolbar Results, April 5, 2010, http://www.dslreports.com/shownews/107744 (accessed 06.05.2010).
[49] Windstream Gives (Sort Of) Explanation For Google Search Hijack, DSL Reports, April 9, 2010, http://www.dslreports.com/shownews/107828 (accessed 06.05.2010).
[50] See list of AVC/H.264 Licensors at http://www.mpegla.com/main/programs/AVC/Pages/Licensors.aspx (accessed 06.05.2010).
[51] See Summary of AVC/H.264 License Terms at http://www.mpegla.com/main/programs/avc/Documents/AVC_TermsSummary.pdf (accessed 09.05.2010).

is free to end users (known as Internet Broadcast AVC Video)" through 2015.[52] This postpones the requirement that Vimeo or YouTube pay license fees for free video, but software developers must pay tribute in order to compatible with the online video. Likewise, what happens in 2015, once this standard is more thoroughly embedded in multiple products and utilized at higher rates, remains to be seen.

But the H.264 codec and its inclusion in HTML 5, has the potential to create a new bottleneck that captures a growing amount of online video traffic and could be utilized as a toll booth to create new "billable moments" for using these services. H.264 is quickly becoming the standard for online video. As of May 1, 2010, an estimated 66% of video content online is available through the H.264 codec, with the majority push coming from YouTube.[53] Constraining video to a particular license scheme is very troubling. Mozilla, for example, would need to pay a reported $5 million license fee in order to play H.264 encoded video on its Firefox Web browser.[54] The license terms for H.264 could also be extended to devices like cameras and video game consoles as well as software where this function is currently available for free.[55]

3. The Need for Open Technology

At its heart, one of the most significant barriers to reform comes down to the differences between *closed* and *open* technologies. These notions often bring to mind issues related to open source and proprietary software (e.g., Linux vs. Windows), but the distinction is more encompassing. Stolterman defines the important attributes thusly:

---

[52] Press Release, *Corrected Version of February 2, 2010 News Release Titled "MPEG LA's AVC License WillContinue Not to Charge Royalties for Internet Video that is Free to End Users."* http://www.mpegla.com/Lists/MPEG%20LA%20News%20List/Attachments/226/n-10-02-02.pdf (accessed 09.05.2010).

[53] YouTube is estimated to account for 40% of all online video content. See Erick Schonfeld, H.264 Already Won—Makes Up 66 Percent Of WebVideos, Tech Crunch, May 1, 2010. http://techcrunch.com/2010/05/01/h-264-66-percent-web-video/ (accessed 09.05.2010).

[54] http://news.cnet.com/8301-30685_3-10440430-264.html (accessed 09.05.2010).

[55] The license for H.264 used in digital cameras such as the Canon 5D or videos edited in Final Cut Pro only allows non-commercial use, see EOS 5D Mark II Instruction Manual. http://gdlp01.c-wss.com/gds/6/0300001676/02/eos5dmkii-im3-en.zip, 241 and http://images.apple.com/legal/sla/docs/finalcutstudio2.pdf (accessed 03.05.2010).

A closed technology is one that does not allow the user to change anything after it has been designed and manufactured. The structure, functionality and appearance of the artifact are permanent.... The technology is a relatively stable variable in social settings.... An open technology allows the user to continue changing the technology's specific characteristics, and to adjust, and/or change its functionality. When it comes to an open technology, changes in functionality pose a question not only of change in the way the existing functionality is used or understood but also of a real change in the artifact's internal manifestation [45].

The Internet, generally speaking, was conceived and remains an open and designable technology. One can "add, embed, contain or surround the artifact with other technology in a way that radically changes it" [45]. This aspect has contributed to the successes of so-called Web 2.0 applications. Unfortunately, this openness is also under attack, as moves by Comcast to block BitTorrent communications, the blocking of pro-choice text messaging by Verizon, and the editing of a live Pearl Jam's concert by AT&T all exemplify [46]. Unfortunately, the "gentlemen's agreements" that have been sold as "solutions" (that these corporations will not engage in these practices again) do nothing to prevent these sorts of anticompetitive, antifree speech, and antidemocratic actions from being repeated at a later date. Thus, a growing list of public interest organizations have grown increasingly worried that by abdicating their responsibility to prevent this sort of corporate malfeasance, the FCC and other regulatory agencies are all but guaranteeing that these sorts of behaviors will continue. As we have shown, discriminatory practices are today being built into the very foundations of next-generation network infrastructure.

In fact, without the advent of the landmark Carterfone decision to allow interconnection of "foreign attachments" to the AT&T telephone network, wireline communications may well have taken a different turn—even preventing the emergence of the Internet in its present form. Prior to Carterfone, the FCC tariff governing interconnecting devices stated, "No equipment, apparatus, circuit or device not furnished by the telephone company shall be attached to or connected with the facilities furnished by the telephone company, whether physically, by induction or otherwise."[56] Even the Hush-a-Phone, a plastic attachment that blocked room noise was originally deemed illegal to affix to telephone handsets [14]. The growth and successes of the Internet are predicated upon an open architecture [47] that facilitates the interconnection of a variety of different devices and technologies [48]. While AT&T may have wanted end-to-end control over every part of their telephone network, the FCC wisely concluded, against the vociferous condemnation of

---

[56] FCC. 1968. FCC 68-661, Carterfone Order.

AT&T, that the best interests of the general public would be achieved by ensuring that innovation could not be stifled by AT&T and that end users could decide for ourselves which devices and technologies we wanted to attach to our telephone lines.

3.1 Limitations on Today's Closed Networks

The limitations to freedom that result from the convergence of networks and devices still exist, as evidenced by the wireless cell phone industry. Many cellular phones are released with exclusivity agreements with carriers; and carriers often place restrictions on the functionality of the phones, turning off key features. For example, when Verizon introduced the Motorola V710, the first U.S. phone with Bluetooth functionality [49], the carrier removed its ability to transfer files over Bluetooth [50]. Customers wanting to transfer photos off of the phone needed to buy another accessory or pay for additional services. The Motorola RAZR was one of the most popular phones of the past decade, shipping 50 million units by July 2006,[57] and many networks offered different models at different times as a differentiator of their services, hoping to win over customers from their competitors' networks.

The problems with this approach are best epitomized by the 2007 deal inked by Apple and AT&T. Apple's iPhone was only available for use on AT&T's network, even though it could be used on any cellular network. Likewise, AT&T only allows certain services and applications to run on the iPhone, even though the iPhone could run many additional programs that would be useful for end users. Innovative iPhone owners and entrepreneurs quickly found ways to unlock their device and install a growing option of after-market applications but the extra work and cost are borne by end users as a result of an anticompetitive business practice that Apple and AT&T refer to as an "exclusive deal." Apple gave some concessions in July 2008, by releasing an application store, but kept tight control over what types of applications would be included in the store. Some applications were rejected because they "duplicate features that come with the phone," such as applications for Google Voice [51]. Others were limited regarding the features they could offer. Thus, while MLB was able to include video streaming over 3G wireless for their applications, Skype was prevented from using streaming technology for their application and was restricted to Wi-Fi use only for this feature. Only in early 2010 did Apple announce that applications that offer VoIP over 3G networks would soon be offered [52]. However, Apple has gone as far to dictate in iPhone Developer Program License Agreement that only certain programming techniques are allowed on their device

---

[57] See http://www.motorola.com/mediacenter/news/detail.jsp?globalObjectId=7031_6980_23 (accessed 22.04.2010).

and "applications that link to Documented APIs through an intermediary translation or compatibility layer or tool are prohibited" [53]. This policy was rescinded following an antitrust investigation by the European Union,[58] carriers and handset manufacturers have collaborated prevent the "jailbreaking," of mobile devices by keeping some software components in read-only memory or designing devices that automatically deactivate if unauthorized software is detected [54]. By comparison, the superiority of open architectures is immediately apparent:

> An open architecture means fewer technological restrictions and, thus, the ability to explore more options. In an open architecture, there is no list of elements and protocols. Both are allowed to grow and change in response to changing needs and technology innovation.... With an open architecture you are not making bets on a specific direction the technology will take in the future. You are not tied to a specific design or a particular vendor or consortium roadmap, so you can evaluate and select the best solution from a broad and energetic competitive field. Competition facilitates innovation and reduces equipment and implementation costs [55].

With data communications networks, the costs of closed architectures are particularly devastating because they impact almost every communications medium. Further, with the dissonance in openness between wireless and wireline networks, and the FCC's push for wireless to provide competition for wireline networks, the need for open networks has never been greater.[59] Although both Verizon and AT&T have declared their intention to run open networks, and the terms of 700 MHz spectrum auction included openness requirements on that new "C-Block" mobile phone band, these details have not yet been clearly defined. However, even these approximations of openness are only baby steps from a fully proprietary infrastructure, but at least the direction of travel is toward a more open, interoperable, and innovation-supporting network.

Within the data communications realm today, most municipal and enterprise 802.11 (Wi-Fi/WiMAX) wireless networks are entirely proprietary. For example, a Motorola 802.11 system will not interoperate directly with a Tropos system, which will not interoperate directly with a Meru system, which will not interoperate directly with a Meraki system, and so on. In fact, most consumers have no idea that the links they rely on to access Internet and Intranet services lock geographical areas into distinct path dependencies with specific vendors (and their specific capabilities and limitations). Disconcertingly, in an era when interoperability of

---

[58] Press Release, Antitrust: Statement on Apple's iPhone policy changes. http://europa.eu/rapid/pressReleasesAction.do?reference=IP/10/1175&format=HTML&aged=0&language=EN&guiLanguage=en.

[59] See footnote 3, Chapter 4.

applications, services, and communications is assumed, and the communities that people participate in are geographically dispersed, the immediate and long-term ramifications of this geospatial lock-in remain almost entirely unexplored. Closed technologies have the potential to constrain the positive potentials of the Internet if their widespread adoption stems more from an emphasis on corporate profits than maximizing wireless networks' public benefits. Today's battles over 802.11n Wi-Fi systems, WiMax, and 4G networking are all indicators of this ongoing tension.

Unlike the Internet, these wireless "last-mile" links can *disallow* users from extending the network (e.g., using bridges and routers), adding applications (e.g., VoIP, P2P, IRC, IM), interconnecting additional services (e.g., streaming servers, distributed file storage, local webhosting), or connecting directly with one another. The wireless medium is a *de facto* throwback to an era paralleling AT&T's control over which devices could be connected to their network and which technologies would thus be developed. For unsuspecting communities and decision makers, the long-term effects of wireless lock-in may be more detrimental than any policy previously witnessed in telecommunications history.

Thus far, regulatory bodies and decision makers remain unwilling to address these fundamental concerns. While Obama's first chairman of the FCC, Julius Genachowski, has proposed to eliminate these discriminatory practices, during the first 2 years of the Obama administration, the FCC has done almost nothing to actually implement these reforms. As this first decade of the twenty-first century draws to a close, this inaction may have profound impacts on the development of feudalistic communications systems in the years to come. Given that all technology is inscribed with social values that foreclose certain possibilities while encouraging others, emphasizing the enclosures that communications technologies facilitate illuminates what is at stake with network neutrality and situates this debate within a larger vision of Internet openness. Today, we sit at a critical juncture for Internet policy and the opportunities that now abound for graceful reforms will soon disappear.

4. Policy Recommendations for the New Critical Juncture in Telecommunications

Comparisons of the Obama administration's use of the Internet to circumvent traditional media with Franklin Delano Roosevelt's use of radio during his fireside chats have become a regular news trope in coverage of the administration. However, what is often forgotten in this news coverage is the more cautionary tale that this historical parallel exemplifies: FDR failed to seize the initiative to set the new media of broadcasting on a democratic course when he had a chance. As a result, we ended up with a broadcast media system that was not only largely commercialized but also largely inoculated against public interest regulation, and therefore never reaching its democratic potential [56]. Lest we repeat this history, the following is a list of policy recommendations that, if implemented, would steer our new digital potentials toward more democratic ends. In earlier work, we laid out a 10-step platform for creating a more democratic Internet [9]. Here, we further flesh out these recommendations to address the range of issues addressed in this chapter.

4.1 Physical Layer Solutions

We recommend overturning the FCC decision following the Brand X Supreme Court case and restoring common carriage provisions to all ISPs. Common carriage ensures that network operators lease their lines to all potential market players, including municipalities, at market (wholesale) rates. This should include universal service provisions and service level agreements for all users (business, residential, municipal, NGO, etc.). As has been seen repeatedly throughout the history of transportation and telecommunications, common carriage protects the general public against price and geographic discrimination and other anticompetitive business practices. From 2000 to 2005, the number of ISPs has nearly halved from 8450 in 2001 to 4417 in 2005. With the demise of common carriage provisions resulting from the Brand X Supreme Court decision, this number will continue to decrease—in the 2010 National Broadband Plan, the FCC revealed that 96% of Americans have access to two or fewer ISPs.[3] Not only is price a primary barrier to adoption of Broadband,[60] but contrary to popular media portrayals, prices are also on the rise [57]. Research from the Pew Internet and American Life project documents that prices are higher when consumers only have one or two providers to choose from [57]. The Harvard Berkman Center has documented the success of open access and common carrier policies in leading broadband nations [4], and the FCC reclassifying broadband as a Title II communications, with Common Carriage regulation, is an essential step to revamping the failed market duopoly in the United States.

---

[60] According to a Federal Communications Commission report 36% of Americans who have not adopted broadband cite cost as the primary reason. A report from the National Telecommunications and Information Administration documents that price is a main reason for 26.3% of nonadopters. See John B. Horrigan, *Broadband Adoption and use in America: OBI Working Papers Series No. 1* (Federal Communications Commission 2010), 5, available at http://hraunfoss.fcc.gov/edocs_public/attachmatch/DOC-296442A1.pdf and National Telecommunications and Information Administration, *Digital Nation: 21st Century America's Progress Towards Universal Broadband Internet Access*, 2010, 13, available at http://www.ntia.gov/reports/2010/NTIA_internet_use_report_Feb2010.pdf.

In addition, current federal spectrum regulation creates a false scarcity of spectrum availability. Allocating blocks of spectrum to exclusive use by single entities ignores the technological strides made over the past 75 years. Similarly, operating digital devices with closer adjacency (or co-channel use with contention protocols to facilitate multiple users on the same frequency—in much the same way that Wi-Fi devices already work) allows for an increased number of devices within a discrete frequency band. Unlicensed spectrum has already proved to be a tremendous boon for innovation and advancing networking technologies and should be dramatically increased. Currently, devices from baby monitors to cordless phones all share the same frequencies with laptops and home computers.

Wi-Fi has proved to be a runaway success, and unlicensing the 2.4 and 5.8 GHz bands has enabled roaming connectivity in homes and businesses, easy Internet access in coffee shops and on airplanes, and mesh networking that is essential to community and municipal broadband networks around the world. Wi-Fi is also an essential component for current cellular phone networks. According to the mobile phone industry, the rapid uptake of smart phones and slow buildout of additional cellular capacity have created network congestion. Cell phone operators have urged the FCC for more spectrum to expand bandwidth, and the FCC has recommended making 500 MHz of spectrum available for broadband access (including cell phones) over the next 10 years.[3] Cell phone networks, as they are currently operated, are becoming oversold and congested. For example, at one time a 30% fail rate for phone calls in New York using an iPhone with AT&T was reportedly considered normal [58]. However, unlicensed bands can offer greater network relief than increased licenses to incumbent carriers. Data from Admob, an advertising company that collects data on traffic use to partner applications and Web sites, reveal that 55% of traffic from Wi-Fi-enabled smart phones is from Wi-Fi connections.[61] Likewise, creating national unlicensed GSM bands would enable anyone to build cellular infrastructure that could utilize today's popular cell phone handsets while relieving congestion on existing networks.

Opportunistic spectrum access (OSA) would allow for secondary use of spectrum. A test in 2004 as part of a National Science Foundation research project found that spectrum efficiency is close to 5% and that even in major metropolitan cities, the highest utilization is under 15%. The current framework for allocating spectrum assumes the need for a single entity to have absolute control over this spectrum. If roadways were distributed like spectrum, cars would be assigned specific lanes for the duration and would never be allowed to change lanes for any reason.

---

[61] AdMob Mobile Metrics Report, November 2009, available at: http://metrics.admob.com/wp-content/uploads/2009/12/AdMob-Mobile-Metrics-Nov-09.pdf.

As commentators have summed up, "Imagine traffic laws which require that each lane in the highway is dedicated to particular makes of car—BMWs and Saabs use lane 1, Toyotas and Fords lane 2, and so on. A Toyota cannot use lane 1 even if that lane is empty!" [59] As we saw so dramatically during 9/11 and Hurricane Katrina, this methodology can easily bring networks to their knees when one of the spectrum "lanes" is destroyed—a far more robust telecommunications system would adapt to changing conditions, allowing devices to change frequencies as necessary. Today, at any given time, 19 out of 20 lanes on the spectrum freeway have no traffic, yet cars would be impeded from driving on them, being forced to all share the single remaining lane. It is time for the FCC to change its rules to allow cognitive radios that are able to detect if a given frequency is in use, and change frequency, power, and modulation in real time to utilize these underused frequencies.

As of this writing, the FCC has plodded along approving rules for TV White Spaces, empty slots and guard bands between TV channels that were originally intended to minimize interference among stations. Radio technology has evolved by leaps and bounds since the current framework of spectrum allocation was first conceived over half a century ago. Currently, broadcasters use less than 20% of the available frequencies in many rural areas [60]. As the FCC's own testing has documented, white space devices (WSDs) can detect TV signals at levels 1/1000 of the signal power needed by televisions to display a picture, thus minimizing the chances of harmful interference. WSDs detect if given frequency is in use and utilize empty bands as needed. In 2008, the FCC has issued rules allowing WSDs, but the rules were contingent on further rules such as the creation of a geolocational database that these devices would query to identify which frequencies they can utilize.[62] An FCC order in September, 2010 approved the use of unlicensed devices operating at 1 W or less, but reservations for wireless microphones limits the potential for so-called Super Wi-Fi in urban markets.[63]

OSA is also valuable and applicable beyond the TV broadcast bands. The federal government makes over 270,000 license allocations and assignments, yet some of these are seasonal or in use only in cases of national emergency or for particular, exceedingly rare occurrences. Maintaining priority for federal users while allowing

---

[62] Federal Communications Commission, *Unlicensed Operation in the TV Broadcast Bands: Additional Spectrum for Unlicensed Devices Below 900 MHz and in the 3 GHz Band*, Second Report and Order and Memorandum Opinion and Order, November 14, 2008. http://hraunfoss.fcc.gov/edocs_public/attachmatch/FCC-08-260A1.pdf. Second Memorandum Opinion and Order

[63] Federal Communications Commission, *Unlicensed Operation in the TV Broadcast Bands: Additional Spectrum for Unlicensed Devices Below 900 MHz and in the 3 GHz Band*, Second Memorandum Opinion and Order, September 23, 2010. http://www.fcc.gov/Daily_Releases/Daily_Business/2010/db0923/FCC-10-174A1.pdf.

secondary access to these bands will preserve the right of the government license holders to use this spectrum, while allowing the frequencies to be used the 95% of the time the airwaves are completely open. Spectrum is essential not only for mobile connectivity but also for fixed wireless networks in low-population density areas like rural regions and Native American Tribal lands. As an essential public resource for communications, in light of today's technological capabilities, the mismanagement of spectrum resources is a national travesty that directly contributes to the digital divide.

4.2 Data Link and Network Layer Solutions

We recommend open architecture and open source driver development, which encourages a digital commons by keeping both the hardware itself and any hardware access layer(s) open. As the Open Source movement gains ground (especially internationally) and hardware prices have plummeted, new business models have arisen to promulgate market capture and path dependence, creating potentials for secondary network enclosures. However, key officials have already begun to challenge governmental over-reliance on proprietary technology. On October 16, 2009, David M. Wennergrem, CIO for the U.S. Department of Defense, issued a memorandum supporting the advantages of open source software.[64] In January 2010, Teri Takai, CTO for the State of California, issued a memorandum "formally establishing the use of Open Source Software" for the state government.[65] Open architectures and access layers help promote competition by creating opportunities for new market entrants and rapid innovation of features and functionality.

We also recommend that private networks do not privilege state security imperatives that compromise individual privacy rights wholesale and that they help ensure a nondiscriminatory environment for content access and information dissemination. Private networking is essential to ensuring the continued expansion of online business, though back doors and other surveillance devices introduce enormous security holes. Likewise, privacy-invasive techniques, when widely deployed, increase impetus for the development and widespread adoption of privacy software that hampers, over the long-term, legitimate law enforcement efforts.

Further, we recommend that ISPs, including wireless carriers, not discriminate against lawful content and applications. Some network management schemes, such

---

[64] Department of Defense, Memorandum for Secretaries of the Military Departments Re: Clarifying Guidance Regarding Open Source Software (OSS), October 16, 2009. http://cio-nii.defense.gov/sites/oss/2009OSS.pdf.

[65] Office of the State CIO, IT Policy Letter, *Subject: Open Source Software Policy*, January 7, 2010. http://www.cio.ca.gov/Government/IT_Policy/pdf/IT_Policy_Letter_10-01_Open_Source_Software.pdf.

as IMS, treat different types of data differently and interfere with normal network operation of the network and transport layers. As included in the FCC's "net neutrality" proceedings, maintaining nondiscrimination is essential to preserving a free and open Internet[66] and preventing a data-obfuscation arms race that will inevitably create additional headaches for future system administrators. Low-latency and first-in/first-out routing protocols help remove the impetus for data packet and application discrimination by requiring that a service providers be responsible for provisioning adequate capacity for its customer base. Service level agreements and minimum speed guarantees help lower over-subscription rates, artificial scarcity and the hoarding of dark fiber and spectrum assets by mandating adequate capacity and providing incentive for network and capacity upgrades.

Broadband Truth-in-Labeling

The Open Technology Initiative of the New America Foundation is calling for Truth-in-Labeling by our nation's broadband operators. Drawn from similar useful disclosure requirements by lenders, these Broadband Truth-in-Labeling disclosure standards will give the marketplace a much-needed tool that clarifies and adds meaning to the terms and conditions of the service being offered. Broadband subscribers are often frustrated that the actual performance of their Internet access service regularly falls far below the advertised speeds. Consumers set their expectations based on phrases like "up to 16 Mbps," and are disappointed to learn that these quotes are worthless as assurances.[67]

Currently, there is no lawful requirement for ISPs to reveal the contents of the broadband services they are providing; customers might be harmed by the invalid or ambiguous languages. Internet Access Providers should disclose the important facts and details of the broadband offering before subscribers sign up. The disclosure should be meaningful, and failing to meet minimum standards should be treated as an important service outage (resulting in a refund or service credit to the consumer). Where there are choices between different products or providers, the disclosure should be made in a way that allows consumers to compare them. Providing clear, meaningful, comparable disclosures ultimately spurs

---

[66] Federal Communications Commission, *Preserving the Open Internet Broadband Industry Practices, Notice of Proposed Rulemaking*: http://hraunfoss.fcc.gov/edocs_public/attachmatch/FCC-09-93A1.pdf.

[67] As developed by Robb Topolski, Benjamin Lennett, Chiehyu Li, Dan Meredith, James Losey, and Sascha Meinrath. Also available at: http://www.newamerica.net/publications/policy/broadband_truth_in_labeling.

competition between ISPs which encourages the future development of broadband technology.

Open Technology Initiative has created a sample Broadband Truth-in-Labeling disclosure. ISPs use a standardized label to notice their customers what broadband services they are subscribing, including Internet speed, service guarantee, prices, service limits, and other related elements. The label aims at educating customers the contents of broadband services and making the broadband services more transparent to spur broadband competition, innovation, and consumer welfare.

To make sure the broadband service is clearly expressed, the Broadband Truth-in-Labeling disclosure should be standardized to comprise several typical elements as indicators of broadband service quality, such as minimum expected speed and latency to the ISP's border router (where the ISP connects to the rest of the Internet) and service uptime. These minimum assurances will be supported by the ISP as guarantees in the delivery of broadband services, backed by technical support and service charge refunds or credits. In addition to the description of minimums being guaranteed of the service, the disclosure should include all applicable fees, a common description of the technology used to provide the services, any service limits such as a bandwidth cap or the application of any traffic management techniques, the length of the contract terms, and a link to all additional terms and conditions.

Requirements should be established for disclosing any highly objectionable or surprising terms such as arbitration restrictions or information or data selling. This Broadband Truth-in-Labeling disclosure must be shown to the consumer as part of the sign-up process and must be assertively presented again any time the ISP decide to alter the terms in such a way that alters the facts on the original Broadband Truth-in-Labeling disclosure.

On a more global level, we recommend replacing and/or dramatically expanding multilateral control over important governance institutions like ICANN in a way that internationalizes control over this vital global resource. As Milton Mueller and others have documented, control over global communications networks and the Internet, in particular, has remained Americentric. Moreover, purportedly representative bodies like ICANN and the regional Internet registries (RIRs) often appear to privilege industry interests. The current United States-controlled ICANN model is unsustainable over the long term and will cause increasing problems as international uptake of the Internet increasingly dwarfs U.S. numbers.

As included in the FCC's open Internet principles and included in the proposed rulemaking,[14] we recommend that ISPs be required to allow any lawful content and

ExampleCom Ultra 15 Mbps
Broadband Truth-in-Labeling

Advertised speed	15 Mbps downstream/2 Mbps upstream
Service guarantees	
Services are measured from and to the border router	
Minimum speed at border router	8 Mbps downstream/384 kbps upstream
Minimum reliability/uptime	96%
Maximum round-trip latency (delay) to border router	50 ms
Service guarantee terms	Daily service credit upon request for any outages or extended periods of under-delivery of service
Prices	$44.99 monthly service
	$19.99 monthly for the first 6 months on promotion
Service limits (list all traffic management techniques)	• Exceeding 100 GB calendar week considered excessive use, subject to disconnect penalties, see http://www.examplecom.invalid/excessive
	• Traffic by heavy users in congested areas is artificially slowed, see http://www.examplecom.invalid/shaping
Other fees (ISPs cannot charge if not listed)	$3 monthly modem rental fee
	$59.99 installation fee
	$19 outlet installation
	$150 early termination during promotion period
	$2 account change fee
	$35 service call fee unless $3 monthly inside wiring maintenance plan is in force
	Sales taxes and franchise fees, vary by location
Contract term	At will, customer may cancel at anytime after first six months. During the first six months, a cancellation results in a $150 fee
Service Technology	DOCSIS 1.1/2.0 HFC
Legal and Privacy Policies	http://www.examplecom.invalid/legal

application. Blocking certain functionality of applications over cellular networks or forging packets to terminate interferes with the core benefits of end-to-end architecture: allowing users to define how the Internet is best used to serve their needs. Blocking lawful transfers of the Bible or content is antithetical to this functionally.[68]

4.3 Transport Layer Solutions

We recommend protecting end-to-end architectures for packet-based data communications networks (E2E). E2E helps remove vulnerabilities to bottlenecks and gatekeeping (e.g., through dynamic routing), illegal surveillance by telcos (e.g., through E2E encryption), and a host of other potential problems. E2E helps speed up network throughput and increases network capacity. In contrast, prioritization schemes, when widely deployed can often create substantial harm to the network throughput as well as users. As Robb Topolski and Chris Riley explain, prioritizing some packets while delaying others can cause packets to be dropped by many applications. As these packets are resent the network generates "greater traffic for the same communication" [28].

An end-to-end architecture helps prevent both governmental and corporate interference in network traffic—an outcome that is especially important at a time when surveillance and DRM techniques that infringe upon our fair use rights are increasingly prevalent. Further, consistent with the proposed FCC rulemaking on preserving the open Internet, we recommend mandating that service providers reveal practices that could interfere with E2E networking.[69] Network management techniques are utilized for a number of reasons. For example, in 2005, Comcast claimed to have stemmed 35% of spam from zombie computers in its network by blocking port 25 [39]. However, blocking SMTP functionality without disclosing to customers the policy,[70] and providing an opportunity for continuing to use e-mail clients with servers of their choice, also permanently removed key functionality for many Comcast users. Furthermore, without verification of the claims made by Comcast, it is quite possible that they have included legitimate e-mail in their spam reduction number. Transparency of these practices helps customers understand the limitations of their connections, whereas the "security through obscurity" that undergirds the argument that these practices should not be discussed has always failed over time.

---

[68] Peter Eckersley, Fred von Lohmann and Seth Schoen, Packet Forgery By ISPs: A Report On The Comcast Affair, November 2007, available at www.eff.org/files/eff_comcast_report2.pdf.

[69] See Federal Communications Commission, *supra* note 133.

[70] User discussions online document user frustration when email services stop working. See http://www.dslreports.com/forum/remark,18116427~days=9999 and http://www.dslreports.com/forum/remark,10460634~mode=flat (accessed 14.10.2010).

4.4 Session, Presentation, and Application Layer Solutions

Interoperability harmonizes different systems and integrates foreign attachments. This is especially important to the continued global expansion of broadband service provision. As Mark Cooper and Barbara van Schewick point out [61], interoperability lowers costs while increasing the collaborative potential of the Internet. Interoperability is critical to ensuring that the 80% of humanity who are not currently online will be able to interconnect with next-generation telecommunications infrastructures. Thus, preventing enclosure of session-level communications means ensuring that "reasonable network management" techniques do not include the ability for providers to harm a particular individual because another user is utilizing substantial network resources. Likewise, any network that limits session-level communications must be viewed with skepticism since there is very little reason to do so if a provider is actually provisioning the speeds and capacity that they have promised to their end users.

We recommend maintaining open protocols and standards that ensure a free-flowing, nonenclosed Internet. This, in turn, facilitates innovation and widespread adoption of new technologies. With the growing pull toward proprietary networking (especially within the wireless medium), it is vitally important to prevent the so-called Balkanization of the Internet. Presentation layer protocols and standards are the building blocks for many of the most widely utilized online applications and are a prime location for potential digital enclosures. Ensuring the continuing democratic potential of the Internet will require continuing vigilance and as the Internet becomes increasingly international and Western dominance of its user base wanes, presentation layer reforms are a key location for much-needed reforms.

We have been vociferous supporters for application neutrality. With application neutrality, Internet television, VOIP, and diverse operating systems and services run unimpeded by any interactions with technologies embedded within the data communications network. Expected convergences in digital communications make this principle increasingly crucial to the long-term growth and health of the Internet. DRM lobbying to protect copyright, irregardless of these technologies' impacts on fair use rights and as exemplified by the Anti-Counterfeiting Trade Agreement (ACTA) proceedings, have also made application neutrality a critical factor if we are to protect the open Internet. In much the same way that telephone systems are neutral transport mediums for voice communications, the Internet must remain free from discriminatory practices that privilege some applications, services, or features over others.

Further, the codices that applications use to give users access to online media, as exemplified by the looming problems of the proprietary H.264 video codec, must be free and open. Enclosing a popular medium behind a licensing bottleneck greatly undermines the future outlook for an open Internet. Continuing down this path will create new digital divides, this time thoroughly embedded into the very heart of the key applications most people use to access online content. Developers and content creators alike should be wary of license fees through this creation of a bottleneck in the end-to-end functionality of the Internet. Finally, we must ensure that new forms of "technological bundling" that create path dependencies we may not even know exist and contain extra costs that may not go into effect for years, must be prevented. Given the extensive and well-documented history of anticompetitive behavior within the high tech industry, agencies like the Federal Trade Commission should investigate how new agreements between content providers and some of the largest application development firms on Earth are detrimentally impacting consumer welfare and prevent them.

5. The Need for a New Paradigm

Taken together, these recommendations support a new paradigm for Internet policy. Whereas the trend in contemporary policy debates has hinged on prioritizing benefits to the major telecommunications companies (of which, both "new investment" and "job creation" have become political code for profit maximizing actions by government officials). This approach assumes that the efficiency and utility of government solutions can be viewed without considering the moral hazards, externalities, and opportunity costs of purposefully supporting the mechanisms that created today's gaping digital divides. Such emphases have had dire consequences and we suggest that the time has come to return to first principles, including a policy framework that limits vertical integration of network layers in order to preserve end-to-end functionality. As we have seen dramatically in recent years, allowing firms to become "too big to fail" has led both to highly distorted markets and the need for massive government intervention that could have been avoided had wiser policies been put in place *a priori*.

We also call for a business model neutral infrastructure that allows for public players such as municipalities and nonprofits, as well as public–private partnerships and private corporations and philanthropies, to provide Internet services. Too often, competition is lessened—and the options for consumers to receive broadband services artificially limited—by shortsighted rules, regulations, and laws crafted to lessen, rather than expand, competition. Maintaining a neutral network requires constant intervention when the providers are limited to specific business models or market players. This also suggests the need to change emphases from the "OSI hourglass model"[71] to a more nuanced approach where the bottleneck can appear at any layer of the network and key enclosures at one layer of the network can be leveraged to control utilization of other layers of the Internet. Announcements by Atheros [62] regarding Wi-Fi/cell phone chip sets are another example of leveraging control over one layer (in their case, the actual wireless chipset) to gain control over other layers (e.g., transport, applications, etc.). Gaining a better understanding of how this new market tactic works will be critical in the coming years since we have already seen anticompetitive behavior by chipset manufacturers in recent years.[72]

Traditionally, many researchers and advocates have focused on Layer 1 monopolies over the physical infrastructure of telecommunications networks. This is very much a hold-over from the Ma Bell and AT&T days that predate the complexity of the Internet. However, this layer may end up being a relatively modest area of concern given the other oligopolies that are developing in plain view—with the potential for corporate deal-making that is both more detrimental and more difficult to understand than anything we have seen previously. The openness battles we have seen thus far could be just a precursor to what is ahead. For example, there has been little debate about the terms of wireless chipset manufacturers even though two companies—Atheros and Broadcom—may control 80% of key Wi-Fi chipset markets. This means that a *de facto* duopoly control could leverage their control into almost every layer of the OSI model (from how physical communications are set up to which applications can run over them). This is an area that has far outpaced communications research and policy debates. As just one other example, the pooling of licenses for the H.264 codec is standardizing not only the technology, but the terms by which the technology itself can be used—thus already impacting the trajectory of innovation for next-generation video services.

Of course, other layers (often called "Layer 8") are not accounted for in the OSI model but should be investigated with equal vigor. These layers (from the "political" to "economic" to "finance") are beyond the scope of this chapter, but are clearly interacting and helping define the parameters of contemporary and future communications networks. Our analysis does not preclude these additional

---

[71] The OSI hourglass model traditionally conceptualizes the main bottleneck within Internet service provision as the transport layer (e.g., the TCP protocol). The notion is that most Internet traffic must go through this one facet and therefore it becomes an essential component for maintaining open communications. See Steve Deering, *Watching the Waist of the Protocol Hourglass*, August 2001, www.iab.org/documents/docs/hourglass-london-ietf.pdf.

[72] Intel has raised anti-trust concerns in both the E.U. (David Lawsky, EU to find Intel anti-competitive: sources, Reuters, May 10, 2009) and the U.S. (Donald Melanson, Intel and FTC settle charges of anticompetitive conduct, Engadget, August 4, 2010. http://www.engadget.com/2010/08/04/intel-and-ftc-settle-charges-of-anticompetitive-conduct/).

complexities (nor is it necessarily meant to supersede them); the key point is that communications systems and their democratic potential are far too precious to leave to the whims of the market.

5.1 Looking Forward

The Internet is not a luxury, it is a necessity. As a lifeline to crucial resources for hundreds of millions of families, businesses, educational institutions, municipalities, and NGOs, high-speed Internet should no longer be considered a commodity, but rather a critical utility on par with water and electricity. However, the social and economic value of the Internet depends on it remaining an open commons. Our national policies have focused on connectivity and universal access with limited discussion on affordable speeds, adoption, and competition. Furthermore, policymakers have taken a timid approach to ensuring the openness of wireless and wireline infrastructures. Not only have policymakers looked on while the United States's international ranking for Internet adoption has plummeted—a reality that is leaving tens of millions of Americans without broadband—but those who are online are facing an increasingly controlled experience that is both far more expensive and far slower than a growing list of countries overseas.

The current path will inevitably lead to a tiered society, one divided along unequal opportunities for education and work, as well as access to arts, culture, and the higher quality of life that these resources make possible. From a National Broadband Plan pushing for largely different speeds between urban populations and the remaining quarter of the population to investing in outdated technology [63], the current policy framework supports a *status quo* that has clearly failed.[73] These asymmetries run counter to the normative ideals embodied by the American Dream. This dream holds that our nation was not designed to maintain an aristocracy and a permanent underclass, but was supposed to be a meritocracy where anyone could succeed and everyone was given the tools they needed to create a better life for themselves and their families.

Broadband connectivity is the new critical infrastructure of the twenty-first century and is the platform on which a growing percentage of all media is transported. Universal broadband should be a national imperative, particularly for rural, low-income, and other underserved constituencies. It is too precious a resource to be solely overseen by an oligopoly of profit-driven corporations who must care for their

---

[73] "The goal is to restore the broadly supported status quo" Chairman Julius Genachowski, *The Third Way: A Narrowly Tailored Broadband Framework*, Federal Communications Commission, May 6, 2010. http://hraunfoss.fcc.gov/edocs_public/attachmatch/DOC-297944A1.pdf.

bottom line first and foremost. Our lack of foresight and attention to ongoing digital divides has already caused great harm to our economy, and it also threatens our future prospects, not just among marginalized constituencies within the United States, but also in relation to our international competitiveness. The United States has thus far failed to grasp the lesson that the past 10 years have been teaching us, but it is not too late to reform our efforts. If the U.S. government elevates affordable Internet access to a top priority and expands open access infrastructure requirements, all Americans will have an opportunity to better their lives and pay prices equivalent to many other countries. The U.S. government must create the same conditions that have fostered broadband competition in other countries—anything less will ensure that the price-gouging and substandard services that many consumers face will continue. Buildout of open access wireline infrastructures and increased unlicensed and opportunistic access to the public airwaves is a logical place to start. In addition to fostering increased competition, an open Internet architecture needs to be protected by maintaining interoperability, network neutrality, and nonproprietary protocols.

While much has been made of the Obama Administration's commitment to the Open Internet, the Genachowski FCC has been far too slow to demonstrate meaningful action. Definitive policy shifts are needed to create a more democratic communications system. Instead, we are faced by a crisis in leadership—in announcing a policy framework to redefine the regulation of the Internet, FCC Chairman Genachowski announced a national goal "to restore the broadly supported *status quo*."[74] Indeed, we stand at a critical juncture, one that may herald a new age of democratic potential. The key question is whether we will harness this untapped promise or allow the U.S. Internet to devolve into an increasingly dysfunctional morass. Taken together, our proposed measures will help create an open, affordable Internet available to all—one that preserves a twenty-first century public sphere as an open commons defined by its users.

REFERENCES

[1] B. van Schewick, Internet Architecture and Innovation, MIT Press, Cambridge, 2010.
[2] Y. Benkler, The Wealth of Networks: How Social Production Transforms Markets and Freedom, Yale University Press, New Haven, 2006.
[3] C. Jeff, Digital Destiny: New Media and the Future of Democracy, New Press, New York, 2007; J. Goldsmith, W. Tim, Who Controls the Internet? Illusions of a Borderless World, Oxford

---

[74] See The Third Way at http://www.broadband.gov/the-third-way-narrowly-tailored-broadband-framework-chairman-julius-genachowski.html.

University Press, New York, 2006; W. Tim, The Master Switch: The Rise and Fall of Information Empires, Knopf Press, New York, 2010.
[4] B. Yochai, Next Generation Connectivity: A Review of Broadband Internet Transitions and Policy from Around the World, Harvard Berkman Center for Internet and Society, 2010. http://cyber.law.harvard.edu/pubrelease/broadband ("Berkman Report").
[5] L. James, L. Chiehyu, Price of the Pipe: Comparing the Price of Broadband Service Around the Globe, New America Foundation, 2010. http://www.newamerica.net/publications/policy/price_of_the_pipe.
[6] B. Lennett, Dis-empowering users vs. maintaining Internet freedom: network management and Quality of Service (QoS), CommLaw Conspectus 18 (2009) 1. http://commlaw.cua.edu/res/docs/articles/v18/18-1/06-lennett-final.pdf, 144.
[7] Federal Communications Commission, OBI Working Paper #4: Broadband Performance. http://www.fcc.gov/Daily_Releases/Daily_Business/2010/db0813/DOC-300902A1.pdf, 2010 (accessed 17.09.2010).
[8] V.W. Pickard, Cooptation and cooperation: institutional exemplars of democratic Internet technology, New Media Soc. 10 (4) (2008) 625–645.
[9] Sascha Meinrath, Victor W. Pickard, The new network neutrality: criteria for Internet freedom, Int. J. CommLaw Policy 12 (2008) 225–243.
[10] V. Pickard, S. Meinrath, Revitalizing the public airwaves: opportunistic unlicensed reuse of government spectrum, Int. J. Commun. 3 (2009) 1052–1084.
[11] J. Habermas, The Structural Transformation of the Public Sphere. Trans. T. Burger, MIT Press, Cambridge, 1989.
[12] J. Habermas, Political Communication in Media Society—Does Democracy Still Enjoy an Epistemic Dimension? The Impact of Normative Theory on Empirical Research, 2006, International Communications Association, Dresden, Germany.
[13] D. Schiller, Digital Capitalism. Networking the Global Market System, MIT Press, Cambridge, 1999.
[14] L. Lessig, Free Culture: How Big Media Uses Technology and the Law to Lock Down Culture and Control Creativity, Penguin Books, New York, 2004.
[15] M. Perelman, Steal This Idea: Intellectual Property Rights and the Corporate Confiscation of Creativity, Palgrave, New York, 2002.
[16] M. Fransman, The New ICT Ecosystem: Implications for Europe, Kokoro, Edinburgh, 2007; J. Zittrain, The Future of the Internet and How to Stop It, Yale University Press, New Haven, 2008, (see Refs. [1,3]).
[17] L. Lessig, The Future of Ideas: The Fate of the Commons in a Connected World, Vintage, New York, 2002.
[18] B. Cammaerts, Critiques on the Participatory Potentials of Web 2.0, Commun. Culture Critique 1 (2008) 358–377.
[19] L. DeNardis, Protocol Politics: The Globalization of Internet Governance, MIT Press, Cambridge, 2009.
[20] M. Beckman, Beware the black market rising for IP addresses, Info World, May 3, 2010. http://www.infoworld.com/print/121729.
[21] J. Gartner, Verizon E-mail Embargo Enrages, Wired, January 10, 2005. http://www.wired.com/techbiz/media/news/2005/01/66226 (accessed 22.04.2010).
[22] N. Anderson, Verizon proposes settlement for class action lawsuit, Ars Tech. April 5, 2006. http://arstechnica.com/old/content/2006/04/6525.ars (accessed 22.04.2010).

[23] M. Delio, When everything was Spam to ISP, Wired, November 11, 2002. http://www.wired.com/science/discoveries/news/2002/11/56235 (accessed 22.04.2010).
[24] J. Cheng, AT&T: 4chan block due to DDoS attack coming from 4chan IPs, Ars Tech. July 27, 2009. http://arstechnica.com/telecom/news/2009/07/att-4chan-block-due-to-ddos-attack-coming-from-4chan-ips.ars.
[25] R. Singel, Wikipedia Bans Church of Scientology, Wired, May 29, 2009. http://www.wired.com/epicenter/2009/05/wikipedia-bans-church-of-scientology/.
[26] K. Zetter, Wiseguys indicted in $25 million online ticket ring, Wired March 1 2010. http://www.wired.com/threatlevel/tag/wiseguys/ (accessed 07.05.2010).
[27] G. Waclawsky John, IMS: a critique of the grand plan, Bus. Commun. Rev. 54 (2005) 55.
[28] M. Chris Riley, R. Topolski, The Hidden Harms of Application Bias, Free Press/New America Foundation Policy Brief, 2009, 6. Available at http://www.newamerica.net/publications/policy/the_hidden_harms_of_application_bias.
[29] A. Odylzko, Network neutrality, search neutrality, and the never-ending conflict between efficiency and fairness in markets, 2009. Available online at http://www.dtc.umn.edu/odlyzko/doc/net.neutrality.pdfA. Odylzko, Pricing and Architecture of the Internet: Historical Perspectives from Telecommunications and Transportation, 2004. Available online at http://www.dtc.umn.edu/odlyzko/doc/pricing.architecture.pdf.
[30] D. Hernandez, Mexico may cut off 30 million cellphones under new registration law, Los Angeles Times, April 8, 2010. http://latimesblogs.latimes.com/laplaza/2010/04/mexico-cell-phones.html.
[31] Shops track customers via mobile phone: signals given off by phones allow shopping centres to monitor how long people stay and which stores they visit, The Times, May 16, 2008. http://technology.timesonline.co.uk/tol/news/tech_and_web/article3945496.ece.
[32] N. Anderson, French anti-P2P law toughest in the World, Ars Tech. March 10 2009. http://arstechnica.com/tech-policy/news/2009/03/french-anti-p2p-law-toughest-in-the-world.ars (accessed 22.04.2010).
[33] N. Anderson, France passes Harsh anti-P2P three-strikes law (again), Ars Tech. September 15, 2009. http://arstechnica.com/tech-policy/news/2009/09/france-passes-harsh-anti-p2p-three-strikes-law-again.ars (accessed 22.04.2010).
[34] N. Anderson, Major labels go Bragh? Irish Judge allows 3 strikes, Ars Tech. April, 2010. http://arstechnica.com/tech-policy/news/2010/04/major-labels-go-bragh-as-irish-judge-allows-3-strikes.ars?utm_source=rss&utm_medium=rss&utm_campaign=rss (accessed 22.04.2010).
[35] C. Doctorow, France's anti-piracy goon squad pirates the font in its logo, Boingboing, January 12, 2010. http://www.boingboing.net/2010/01/12/frances-anti-piracy.html.
[36] J. Cheng, Comcast settles P2P throttling class-action for $16 million, Ars Tech. December 22, 2009. http://arstechnica.com/tech-policy/news/2009/12/comcast-throws-16-million-at-p2p-throttling-settlement.ars; N. Anderson, Just like Comcast? RCN accused of throttling P2P, Ars Tech. April 2010. http://arstechnica.com/tech-policy/news/2010/04/just-like-comcast-rcn-accused-of-throttling-p2p.ars.
[37] E. Lin, SFN report: Japan leads with 3G penetration, SFN Blog May 10 2010.http://www.sfnblog.com/industry_trends/2010/05/sfn_report_japan_leads_with_3g_penetrati.php (accessed 17.09.2010).
[38] S. Lawson, Vonage says ISP blocked its calls: a broadband provider prevented customers from using its service, company says, PC World, February 16, 2005. http://www.pcworld.com/article/119695/vonage_says_isp_blocked_its_calls.html.
[39] J. Hu, Comcast takes hard line against spam, Cnet, June 10, 2004. http://news.cnet.com/2100-1038_3-5230615.html.

[40] J. Hu, Comcast reports 35 percent decline in spam, Cnet, June 29, 2004. http://news.cnet.com/2100-1038_3-5251909.html.
[41] D. Roth, The Dark Lord of Broadband Tries to Fix Comcast's Image, Wired, January 19, 2009. http://www.wired.com/techbiz/people/magazine/17-02/mf_brianroberts.
[42] B. Stone, Comcast: we're delaying, not blocking, BitTorrent traffic, New York Times October 22, 2007. http://bits.blogs.nytimes.com/2007/10/22/comcast-were-delaying-not-blocking-bittorrent-traffic/.
[43] J. Cheng, Say hello to.كوم as domain names go truly global, Ars Tech. October 30 2009. http://arstechnica.com/web/news/2009/10/domain-extensions-go-global-goodbye-com-welcome.ars.
[44] C. Griffiths, Domain helper service: here to help you, Comcast Voices July 9 2009. http://blog.comcast.com/2009/07/domain-helper-service-here-to-help-you.html (accessed 06.05.2010).
[45] E. Stolterman, Creating community in conspiracy with the enemy, in: L. Keeble, B. Loader (Eds.), Community Informatics, Routledge, New York, 2002, p. 45.
[46] G. Kaufman, AT&T Admits It Edited Webcasts Before Pearl Jam's, MTV, August 13, 2007. http://www.mtv.com/news/articles/1566946/20070813/pearl_jam.jhtml (accessed 10.05.2010); A. Liptak, Verizon Blocks Messages of Abortion Rights Group, New York Times September 27 2007. http://www.nytimes.com/2007/09/27/us/27verizon.html (accessed 10.05.2010).
[47] M. Cooper, Open Architecture as Communications Policy, Center for Internet and Society, Stanford, 2004; B. Kahin, J. Keller, Coordinating the Internet, MIT Press, Cambridge, 1997.
[48] P.J. Louis, Telecommunications Internetworking, in: McGraw-Hill Professional, New York, 2000.
[49] C. Nobel, Verizon to launch its first Bluetooth phone, eWeek July 26 2004. http://www.eweek.com/c/a/Mobile-and-Wireless/Verizon-to-Launch-Its-First-Bluetooth-Phone/.
[50] S. Segan, Motorola V710, PC Mag. August 26, 2004. http://www.pcmag.com/article2/0,2817,1639784,00.asp.
[51] J. Kincaid, Apple is growing rotten to the core: official Google Voice App Blocked from App Store, Tech Crunch, July 27, 2009. http://techcrunch.com/2009/07/27/apple-is-growing-rotten-to-the-core-and-its-likely-atts-fault/.
[52] S. Higginbotham, Apple brings 3G VoIP to the iPhone, GigaOM, January 28, 2010. http://gigaom.com/2010/01/28/apple-brings-3g-voip-to-the-iphone/.
[53] B.X. Chen, Adobe Apps: easier to pass through the 'i' of a needle? Wired, April 8, 2010. http://www.wired.com/gadgetlab/2010/04/iphone-developer-policy/ (accessed: 10.05.2010).
[54] D. Meredith, J. King, S. Meinrath, J. Losey, Mobile Devices are Increasingly Locked Down and Controlled by the Carriers: How Cell Phone "Customization" Undermines End-Users by Redefining Ownership, New America Foundation, 2010. http://oti.newamerica.net/blogposts/2010/mobile_devices_are_increasingly_locked_down_and_controlled_by_the_carriers-38418.
[55] J.C. Waclawsky, Closed systems, closed architectures and closed minds, Bus. Commun. Rev. (2004) 61.
[56] V. Pickard, Media Democracy Deferred: The Postwar Settlement for U.S. Communications, 1945–49, 2008, PhD Dissertation, University of Illinois.
[57] J. Horrigan, Home Broadband Adoption 2009: broadband adoption increases, but monthly prices do too, Pew Internet and America Life Project, June 2009.
[58] M. Buchanan, Apple Genius Bar: iPhones' 30 percent call drop is "normal" in New York, Gizmodo, September 29, 2009. http://gizmodo.com/5370493/apple-genius-bar-iphones-30-call-drop-is-normal-in-new-york, (accessed 29.09.2010).
[59] C. Santivanez, R. Ramanathan, C. Partridge, R. Krishnan, M. Condell, S. Polit, Opportunistic Spectrum Access: Challenges, Architecture, Protocols, WiCon'06, August 2–5, 2006. http://www.ir.bbn.com/ramanath/pdf/osa-wicon06.pdf, Boston, MA, USA.
[60] B. Scott, M. Calabrese, Measuring the TV "White Space" Available for Unlicensed Wireless Broadband, Free Press and New America Foundation, 2005.

[61] M. Cooper, Open Communication Platforms: Cornerstone of Innovation and Democratic Discourse in the Internet Age, J. Telecommun. High Technol. Law 1 (2003), (see Ref. [1]).
[62] M. Ricknäs, Atheros Turns Cell Phone Into Access Point, PC World, January 10, 2009. http://www.pcworld.com/printable/article/id,156873/printable.html.
[63] J. Losey, C. Li, S. Meinrath, Broadband Speeds in Perspective: A Comparison of National Broadband Goals from Around the Globe, New America Foundation, 2010N. Anderson, 4Mbps broadband for all to cost $23 billion, won't use fiber, Ars Tech., May, 2010. http://arstechnica.com/tech-policy/news/2010/05/4mbps-broadband-for-all-to-cost-23-billion-wont-use-fiber.ars.

Online Advertising

AVI GOLDFARB

Rotman School of Management, University of Toronto, Toronto, Ontario, Canada

CATHERINE TUCKER

MIT Sloan School of Management, Cambridge, Massachusetts, USA

Abstract

This chapter explores what makes online advertising different from traditional advertising channels. We argue that online advertising differs from traditional advertising channels in two important ways: measurability and targetability. Measurability is higher because the digital nature of online advertising means that responses to ads can be tracked relatively easily. Targetability is higher because data can be automatically tracked at an individual level, and it is relatively easy to show different people different ads. We discuss recent advances in search advertising, display advertising, and social media advertising and explore the key issues that arise for firms and consumers from measurability and targetability. We then explore possible public policy consequences, with an in depth discussion of the implications for consumer privacy.

1. Introduction 290
2. What Is Different About Online Advertising? 292
    2.1. Measurability 292
    2.2. Targetability 294
    2.3. Pre-Internet Targeting 294
    2.4. How Firms Target Online 295
3. Different Types of Online Advertising 298
    3.1. Display 298

3.2. Search Advertising 300
3.3. Social Media Advertising 302
3.4. Other Forms of Online Advertising 303
4. Consequences of Measurability and Targetability on the
Online Advertising Industry 303
   4.1. Accountability 304
   4.2. Two-Sided Platforms and Pricing 304
   4.3. The Effect of Online Advertising on Traditional Forms of Advertising . . . 305
5. Regulation 306
   5.1. Privacy 308
6. Conclusions and Avenues for Future Research 310
References 311

1. Introduction

Since the first banner ad was shown in October 1994, online advertising has grown at a furious rate. By 2009, online ads accounted for $22 billion in spending. Online advertising is also important for what it enables. In the United States alone, Web sites supported by advertising represent 2.1% of the total U.S. gross domestic product (GDP) and directly employ more than 1.2 million people [1].

Figure 1 from IAB [2] compares the growth of internet advertising for its first 15 years (1995–2009) relative to cable television (1980–1994) and broadcast television (1949–1963), in current inflation-adjusted dollars. It is clear that online advertising has grown much faster, especially in the past 5 years, than traditional advertising channels did in their first 15 years. However, it is less clear whether and how online advertising differs substantively from other, more established advertising channels. It is this question we explore in this chapter.

Specifically, we draw on a wide literature to argue that online advertising differs from traditional advertising channels in two important ways: measurability and targetability. Measurability is higher because the digital nature of online advertising means that responses to ads can be tracked relatively easily. Targetability is higher because data can be automatically tracked at an individual level, and it is relatively easy to show different people different ads. Measurability and targetability are key features of all kinds of online ads, including display ads, search ads, and social media ads. These features have important implications for the nature of the online ad market. For example, measurability means that advertising agencies and platforms

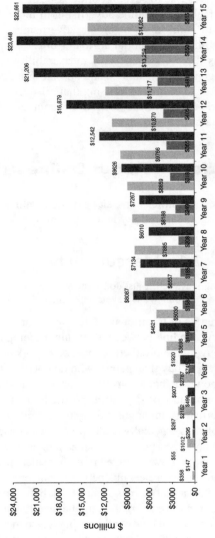

FIG. 1. Ad revenue for the first 15 years of online media, cable TV, and broadcast TV. Source: IAB [2].

are more accountable. More generally, we will argue that the combination of measurability and targetability has implications for how companies behave in online advertising markets and gives rise to several policy issues.

We discuss these two features of online advertising in more depth in Section 2. Section 3 discusses the different kinds of online advertising and documents how measurability and targetability are essential attributes of each type of online advertising. Section 4 explains the consequences of measurability and targetability on advertisers, advertising agencies, consumers, and ad platforms. Section 5 discusses several policy issues related to online advertising that arise from measurability and targetability, including privacy concerns. Section 6 concludes with some areas that seem ripe for further research.

2. What Is Different About Online Advertising?

We discuss in turn the two dimensions along which online advertising differs from traditional advertising: measurability and targetability.

2.1 Measurability

I know half my advertising is wasted, I just don't know which half
John Wanamaker, department store innovator, 1838–1922.

Measuring the effectiveness of traditional advertising is hard, for two reasons.

First, it is hard to observe the link between a consumer seeing an ad and the same consumer subsequently buying the product. It appears to work (i.e., people who see the ads might be more likely to buy than people who do not), but the firm cannot see how. The firm does not know whether a consumer was motivated to buy because of a particular newspaper ad, or because of a TV ad, a specific billboard, or their new radio jingle. For long-term advertising campaigns that try to build affection over time for a particular brand, this problem is especially acute. Macy's can observe who uses their 20% off coupon in the Sunday paper, but Budweiser cannot observe whether their Bud Light ad shown during the Superbowl is linked to higher sales in the long run.

Second, even if firms can observe a clear link between someone seeing an ad and then buying the product, it is not clear that there is a causal link between the two. This is the classic endogeneity problem of advertising. It could be merely that the kind of person who was chosen to be exposed to that kind of advertising is also more likely to purchase the product. For example, even with the coupon, Macy's cannot observe whether the people who used the coupon would have bought anyway. Both these problems are well recognized as limitations of empirical studies in offline advertising [3].

By contrast, online advertising is inherently measurable. The digital nature of online advertising means that individual responses to ads can be easily recorded. For example, the effectiveness of many forms of online advertising can be measured by whether someone clicks on an ad. Often, through the use of cookies, IP addresses, and other tracking technologies, advertisers can go beyond this simple click metric and observe directly whether users engage in a certain online action (such as an online purchase or subscribing to receive more information) after being exposed to an ad.

In addition to the digital nature of online advertising, another important feature of internet advertising which facilitates measurability is the way it permits randomized field tests. These are known in the industry as "a/b tests." These tests randomly expose one group of consumers to a particular online ad, while another group of consumers is not exposed to the ad, and compare their subsequent responses. This randomization allows precise estimates of the effect of advertising without the usual endogeneity concerns that have bedeviled previous work, because the consumers in the test are in expectation identical. Measuring effectiveness is straightforward for ads whose call to action involves clicking through on the ad. However, combined with surveys, such tests also allow for examination of brand-focused advertising. For example, Goldfarb and Tucker [25] use detailed survey data from online randomized field tests to examine how well online ads work. These online surveys (which appeared in pop-up windows) asked questions about brand favorability and purchase intent *immediately* upon a person having been exposed (or not) to a particular piece of display advertising.

While these surveys have the advantage of experimental design, the response rates are low. Only a small fraction of Web site visitors are willing to click on the survey pop-up, and these are likely to be the Web site visitors that are more susceptible to online advertising (because they responded to the ad for the survey). This means that overall advertising effectiveness measures may be biased upward. Fortunately, this bias is broadly consistent across campaigns and therefore these field studies do a good job of measuring the relative performance of advertising campaigns. Such tests therefore are a substantial improvement over the offline state-of-the-art in benchmarking advertising campaign effectiveness.

One interesting side note is that for the first time, firms can also measure directly the extent to which online advertising is *ineffective* and the extent to which consumers try to avoid it. As discussed by Dreze and Hussherr [5], many consumers purposely avoid directly looking at digital ads when confronted with them. This lack of attention is not unique to online advertising. Ritson [6] documents ethnographic research that explores the extent to which people ignore television advertising. What is unique to online advertising is that researchers can cheaply and easily access quantitative measures of how ineffective advertising is. One of the challenges of any measurement of online advertising is that positive actions and effects are reasonably sparse. Few people, on seeing a pop-up ad click-through. Even fewer click-through and buy. Trying to measure what ad features affect click-through rates that are less than one-tenth of 1% presents unique challenges for researchers [7] because current discrete choice models generally do not handle data sparsity well. This challenge is particularly acute when measuring the effects of brand advertising. Researchers like Reiley and Lewis [8] have had to collect millions of incidents of ad exposure to be able to measure any effects precisely.

2.2 Targetability

Ad targeting occurs when an advertiser selects a particular subset of potential viewers of the ad to show the ad to, and displays the ad to that subset rather than to everyone using the media platform. An example would be choosing to advertise, not to the hundreds of millions of Facebook users in general, but only to those Facebook users who are female *and* aged between 26 and 54 *and* list on their profile that they like the poet Maya Angelou. No newspaper can offer this level of targeting. The targetability of online advertising can be thought of as reducing the search costs for advertisers of identifying consumers. Targeting ads has always been known to be desirable, but internet advertising has two primary advantages over offline advertising. First, the internet has made it virtually costless for advertisers to collect huge amounts of customer data. In contrast, the costs of collecting detailed enough individual-level data to target offline, for example, through "direct response mail," have generally been prohibitive except in a very few circumstances. Second, internet technology makes it relatively easy to serve different customers different ads because packets are sent to individual computers. In contrast, with current technology, splitting cable TV ads across consumer types or sending hundreds of different newspaper ads to different households is prohibitively costly.

Between the data collection and the serving of ads, targeting requires sophisticated algorithms and data processing capabilities to serve the right ads to the right people. There is nothing inherent in internet technology that makes online advertising better at this computational step, but the data collection and individual-level sending of ads make this computational step particularly fruitful.

2.3 Pre-Internet Targeting

The focus on targeting of online advertising revives a theoretical literature on the potential benefits of targeted advertising. In their classic model of informative advertising, Grossman and Shapiro [9] model both the need for and the ease of targeting. In their model, the potential customers for a product can be interpreted as the degree of differentiation (t) in the market and the ability to target can be interpreted as the fraction of messages sent to consumers interested in the product group $(1 - \alpha)$. Perhaps owing to the limitations of targeting technology when the paper was written, they interpreted this parameter as the cost of advertising. However, the increasing sophistication of advertising technologies has led marketing researchers to revisit the importance of targeting. For example, Iyer et al. [10] describe how targeting can improve equilibrium profits in the industry.

Chandra [11], in a study of pricing of newspapers, shows that in old media the ability to price ads was a function of the similarity of user characteristics. By contrast, in online media platforms, users do not have to be similar to take advantage of targeting. Instead, the electronic automation of the serving of online ads means that users for a media platform can be very different and be shown very different ads.

Though, as described by Gal-Or et al. [12], there have been a few attempts in older media models to target advertising based on demographics, we argue in the next section that what makes internet advertising unique is the ability to target based on *actions* or *intent*.

2.4 How Firms Target Online

Many different types of online targeting are used today. Table I summarizes the broad categories of targeting used by advertisers in 2010. There are of course many hybrids of these techniques. The targeting techniques used vary across different

TABLE I
DIFFERENT TYPES OF ONLINE TARGETING

Name	Description
Contextual targeting	Ad is matched to content it is displayed alongside
Behavioral targeting	Use prior click-stream data of customer to determine whether they are a good match for the ad. Scope generally depends on whether ad network or Web site publisher controls which ads get displayed
Retargeting (search)	Online ad is shown to user who previously searched using a particular search term
Retargeting (Web site)	Online ad is shown to user who previously visited a Web site but did not "convert"
Real-time targeting	Advertiser has power to decide in "real time" whether to serve an ad to a customer based on data the Web site shares with them about that user.
"Look-alike" targeting	Targeting based on users having similar characteristics to current customers
"Act-alike" targeting	Targeting based on users having click-through paths which resemble successful conversions
Demographic targeting	Publisher uses data that customer has volunteered such as age, gender, location, and interests to choose whom to display ads to

advertising formats. Generally, search engines can rely on context-based advertising where ads are displayed based on the search term typed into the search engine. This means that if someone searches for information on Aviation Accident Attorneys, they see only ads that were bid on by advertisers who are advertising some form of aviation accident legal service. Search engine ads are therefore placed in the context of the stated intent of the user. Other Web sites, for example, theknot.com, whose services are focused on a very specific topic (in this case weddings) attract advertisers who rely primarily on the fact that the user base of theknot.com will be interested in buying wedding-related products and services. By contrast, display advertising on Web sites where the content is not easily monetizable or matched to consumers (such as news, Web services, and entertainment) generally use various forms of targeting based on user behavior rather than content to try to match ads with consumers.

Behavioral targeting in its most basic form uses data from a user's clickstream to try to work out whether the user is interested in a particular service. For example, if Yahoo! observes a person looking at Infiniti cars on Yahoo! Autos, they can use this information to then serve them ads for Infiniti cars when they look at news stories on Yahoo! News. This kind of targeting can be conducted either by a single content provider or by an ad network that provides ads to many different Web sites. An ad network can use the information that someone is researching a certain Caribbean vacation destination on one Web site, to serve ads for resorts in that destination when a user starts reading celebrity gossip on a separate Web site. Murthi and Sarkar [13] were among the first to highlight research questions related to how internet technology would enable more personalization of the ads we see.

Another increasingly important form of targeting is "retargeting." This means that a Web site tracks whether a customer has expressed in interest in a particular product or service. If the customer fails to purchase the product, then the advertiser shows new ads to those specific customers in an effort to get them to return to the Web site. This kind of targeting is also possible in search advertising, where search ads can be replayed after the initial search and the user is searching using different search terms.

The theoretical and empirical literature has in general not distinguished between these different forms of targeting. One exception is Goldfarb and Tucker [14], which explicitly looks at the benefits of contextual targeting in search engine advertising when offline targeting alternatives are not available. Another exception is Lambrecht and Tucker [15], which looks at the benefits of ad retargeting at different points in the customers' purchase decision process. However, other categories of online ad targeting such as "behavioral targeting," "look-alike," "act-alike," and demographic targeting have not been studied explicitly in the academic literature, to the authors' knowledge, perhaps because of their novelty.

An important development in ad targeting has been the advent of "real-time targeting." The distinction between this and the two methods discussed above is that real-time targeting puts the anonymous data about users into advertisers' hands directly rather than the media platform deciding who to serve an ad to at any one time. The label "real-time" emphasizes the fact that this type of targeting is implemented by real-time auctions based on recent movements by a certain user.

These different targeting methods, generally of course, require media platforms to collect comprehensive data on the Web pages that customers have previously browsed. This data is commonly called click-stream data. Typically advertisers and Web site owners track and identify users in this click-stream data using a combination of cookies, flash cookies, and Web bugs.

Web bugs are 1 × 1-pixel pieces of code that allow advertisers to track customers remotely.[1] Web bugs are different from cookies, because they are designed to be invisible to the user and also are not stored on a user's computer. This means that without inspecting a Web page's underlying html code, a customer cannot know they are being tracked. Web bugs allow advertisers to track customers as they move from one Web page to another. They also allow advertisers to document how far a Web site visitor scrolls down a page. Combined, this means they are very helpful in determining Web site visitor interests. Web bugs are very widely used on commercial Web sites [17]. Murray and Cowart [18] found that 96% of Web sites that mentioned a top 50 brand (as determined by the 2000 FT rankings) had a Web bug.

A cookie is simply a string of text stored by a user's Web browser. These allow firms to track customers' progress across browsing sessions. This can also be done using a user IP address but cookies are generally more precise, especially when IP addresses are dynamic as is the case for many residential internet services.

Advertisers tend to use cookies and Web bugs in conjunction because of the challenge of customer deletion of cookies. For example, 38.4% of respondents in a recent survey said they deleted cookies each month.[2] Therefore, Web bugs (which a user cannot avoid) have been increasingly used in conjunction with, or even in place of, cookies in targeting advertising [8]. Web bugs also have greater reach in terms of tracking ability than cookies, because they can be used to track consumers' scrolling within a Web page. Advertisers may also use a flash cookie is an alternative to a regular cookie. A flash cookie differs from a regular cookie in that is saved as a

---

[1] These are also sometimes referred to as "beacons," "action tags," "clear GIFs," "Web tags," or "pixel tags" [16].

[2] Burst Cookie Survey: Consumers don't understand, say maybe useful, but some delete anyhow, marketing vox, June 2003.

"Local Shared Object" on an individual's computer, making it harder for users to delete using regular tools on their browser.

There are also other even more comprehensive ways of obtaining user browsing behavior. For example, Phorm, an advertising agency in the United Kingdom, used deep-packet inspection to evaluate the content of packets sent between the user and the ISP, at the ISP level. This is different from most behavioral targeting where users are tracked across a subset of Web sites only. Researchers such as Clayton [19] argued that this is akin to "warrantless wiretapping" because theoretically the firm can observe the content of private communications.

All these data types, whether cookies, clickstreams, Web bugs, and so on are used in several ways to better target ads to customers. By examining past surfing and click behavior, firms can learn about current needs as well as general preferences. Web sites and ads can even morph to fit their users' cognitive styles [20,21]. For example, advertisers can use the information given by user clickstreams to assess whether users prefer visual images or text information.

3. Different Types of Online Advertising

There are many different types of online advertising. Table II summarizes how spending was distributed across the different types in the United States in 2009.

3.1 Display

"Display advertising" includes display banner ads, media-rich ads, and digital video ads. Display advertising is the major mechanism by which Web pages that provide nonsearch content finance their Web sites.

TABLE II
HOW $22.7 BILLION WAS DISTRIBUTED ON ONLINE ADVERTISING IN 2009

Type of ad	Percentage spent
Search	47
Display banner ads	22
Classifieds	10
Rich media	7
Lead generation	6
Digital video	4
Sponsorship	2
Email	1

Source: IAB [2].

In the early years of the World Wide Web, banner ads were the predominant advertising medium. Early marketing research, such as Chatterjee et al. [22], focused on modeling "click-through rates." Manchanda et al. [23] used detailed data from a single health and beauty firm to show the effect of clicks on purchase behavior. Click-through rates were a substantial improvement over measures of offline advertising effectiveness that tried to tease out causal effects by examining time series data on advertising and sales.

However, click-through rates for banner ads have fallen sharply since the early 2000s, perhaps because they are less novel to the consumer, and because there are many more banner ads. The fall in click-through rates has led researchers to try and understand the role of banner-based brand advertising on customer behavior more generally, without relying on analysis of click-through rates. For example, Danaher and Mullarkey [24] show that the longer a person is exposed to a Web page containing a banner ad, the more likely they are to remember that banner ad. Reiley and Lewis [8] use a large field experiment with over 1 million ad views to assess whether banner-based brand advertising where consumers are just exposed to the ad but do not click-through can affect store purchases offline. They find small but positive effects, particularly for older consumers. Offline, running such field experiments, even at a much smaller scale, is extremely difficult. Internet technology meshed with an offline point of sales system facilitated the experimental design and the tracking of people after they saw the ad.

In response to falling effectiveness (as measured by, e.g., click-through rates), display advertising has evolved substantially beyond the "electronic billboard." Display advertising (which encompasses plain banner ads as well as new rich media and video) is now a multibillion-dollar market where ads include many sophisticated visual and auditory features that make ads more obtrusive and harder to ignore. Another orthogonal but noteworthy development has been Google's development of a highly profitable nonsearch display advertising division (called "AdSense") that generates an estimated $6 billion in revenue by displaying plain content-targeted text ads. For both highly visible ads and plain content-targeted ads, advertisers carefully measure consumer response through click-throughs, through purchase intention, and through other online behavior. Goldfarb and Tucker [25] explore how well these divergent strategies work for online advertising, and how consumer perceptions of obtrusiveness and privacy influence their success or lack of it, both independently and in combination. They find that there is a negative relationship between targeting and the use of "media-rich" features in advertising. In other words, ads that are both targeted and obtrusive are less effective at increasing purchase intent than ads that are either just targeted (like AdSense ads) or just obtrusive. That may explain why these separate advertising strategies have developed.

Overall, both measurability and targetability are important influences on the development and effectiveness of online display advertising.

3.2 Search Advertising

Search engine advertising is enormously important because of the extent to which consumer search on the internet across Web sites is reasonably limited [26]. Search engines are a key gateway to all Web sites on the internet, and as such provide a crucial advertising venue. "Search ads" or "paid search" are the ads that appear alongside search results after a consumer types a search term or "keyword" into a search engine. Typically, these ads are placed above or to the right hand side of the "organic" or main search results.

In 1998, Goto.com introduced two features that set search engine advertising apart from other advertising markets[3]: (a) atomized pricing and display of ads based on search terms or "keywords"; and (b) automated, electronic "position auctions" to price the advertising slots for these keywords.

Because each search is a statement of intent, ads that are targeted and priced to these statements of intent are highly effective. A search engine ad is a direct response to a statement of intent and therefore targets potential customers at exactly the time they are looking for something. Goldfarb and Tucker [14] show that targeting plays an important role in generating the final prices of the auctions.

While targeting is a key aspect of search engine advertising, much of the academic literature has focused on auctions. This is likely because there was already a deep economic literature on auction pricing. There have been a variety of studies of specific features of the search advertising auction mechanism, such as the importance of position given the limited number of slots [27,28]; the interaction between the list of search results and the list of sponsored links on the search page [29]; and whether advertisers pay for clicks or impressions [30] or clicks or actions [31]. Generally, auctions enable search engines to target ad prices narrowly without active involvement in setting the price for each specific keyword. The auction ensures that prices reflect the demand in the narrow advertising segment targeted by the keyword. Athey and Ellison [32] provide a good summary of the literature.

The economic auctions literature made several early contributions to the development of search engine advertising auctions that greatly improved the efficiency of, and revenue from, these auctions. For example, Goto.com's auctions were "first-price" auctions, where bidders paid the amount that they bid. These first-price

---

[3] Goto.com was renamed Overture in 2001 and purchased by Yahoo in 2003. Before 1998, search engine ads were priced by impressions and demographics.

auctions meant that advertisers would benefit by continuously updating their bids to be 1 penny higher than their competitors. Advertising became a time-intensive activity where advertisers would need to check the site regularly. Economic theory shows that first-price auctions lack an equilibrium, and therefore we see continuously changing prices and bids. This led to "price cycles" similar to those observed in retail gasoline markets [33]. Instead, a "generalized second price" auction was developed (discussed in, for e.g., Edelman et al. [27] and Varian [28]) that had a stable equilibrium in which advertisers would pay the bid of the next highest bidder. Edelman et al. [27] and Varian [28] also showed that ranking advertisers by the expected revenue (the price times a measure of the expected click-through rate or the "quality score") would generate higher revenues for the advertising platform (i.e., Google, Yahoo, or Bing).

There has been relatively little empirical work on search engine advertising and auctions because of the very complex nature of the position auction mechanism. One exception is Yao and Mela [34], who make the problem tractable by looking at a hybrid auction mechanism developed for an individual Web site and suggest that dynamics are important in understanding firm bidding behavior. Another question addressed by academics is how the position of a search ad affects its performance. Agarwal et al. [35] show that it is not always most profitable to be placed at the top of the position auction, but instead it is often better to be placed in the middle. This was confirmed by Ghose and Yang [36] using a Bayesian model.

Also of interest is the relationship between organic search results (sometimes called "algorithmic search") and search ads (or "sponsored search"). Before the dot-com shakeout, as described by Bradlow and Schmittlein [37], there were six search engines with different algorithms and funding strategies. A decade later there are two major search engines, Google and Bing, with Google the much larger player. For both search engines, ads appear at the same time as the organic search results that are ordered by the relevance of the results to the search. The display of both organic and sponsored listings on the same page raises the question of whether organic and paid listings are complements or substitutes. To answer this question, Yang and Ghose [38] analyze the relationship between organic search and paid search rankings by a particular advertiser and find a positive interdependence between paid and organic search listings. Chiou and Tucker [39] examine this idea at a finer level and find that there is a positive relationship for search queries about generic products, but a negative relationship for searches that use brand names. Rutz and Bucklin [40] use a dynamic model to show that paid ads can actually have positive spillovers for later search behavior (i.e., they are complements to future search). It is important to resolve whether organic and paid search results are complements or substitutes because, as White [41] discusses, search advertising may create incentives for search engines to reduce the quality of the organic listings if paid and organic links are substitutes. In each of the empirical studies, measuring ad effectiveness was possible due to the digitally automated nature of the advertising medium.

3.3 Social Media Advertising

Marketers have always recognized that word-of-mouth is important. However, before online media evolved it was simply very expensive and difficult to measure it or for firms to proactively encourage it. Recent work by Trusov et al. [42] has suggested that online word-of-mouth can be more effective than traditional marketing campaigns. This may be because when advertising messages are transmitted over social networks there is a potential for amplification of the message because social networks are so diffuse [43,44].

Online social media advertising also allows firms explicitly to direct word-of-mouth [45,46]. New advertising agencies, such as Bzz Agent and Tremor, now exist to encourage positive online word-of-mouth. Of particular importance here are changes in regulation in October 2009 concerning Federal Trade Commission Guidelines Concerning the Use of Endorsements and Testimonials in Advertising. These were directed specifically at clarifying what constitutes an endorsement when the message is conveyed by bloggers or other "word-of-mouth" marketers. Such regulatory changes may allow researchers to start to estimate the effect of authentic compared to inauthentic word-of-mouth.

A specific area of online social media marketing that is growing in importance is advertising on social networks such as Facebook. Table III summarizes the average time spent each month by users on social network Web sites. What is striking is just

TABLE III
GLOBAL REACH OF SOCIAL NETWORKING SITES

Country	Unique audience (000)	Time each month per person (h:min)
United States	142,052	6:09
Japan	46,558	2:50
Brazil	31,345	4:33
United Kingdom	29,129	6:07
Germany	28,057	4:11
France	26,786	4:04
Spain	19,456	5:30
Italy	18,256	6:00
Australia	9895	6:52
Switzerland	2451	3:54

Source: The Nielsen Company, January 2010.

how long users are spending on these sites relative to other kinds of Web sites. For example, in the United State people spend on average over 6 h a month on these sites, which is more than double the time spent on other popular types of Web sites such as portals or search engines. Befitting the newness of this kind of media, empirical research is limited. Aral and Walker [47] use data from a field experiment on Facebook concerning the distribution of a new application, and find that mandating a word-of-mouth response from consumers upon its adoption was more effective than making it discretionary. Tucker [48] looks at the effect of advertising on Facebook, using a field experiment conducted by a nonprofit to measure the extent to which information that users provide on social networks can be used to both target ads and personalize the content of ads.

3.4 Other Forms of Online Advertising

There are four other categories described in Table II: classifieds, lead generation, sponsorship, and email. Generally, there has been less academic research into these forms of advertising. One reason is that there is less data. For example, Craiglist.org is a powerhouse in the online classifieds world, receiving around 9 billion pageviews per month [49]; however, Craigslist is notoriously protective of its users' privacy and therefore it does not share data. Two papers that do explore other types of online advertising are Tucker and Zhang [50], who look at how Web sites devoted to online classifieds can grow their user base by advertising how many people are using the Web site, and Ansari and Mela [51], who emphasize the role of email in facilitating customization.

4. Consequences of Measurability and Targetability on the Online Advertising Industry

In this section, we discuss the consequences of measurability and targetability on the online advertising market. First, we discuss how measurability implies accountability and the consequences of accountability for participants in online advertising markets. Second, we discuss the role of targeting in two-sided markets such as advertising. Third, we explore substitution between online and offline advertising and document the key role that targetability plays in this substitution.

4.1 Accountability

Accountability is the ability of the advertiser to now evaluate the performance of advertising media and advertising agencies on the basis of whether they have "measured" success. This is a direct consequence of measurability. It is a new characteristic of online advertising because for other broadcast media, measurement is so problematic that it is difficult to hold an advertising platform or an ad agency to account if an advertising campaign fails because it is not clear it has failed. Using once again the newspaper example, newspaper advertisers find it hard to reduce their ad buys in response to a lower response rate, because only a few of the responses they receive can be directly and exclusively attributed to the newspaper ad.

This change in accountability may be an important factor in the rise of disintermediation for some forms of online advertising campaigns [52]. Previously, most advertisers outsourced their media buys to a traditional advertising agency. However, new online advertising models have led advertisers to be able to purchase advertising directly from content providers in ways that were not previously possible. Google and other search engines have faced resistance from traditional advertising agencies, who resent the fact that Google makes it easy for advertisers to buy ads for search engines directly rather than going through the advertising agency and paying a traditional commission.

Last, online advertising platforms may, in the future, be held to different standards of accountability by governments than other forms of advertising, simply because of the amount of data that is collected and the extent to which this threatens users' privacy.

There are also other key issues concerning liability that stem from this increased accountability. There is, for example, the matter of the liability of portals for materials advertised on their sites—the key litigation here is *United States Court of Appeals, Ninth Circuit—433 F.3d 1199, Yahoo! Inc. v. LICRA and UEJF, January 12, 2006*. Further, the fact that with "ad network" forms of online advertising, owners of individual Web sites do not necessarily know and cannot be held legally liable for the materials advertised on their Web site. Nonnetwork ads provide more accountability; "ad network" ads may provide less accountability than nonnetwork ads do, by adding back in an intermediary in the form of the organizer of the network.

4.2 Two-Sided Platforms and Pricing

Online advertising markets have some of the typical features of a two-sided media platform [53–56]. Quite simply, a two-sided platform is anything that facilitates the exchange of a product, service, or information between two disparate groups of users. What makes such platforms interesting from an economics perspective is that they generate network effects because each group only wants to use the platform because of the presence of the users from the other group. For example, no advertiser would want to place a search ad on Yahoo!'s front page if there were not millions of consumers visiting it.

The two-sided structure with advertisers and viewers, combined with an ability to target to narrow audiences, means that the structure of online advertising-supported markets has some commonalities with the magazine industry [57]. However, online advertising platforms affect the relationship between advertisers and content providers in some unusual ways. Hu [58] points out that the measurability of online advertising means that online advertising platforms can offer performance-based pricing in ways impossible with the traditional types of advertising discussed by Anderson and Coate [53]. It is not clear empirically how this has affected the advertiser–media platform relationship.

The fragmentation of the Web has also destabilized traditional relationships between advertisers, advertising agencies, and media content providers. Advertising agency services have become more unbundled [52]. Television advertising buys used to be characterized by a strict set of procedures which tended to shut out individual advertisers from buying their advertising directly. Generally, it was preferable for firms to be fronted by an advertising agency, so that they could take advantage of both the bargaining power and the relationships that agency had with the television stations. However, online advertising is frequently sold remotely and through rigid pricing mechanisms such as online auctions, which are not often initially set up to reflect a media buyer's bigger marketing clout. This means that firms have less need online than offline to buy their advertising through agencies.

In addition, some research such as Wilbur and Zhu [59] has highlighted how these new models of accountability can lead to tension when advertisers can assess the extent that they are paying for ads shown to users they did not want the ads shown to. This phenomenon is called "click-fraud," and occurs when firms pay for clicks that are fraudulent and are placed either by the Web site hosting the ad or by rivals who are attempting to raise their competitors' advertising costs.

4.3 The Effect of Online Advertising on Traditional Forms of Advertising

There is a growing literature in marketing that explores the relationship between offline and online environments for customer acquisition [60], brands [61], word-of-mouth [62,63], purchases [64,65], customized promotions [66], ad pricing [14], search behavior [67], and price sensitivity [68]. Kempe and Wilbur [69] discuss what television networks can learn from the best practices for the selling and pricing online advertising.

However, there has been less work that explicitly deals with the general relationship between online and offline media. Silk et al. [70] point out that it is not clear whether internet advertising should be viewed as a threat to traditional forms of advertising. Goldfarb and Tucker [71] explore this in the alcohol advertising industry. They find that when states ban the advertising of alcohol on outdoor media (such as billboards and signage), the effectiveness of online advertising increases. Similarly, Goldfarb and Tucker [14] show that search ad prices rise when offline ads are banned. These two studies suggest that online and offline advertising should be viewed as substitutes.

Two recent theory papers have emphasized how highly targeted online ads might affect the offline advertising industry. Specifically, Athey and Gans [72] and Bergemann and Bonatti [73] investigate how the evolution of targeting technologies will affect competition in the advertising industry, and show that the effects are not necessarily negative for "old media firms," as advertising becomes bifurcated between targeted and nontargeted ads. These models also suggest that targeting is most valuable in the absence of channel competition.

5. Regulation

It is striking how unregulated online advertising is, given its sizable economic clout and social influence. There are five major dimensions along which online advertising has been regulated or looks likely to be regulated in the United States.

First, there have been attempts to regulate the content and placement of ads for contentious products. For example, Chiou and Tucker [39] study the FDA's recent attempt to limit the extent to which pharmaceutical companies can use search ads. The space constraints of the ad format mean that pharmaceutical firms cannot adequately display information about side effects. The study finds that there is a potential downside to attempts to curb pharmaceutical online advertising, as users became more likely to visit non-FDA-regulated sites, such as Canadian pharmacies and herbal alternative remedies, when pharmaceutical ads are taken down. Interestingly, there appears to have been substantial self-regulation by media platforms when it comes to other forms of contentious advertising. For example, Google does not accept hard liquor search ads, nor ads for its content networks that contain images that are not "family safe."

Second, online advertising can raise issues of copyright and digital rights protection. The internet has undoubtedly facilitated the reuse of media and other content, potentially without copyright holders' permission, in ways that were not possible before. Chiou and Tucker [74] have analyzed how regulation about use of trademarks has affected online advertising. Generally, upstream firms have resisted the use of the trademarks and slogans by downstream resellers of their products. However, this research uses a natural experiment where Google loosened its policy regarding the use of trademarks in advertising copy, and found that firms' resistance is perhaps not warranted in all situations. They found the use of trademarks by resellers of hotel rooms in search advertising actually increases primary demand for the hotel's own Web site.

Third, there is concern that the largest companies in the online advertising market are getting too big [75]. An important aspect of this argument is that the importance of data to targeting technology might lead to a natural monopoly. Because targeting technologies rely on data from past behavior and on experiments run on current users, the more data a company has, the better it will be able to target its customers. This means that it might be extraordinarily difficult to overcome an initial market share lead because the initial lead means more data and therefore better targeting. Consequently, the European Union has opened an antitrust investigation into Google.

Unfortunately, there is little research to inform such an investigation. Even the market definition is not well defined. For example, while the European Union declared that online and offline advertising are separate markets in its decision on Microsoft's acquisition of Yahoo's search business, the only empirical work on the subject shows that online competes with offline [14,71]. In the theory literature, White and Jain [76] note that it is an open question whether search and display advertising are substitutes or complements, and they show that the answer has important implications for any antitrust case. More generally, there is still little theoretical or empirical understanding of whether a dominant position in the algorithmic search market can be extended to generate a dominant position in other markets such as advertising, maps, or email.

Fourth, another area which has received some policy attention is online advertising directed at children. The Children's Online Privacy Protection Act (COPPA) was implemented in April, 2000. This rule stated that commercial Web sites directed to children under 13 years old or general audience sites that have actual knowledge that they are collecting information from a child must obtain parental permission before collecting such information. However, as pointed out by Montgomery and Chester [77], given the extent to which advertisers use social media to target ads, and given the lack of clarity of current laws for these new media, it is not clear that current law is achieving this aim.

The last area, requiring a more detailed discussion, is proposed regulation concerning user privacy and the use of customer data by online advertisers.

5.1 Privacy

What is obvious in Section 2.4 is the extent to which advertisers are able to collect detailed user-level data which they can use to optimize their advertising. Online advertising has enabled firms to track users electronically and store the data almost costlessly. The collection of such data is often argued to be harmless because it typically involves a series of actions linked by an IP address or otherwise anonymous cookie-id number. However, attempts by advertisers to use such information have met resistance from consumers due to privacy concerns. In a well-publicized survey, Turow et al. [78] found that 66% of Americans do not want marketers to tailor advertisements to their interests. This customer resistance to tailored advertising is a major problem for advertisers. Theoretically, they would like to use social network data to *target* the users who see an advertisement and to *tailor* the content of advertising appeals. Fear that users may react unfavorably because of privacy concerns has led advertisers to limit their tailoring of ads. A recent survey suggested that concerns about consumer response have led advertisers to reduce the targeting of advertising based on online behavior by 75% [79].

As discussed by Hui and Png [80], it is not straightforward to incorporate notions of privacy into economic models. However, there are reasons to think that such collection of data may not be entirely harmless.

First, it is not clear that such data are strictly anonymous. For customers who browse using a static IP address, it is reasonably straightforward to trace back their movements to an offline identity. It is also possible to trace back user identities by various click-stream actions. For example, in *Pharmatrak, Inc. Privacy Litig., 329 F.3d 9, 15 (1st Cir. 2003)*, the defendant was accused of having collected personal data because the plaintiffs were able to construct individually identifiable profiles for 232 users out of the 18.7 million profiles (0.001%) in the defendant's data set. In this case, the defendant (Pharmatrak) collected data about users who browsed multiple pharmaceutical company Web sites, in order to compare traffic on and usage of different parts of these Web sites. The plaintiffs were able to construct these individual profiles largely because the Web server recorded the subject, sender, and date of the Web-based email message a user was reading immediately prior to visiting the Web site.

Second, potentially the collection of such data could lead to a form of behavioral price discrimination, that may harm consumer surplus. This has been discussed in a purely pricing context by Acquisti and Varian [81], Fudenburg and Villas-Boas [82], and Hui and Png [80]. Since pricing promotions are often an important feature of online advertising campaigns, consumers could be offered very different prices based on their click-stream data without their knowledge.

Third, at the moment it is not clear that property rights over such clickstream data used for targeting have been assigned in a transparent way which facilitates the postassignment type of bargaining essential for Coasian efficiency. Instead, firms collect information on customers' clickstreams, often without informing consumers in an upfront way that they are doing so, and use that to conduct profitable advertising campaigns. Consumers are not given a chance to either profit or negotiate with advertising networks to take advantage of the profitable use that their data are being put to.

These three areas of potential negative consequences of ad targeting and the collection of unprecedented amounts of data, have led governments to both enact and contemplate regulation to govern the collection of such data. As discussed by Baumer et al. [83] and Debussere [84], generalized privacy regulation in Europe has curtailed to some extent to the use of Web bugs and cookies if consumers are not informed about their use. By contrast, the United States has lagged behind in terms of the strictness of online privacy laws. Previously in the United States, behavioral targeting was governed by self-regulation. In 2009, the Federal Communications Commission released four principles of self-governance. These are summarized in Table IV.

In the United States, there have been calls from elected officials for explicit privacy regulation as opposed to relying on self-regulation. For example, Congressman Rick Boucher, chair of the subcommittee on "Communications, Technology,

TABLE IV
FTC PRINCIPLES OF SELF-REGULATION FOR BEHAVIORAL TARGETING

Principle	Description
Transparency and consumer control	Every Web site that uses behavioral targeting should clearly and concisely spell out what they are doing
Reasonable security, and limited data retention, for consumer data	Firms should retain data only as long as is necessary to fulfill a legitimate business or law enforcement need
Affirmative express consent for material changes to existing privacy promises	A firm must keep its promises to consumers regarding protecting their data. If they get bought or merged with another company, those pledges still hold, unless consumers agree to the changes. If the company revises its policies on privacy, they must receive users' consent before implementing the new rules
Affirmative express consent for (or prohibition against) using sensitive data for behavioral advertising	A firm wishing to collect "sensitive" personal data must get users' permission before, not after, it starts collecting

FTC staff report: February 2009 self-regulatory principles for online behavioral advertising.

and the Internet," has proposed a draft bill to try and regulate behavioral advertising in the United States. The content of the bill has been marketed as codifying best practices in the industry [85].

However, such regulation may have costs. As set out by Evans [86] and Lenard and Rubin [87], there is a trade-off between the use of online customer data and the effectiveness of advertising. Goldfarb and Tucker [88] examined responses of 3.3 million survey-takers who had been randomly exposed to 9596 online display (banner) advertising campaigns to explore how strong privacy regulation in the form of the 2002/58/EC Privacy Directive in the European Union has influenced advertising effectiveness. They find that display advertising became far less effective at changing stated purchase intent after the laws were enacted relative to other countries. The loss in effectiveness was more pronounced for Web sites that had general content (such as news sites), where nondata-driven targeting is particularly hard to do. The loss of effectiveness was also more pronounced for ads with a smaller page presence and for ads that did not have additional interactive, video, or audio features.

An alternative approach to addressing user privacy concerns regarding advertising, rather than explicit regulation, is to empower users to control what information is used. Tucker [48] uses field experiment data to evaluate the effect of Facebook giving users increased control over their privacy settings. She finds that after Facebook allowed users more transparent control over their privacy settings, personalized advertising, or mentioning specific details about a user in the ad copy, became more effective. This has relatively optimistic implications for future regulation which rather than focusing on banning specific practices, could instead focus on allowing users control over their privacy settings, thereby reducing the potential to harm the online advertising industry.

6. Conclusions and Avenues for Future Research

In this chapter, we have outlined the key aspects of the online advertising market. We argued that two dominant themes emerge in studying online advertising: measurability and targetability. These features have affected the way firms behave in these markets by increasing accountability and changing the nature of competition. The rise of online advertising has also generated a number of policy questions.

As this chapter has highlighted, there are many unexplored avenues for future academic research into online advertising. We categorize these into four major topics.

First, with respect to measurability, academic researchers have taken advantage of the evolution for the first time of reliable measures of advertising effectiveness to answer longstanding questions about advertising in general, but there has been little work that studies how "measurability" has changed advertisers' and media platforms' relationships. There has also been little work on what are the optimal pieces of information for internet platforms to provide to potential advertisers and whether there are ever situations where platforms benefit from not sharing information with advertisers.

Second, online advertising is unique compared to other media in terms of the ease of targeting. However, little research has been done to explore this important feature beyond simply documenting that targeting improves ad performance. However, it is important to know whether targeting is always empirically desirable or whether there are occasions when firms can miss important segments by being too targeted. Further, there has been little comparison of which methods of targeting are most effective and when they are most effective. It is also important to understand whether there are ways that online ad targeting can be conducted which simultaneously reassure users about their online privacy.

Third, the evolution of online advertising has led to a whole new ecosystem of online advertising agencies and online media metrics agencies. However, we know little about how online advertising agencies operate and how they differ in their operations from traditional advertising agencies. We also know little about relative levels of control over both advertising content and placement for advertisers and media agencies have changed with the evolution of the internet. There, has also been no empirical work on how the ability of advertisers to accurately measure what part of advertising is "wasted" and consequently potentially fraudulent, has changed industry structure.

Fourth, there are increasing calls for regulation in the online advertising industry; however, we know little about the potential costs and benefit of such regulations. An explicit research agenda that expands on the research cited above to examine the impact of privacy regulation, copyright law, and perhaps merger analysis would help inform policymakers.

References

[1] J. Deighton, J. Quelch, Economic Value of the Advertising-Supported Internet Ecosystem, IAB Report, 2009, June.
[2] IAB, IAB Internet Advertising Revenue Report: 2009 Full-Year Results, IAB and Price waterhouse Coopers, 2010 (Technical Report).
[3] G. Assmus, J.U. Farley, D.R. Lehmann, How advertising affects sales: meta-analysis of econometric results, J. Mark. Res. 21 (1) (1984) 65–74.
[4] A. Goldfarb, C. Tucker, Boom to Bust Advertising, MIT, mimeo, 2010.

[5] X. Dreze, F.-X. Hussherr, Internet advertising: is anybody watching? J. Interact. Mark. 17 (4) (2003) 8–23.
[6] M. Ritson, Creative Business: Talking, Reading, Tasking, Financial Times, 2003
[7] A. Lambrecht, C. Tucker, Paying with Money or with Effort: Pricing When Customers Anticipate Hassle, LBS, mimeo, 2009.
[8] D. Reiley, R. Lewis, Retail advertising works! measuring the effects of advertising on sales via a controlled experiment on yahoo!, Working Paper, Yahoo! Research, 2009.
[9] G.M. Grossman, C. Shapiro, Informative advertising with differentiated products, Rev. Econ. Stud. 51 (1) (1984) 63–81.
[10] G. Iyer, D. Soberman, M. Villas-Boas, The targeting of advertising, Mark. Sci. 24 (3) (2005) 461.
[11] A. Chandra, Targeted advertising: the role of subscriber characteristics in media markets, J. Ind. Econ. 57 (1) (2008) 58–84 (March 2009).
[12] E. Gal-Or, M. Gal-Or, J.H. May, W.E. Spangler, Targeted advertising strategies on television, Manage. Sci. 52 (5) (2006) 713–725.
[13] B. Murthi, S. Sarkar, The role of the management sciences in research on personalization, Manage. Sci. 49 (10) (2003) 1344–1362.
[14] A. Goldfarb, C. Tucker, Search Engine Advertising: Channel Substitution when Pricing Ads to Context. Manage. Sci. 2011 (Forthcoming).
[15] A. Lambrecht, C. Tucker, Behavioral Targeting, Retargeting and Customer Decision Making, MIT, mimeo, 2010.
[16] F. Gilbert, Beacons, bugs, and pixel tags: do you comply with the FTC behavioral marketing principles and foreign law requirements? J. Internet Law 2008.
[17] D. Martin, H. Wu, A. Alsaid, Hidden surveillance by web sites: web bugs in contemporary use, Commun. ACM 46 (12) (2003) 258–264.
[18] B.H. Murray, J.J. Cowart, Webbugs a Study of the Presence and Growth Rate of Web Bugs on the Internet, Cyveillance, Inc., 2001 (Technical Report).
[19] D.R. Clayton, Problems with Phorm, University of Cambridge, 2008 (Technical report).
[20] J.R. Hauser, G.L. Urban, G. Liberali, M. Braun, Website morphing, Mark. Sci. 28 (2) (2009) 202–223.
[21] E. MacDonald, R. Bordley, J.-M. Kim, G. Urban, Improving click-through with web advertisements designed for cognitive style, Marketing Science Presentation, 2010.
[22] P. Chatterjee, D.L. Hoffman, T.P. Novak, Modeling the clickstream: implications for web-based advertising efforts, Mark. Sci. 22 (4) (2003) 520–541.
[23] P. Manchanda, J.-P. Dube, K.Y. Goh, P.K. Chintagunta, The effect of banner advertising on internet purchasing, J. Mark. Res. 43 (1) (2006) 98–108.
[24] P.J. Danaher, G.W. Mullarkey, Factors affecting online advertising recall: a study of students, J. Advert. Res. 43 (03) (2003) 252–267.
[25] A. Goldfarb, C. Tucker, Online display advertising: targeting and obtrusiveness, Mark. Sci. 2011 (Forthcoming).
[26] E.J. Johnson, W.W. Moe, P.S. Fader, S. Bellman, G.L. Lohse, On the depth and dynamics of online search behavior, Manage. Sci. 50 (3) (2004) 299–308.
[27] B. Edelman, M. Ostrovsky, M. Schwarz, Internet advertising and the generalized second-price auction: selling billions of dollars worth of keywords, Am. Econ. Rev. 97 (1) (2007) 242–259.
[28] H. Varian, Position auctions, Int. J. Ind. Org. 25 (6) (2007) 1163–1178.
[29] Z. Katona, M. Sarvary, The race for sponsored links: bidding patterns for search advertising, Mark. Sci. 29 (2) (2010) 199–215.

[30] Y. Zhu, K.C. Wilbur, Strategic bidding in hybrid cpc and cpm auctions, NET Institute Working Paper (08–25), 2008.
[31] N. Agarwal, S. Athey, D. Yang, Skewed bidding in pay-per-action auctions for online advertising, Am. Econ. Rev. 99 (2) (2009) 441–447.
[32] S. Athey, G.D. Ellison, Position Auctions with Consumer Search, NBER Working Paper No. w15253, 2009.
[33] X. Zhang, J. Feng, Price Cycles in Online Advertising Auctions, MIT, mimeo, 2005.
[34] S. Yao, C.F. Mela, A Dynamic Model of Sponsored Search Advertising, Kellogg, mimeo, 2010.
[35] A. Agarwal, K. Hosanagar, M.D. Smith, Location, Location, Location: An Analysis of Profitability of Position in Online Advertising Markets, Wharton, mimeo, 2008.
[36] A. Ghose, S. Yang, An empirical analysis of search engine advertising: sponsored search in electronic markets, Manage. Sci. 55 (10) (2009) 1605–1622.
[37] E.T. Bradlow, D.C. Schmittlein, The little engines that could: modeling the performance of World Wide Web search engines, Mark. Sci. 19 (1) (2000) 43–62.
[38] S. Yang, A. Ghose, Analyzing the relationship between organic and paid search advertising: positive, negative or zero interdependence? Mark. Sci. 29 (4) (2010) 602–623.
[39] L. Chiou, C. Tucker, How Does Pharmaceutical Advertising Affect Consumer Search? MIT, mimeo, 2010.
[40] O.J. Rutz, R.E. Bucklin, From Generic to Branded: A Model of Spillover Dynamics in Paid Search Advertising, UCLA, mimeo, 2008.
[41] A. White, Search Engines: Left Side Quality Versus Right Side Profits, Toulouse School of Economics, mimeo, 2009.
[42] M. Trusov, R.E. Bucklin, K. Pauwels, Effects of word-of-mouth versus traditional marketing: Findings from an internet social networking site, J. Mark. 73 (2009) 90–102.
[43] J. Campbell, Marketing to a Network of Consumers, University of Toronto, mimeo, 2010.
[44] P.P. Zubcsek, M. Sarvary, Direct Marketing on a Social Network, Insead, mimeo, 2009.
[45] D. Godes, D. Mayzlin, Firm-created word-of-mouth communication: evidence from a field test, Mark. Sci. 28 (4) (2009) 721–739.
[46] D. Mayzlin, Promotional chat on the internet, Mark. Sci. 25 (2) (2006) 155–163.
[47] S. Aral, D. Walker, Creating Social Contagion Through Viral Product Design: Theory and Evidence from a Randomized Field Experiment, NYU, mimeo, 2010.
[48] C. Tucker, Social Networks, Personalized Advertising, and Privacy Controls, MIT, mimeo, 2010.
[49] H. Blodget, Craigslist valuation: 80 million in 2008 revenue, worth 5 billion, Business Insider, 2008, April 3.
[50] C. Tucker, J. Zhang, Growing two-sided networks by advertising the user-base: a field experiment, Mark. Sci. 29 (5) (2010) 805–814.
[51] A. Ansari, C. Mela, E-customization, J. Mark. Res. 40 (2) (2003) 131–145.
[52] M. Arzaghi, E.R. Berndt, J.C. Davis, A.J. Silk, Economic Factors Underlying the Unbundling of Advertising Agency Services, NBER Working Paper, 2010.
[53] S.P. Anderson, S. Coate, Market provision of broadcasting: a welfare analysis, Rev. Econ. Stud. 72 (4) (2005) 947–972.
[54] M.R. Baye, J. Morgan, Information gatekeepers on the internet and the competitiveness of homogenous product markets, Am. Econ. Rev. 91 (3) (2001) 454–474.
[55] D.S. Evans, The economics of the online advertising industry, Rev. Netw. Econ. 7 (3) (2008) 359–391.
[56] K.C. Wilbur, A two-sided, empirical model of television advertising and viewing markets, Mark. Sci. 27 (3) (2008) 356–378.

[57] A. Goldfarb, Concentration in advertising-supported online markets: an empirical approach, Econ. Innovat. New Technol. 13 (6) (2004) 581–594.
[58] Y.J. Hu, Performance-based Pricing Models in Online Advertising, Purdue University—Krannert School of Managemen, mimeo, 2004.
[59] K.C. Wilbur, Y. Zhu, Click fraud, Mark. Sci. 28 (2) (2009) 293–308.
[60] D. Bell, J. Choi, Preference Minorities and the Internet. J. Mark. Res. 2011 (Forthcoming).
[61] P.J. Danaher, I.W. Wilson, R.A. Davis, A comparison of online and offline consumer brand loyalty, Mark. Sci. 22 (4) (2003) 461–476.
[62] D. Bell, S. Song, Neighborhood effects and trial on the internet: evidence from online grocery retailing, Quant. Mark. Econ. 5 (4) (2007) 361–400.
[63] C. Forman, A. Ghose, B. Wiesenfeld, Examining the relationship between reviews and sales: the role of reviewer identity disclosure in electronic markets, Inf. Syst. Res. 19 (3) (2008) 291–313.
[64] E. Brynjolfsson, Y. Hu, M. Rahman, Battle of the Retail Channels: How Product Selection and Geography Drive Cross-Channel Competition, MIT, mimeo, 2009.
[65] C. Forman, A. Ghose, A. Goldfarb, Competition between local and electronic markets: how the benefit of buying online depends on where you live, Manage. Sci. 55 (1) (2009) 47–57.
[66] J. Zhang, M. Wedel, The effectiveness of customized promotions in online and offline stores, J. Mark. Res. 46 (2) (2009) 190–206.
[67] D. Lambert, D. Pregibon, Online effects of offline ads, in: ADKDD '08: Proceedings of the 2nd International Workshop on Data Mining and Audience Intelligence for Advertising, ACM, New York, NY, 2008, pp. 10–17.
[68] J. Chu, P. Chintagunta, J. Cebollada, Research note—a comparison of within-household price sensitivity across online and offline channels, Mark. Sci. 27 (2) (2008) 283–299.
[69] D. Kempe, K.C. Wilbur, What Can Television Networks Learn from Search Engines? How to Select, Order, and Price Advertisements to Maximize Advertiser Welfare, Duke, mimeo, 2009.
[70] A.J. Silk, L.R. Klein, E.R. Berndt, The emerging position of the internet as an advertising medium, Netnomics 3 (2) (2001) 129–148.
[71] A. Goldfarb, C. Tucker, Advertising bans and the substitutability of online and offline advertising, J. Mark. Res. 2011 (Forthcoming).
[72] S. Athey, J.S. Gans, The impact of targeting technology on advertising markets and media competition, AER Paper and Proceedings, 2010
[73] D. Bergemann, A. Bonatti, Targeting in Advertising Markets: Implications for Offline vs. Online Media, MIT, mimeo, 2010.
[74] L. Chiou, C. Tucker, How Does the Use of Trademarks by Intermediaries Affect Online Search? MIT, mimeo, 2010.
[75] E.K.N.M. Clemons, Regulation of Digital Businesses with Natural Monopolies or Third Party Payment Business Models: Antitrust Lessons from the Analysis of Google, Wharton, mimeo, 2010.
[76] A. White, K. Jain, The attention economy of search and web advertisement, Toulouse School of Economics, mimeo, 2010.
[77] K. Montgomery, J. Chester, Interactive food and beverage marketing: targeting adolescents in the digital age, J. Adolesc. Health 45 (2009) 518–529.
[78] J. Turow, J. King, C.J. Hoofnagle, A. Bleakley, M. Hennessy. Americans Reject Tailored Advertising and Three Activities That Enable It, Berkeley, mimeo, 2009.
[79] S. Lohr, Privacy concerns limit online ads, study says, New York Times, 2010, April 30.
[80] K. Hui, I. Png, Chapter 9: The Economics of Privacy, Economics and Information Systems, Handbooks in Information Systems, vol. 1, Elsevier, 2006.

[81] A. Acquisti, H.R. Varian, Conditioning prices on purchase history, Mark. Sci. 24 (3) (2005) 367–381.
[82] D. Fudenburg, J.M. Villas-Boas, Chapter 7: Behavior based price discrimination and customer recognition, Handbooks in Information Systems, vol. 1, Emerald Group Publishing, 2006, pp. 377–435.
[83] D.L. Baumer, J.B. Earp, J.C. Poindexter, Internet privacy law: a comparison between the United States and the European Union, Comput. Secur. 23 (5) (2004) 400–412.
[84] F. Debussere, The EU e-privacy directive: a monstrous attempt to starve the cookie monster? Int. J. Law Inf. Technol. 13 (1) (2005) 70–97.
[85] R.R. Boucher, Behavioral Ads: The Need for Privacy Protection, The Hill, 2009.
[86] D.S. Evans, The online advertising industry: economics, evolution, and privacy, J. Econ. Perspect. 23 (3) (2009) 37–60.
[87] T.M. Lenard, P.H. Rubin, In Defense of Data: Information and the Costs of Privacy, Technology Policy Institute Working Paper, 2009.
[88] A. Goldfarb, C. Tucker, Privacy regulation and online advertising, Manage. Sci. 57 (1) (2011) 57–71.

Author Index

A

Acquisti, A., 308
Agarwal, A., 301
Agarwal, N., 300
Agbota, H., 119
Alexandersson, M., 156
Allen, C., 42
Alsaid, A., 297
Amendolare, V., 110
Amin, A.S., 53–54
Anderson, N., 256
Anderson, S.P., 304–305
Andreasen, F., 5
Angermann, M., 115
Ansari, A., 303
Aral, S., 303
Arkko, J., 42
Arzaghi, M., 304–305
Assmus, G., 292
Athey, S., 300, 306
Atkinson, R., 15, 42

B

Bagchi, S., 42
Baglietto, P., 52
Baiker, S., 112
Bailey, T., 112–114
Balduzzi, M., 20
Barnard, P., 3
Baset, S.A., 53
Baugher, M., 15
Baumer, D.L., 309

Baye, M.R., 304
Beauregard, S., 111, 117, 119
Beckman, M., 252
Beijar, N., 53
Bell, D., 305
Bellman, S., 300
Benini, M., 53–54
Benkler, Y., 239
Bergemann, D., 306
Berndt, E.R., 304–306
Berry, R., 84
Bhargava, V., 109
Blanco, J.-L., 114
Bleakley, A., 308
Blodget, H., 303
Bonatti, A., 306
Bonfiglio, D., 53
Bordley, R., 298
Borella, M., 84–85
Borriello, G., 121
Boucher, R.R., 310
Bradlow, E.T., 301
Braeckel, P., 161
Braun, M., 298
Brooks, A., 113
Brynjolfsson, E., 305
Buchanan, M., 272
Bucklin, R.E., 301–302

C

Calabrese, M., 273
Camarillo, G., 5, 14, 42
Cammaerts, B., 244

Campbell, J., 302
Carrara, E., 15
Casner, S., 9
Cavallaro, L., 20
Cebollada, J., 305
Chae, M.J., 53–54
Chandra, A., 295
Chan, S.H.G., 84
Chatterjee, P., 299
Chen, B.X., 269
Cheng, J., 253, 257, 263
Chen, H., 43
Chen, K.T., 53, 82
Chester, J., 307
Chiehyu, L., 240
Chintagunta, P.K., 299, 305
Chiou, L., 301, 306–307
Cho, D.H., 84
Choi, J., 305
Chong, H.M., 52
Chow, W.Y., 53–55, 59, 82, 87–88, 90, 97
Chris Riley, M., 254, 278
Chu, J., 305
Cicco, L.D., 53
Cinaz, B., 115
Clark, S., 111
Clark, W.E., 107
Clayton, D.R., 298
Clemons, E.K.N.M., 307
Coate, S., 304–305
Coates, M., 58
Condell, M., 273
Cooper, M., 267, 279
Corke, P., 112, 114
Cowart, J.J., 297
Coyne, J., 110
Csorba, M., 111
Cyganski, D., 110

D

Daempfling, H., 110
Dagiuklas, T., 53–54
Danaher, P.J., 299, 305

Daniilidis, K., 114
Das, A., 109
Davis, J.C., 304–305
Davis, R.A., 305
Davis, W.D., 107–108
Deans, M., 114, 127
Debussere, F., 309
Deighton, J., 290
Delio, M., 244, 253
Dell'Amico, M., 20
De Marco, A., 52
DeNardis, L., 250
Dierks, T., 14, 42
Djugash, J., 109, 111–114, 127
Doctorow, C., 256
Donnelly, M.K., 107–108
Draughorne, R., 53–54
Dreze, X., 293
Drezner, Z., 58
Dube, J.-P., 299
Duckworth, R.J., 110
Durrant-Whyte, H.F., 111–114

E

Earp, J.B., 309
Edelman, B., 300–301
Eisenberg, Y., 84
Ellison, G.D., 300
El-Sheikh, H.M., 53–54
Enck, W., 43
Evans, D.S., 304, 310

F

Fader, P.S., 300
Fahy, R.F., 105
Farley, J.U., 292
Farley, R., 47
Fattori, A., 20
Feder, H.J.S., 113
Feng, J., 301
Fernandez-Madrigal, J.-A., 114
Fischer, C., 103, 109, 133

Forman, C., 305
Fors, K., 156
Foster, B., 5
Fox, D., 121
Foxlin, E., 117–118, 135
Franks, J., 13, 42
Fransman, M., 243
Frederick, R., 9
Fudenburg, D., 308

G

Galiotos, P., 53–54
Gal-Or, E., 295
Gal-Or, M., 295
Gamini Dissanayake, M.W.M., 111
Gandikota, V.R., 94
Gans, J.S., 306
Garey, M.H., 62–64
Garg, S., 42
Gartner, J., 253
Gellersen, H., 103, 109, 119, 133
Geneiatakis, T.D.C.L.D., 42
Ghafarian, A., 53–54
Ghahramani, Z., 113
Ghosal, D., 42
Ghose, A., 301, 305
Gilbert, F., 297
Glasmann, J., 53
Glitho, R.H., 52
Godes, D., 302
Goh, K.Y., 299
Goldfarb, A., 289, 293, 296, 299–300, 305–307, 310
Gonzalez, J., 114
Goode, B., 52
Graham-Rowe, D., 110
Grainger, S., 53–54
Griffiths, C., 264
Gritzalis, S., 42
Grossman, G.M., 294
Guivant, J., 112–113
Gutmann, J.-S., 156

H

Habermas, J., 241–242
Hagirahim, H., 53–54
Hahn, M., 53–54
Hallam-Baker, P., 13, 42
Hamacher, H.W., 58
Handley, M., 5, 9, 55, 82, 95
Hardman, V., 55, 82–83, 94–95, 97
Hargraves, S., 53–54
Harle, R., 111
Haukka, T., 42
Hauser, J.R., 298
Hazas, M., 103, 109, 119, 131, 133
Hebert, M., 114, 127
Hennessy, M., 308
Hernandez, D., 255
Higginbotham, S., 268
High, S., 53–54
Hightower, J., 121
Hillier, F.S., 58
Hodson, O., 55, 82–83, 94–95, 97
Hoffman, D.L., 299
Hollingworth, J., 11
Hoofnagle, C.J., 308
Horrigan, J., 271
Hosanagar, K., 301
Hostetler, J., 13, 42
Huang, P., 53, 82
Huang, T.Y., 53, 82
Hui, K., 308
Hu, J., 260, 278
Hussherr, F.-X., 293
Hu, Y.J., 305
Hwang, E., 53–54

I

Iyer, G., 295

J

Jackson, C., 53–54
Jacobson, V., 5, 9
Jain, K., 307

Jajodia, S., 42, 53
James, L., 240
Jeff, C., 239
Jiang, X., 43, 47
Johnson, D.S., 62–64
Johnson, E.J., 300
Johnston, A., 5, 14
Jo, M., 53–54

K

Kambourakis, G., 42
Kang, I., 53–54
Kang, M.G., 84
Kantor, G., 109, 111, 113, 127
Karapantazis, S., 52–54
Katona, Z., 300
Katsaggelos, A.K., 84
Kaufman, G., 267
Kellerer, W., 53
Kempe, D., 305
Kenn, H., 115
Kent, S., 15, 42
Khendek, F., 52
Kim, H., 53–54
Kim, J.G., 53–54, 112, 114
Kim, J.-M., 298
Kincaid, J., 268
King, J., 269, 308
Klann, M., 107, 109
Kleiner, A., 113
Klein, L.R., 306
Klepal, M., 111
Koller, D., 113
Konolige, K., 156
Kortuem, G., 119
Krach, B., 115
Kray, C., 119
Kretkowski, P.D., 20
Krishnan, R., 273
Krohn, A., 119
Kumar, V., 109
Kurose, J.K., 55, 82–83, 95, 97

L

Lambert, D., 305
Lambrecht, A., 294, 296
Lavallee, A., 18
Lawrence, S., 13, 42
Lawson, J.R., 107–108
Lawson, S., 260
Leach, P., 13, 42
Lee, B., 53–54
Lee, D.H., 53–54
Lee, H., 53–54
Lee, J.Y.B., 53–54
Lee, Y.B., 94
Lehmann, D.R., 292
Lenard, T.M., 310
Lennett, B., 241
Leonard, J.J., 112–114
Lessig, L., 243–244, 267
Leung, Y.-W., 51, 53–55, 57, 59, 82, 84, 86–88, 90, 97
Lewis, R., 294, 297, 299
Liao, L., 121
Liberali, G., 298
Li, C., 282
Lieberman, G.J., 58
Liew, S.C., 53–54, 94
Lim, L., 108
Lin, E., 259
Ling, N., 94
Lin, X., 94
Liu, Y., 113
Li, Z.G., 94
Lohr, S., 308
Lohse, G.L., 300
Losey, J.W., 237, 269, 282
Louis, P.J., 267
Luotonen, A., 13, 42

M

MacDonald, E., 298
MacFarlane, A., 108
Ma, D., 43

Mahler, J., 84–85
Mahy, R., 46
Makarov, S., 110
Manchanda, P., 299
Manoj, B.S., 58
Maresca, M., 52
Martin, D., 297
Mascolo, S., 53
Mathar, R., 58
Matthews, H.S., 52
Matthews, P., 46
May, J.H., 295
Mayzlin, D., 302
McDaniel, P., 43
McGann, S., 42
McGrew, D., 15
Meinrath, S.D., 237, 241, 250, 269, 271, 282
Mela, C.F., 301, 303
Mellia, M., 53
Meo, M., 53
Meredith, D., 269
Merminod, B., 111
Miller, L.E., 111
Moe, W.W., 300
Montemerlo, M., 113, 139
Montgomery, K., 307
Morgan, J., 304
Mullarkey, G.W., 299
Muller, H., 53
Murray, B.H., 297
Murthi, B., 296
Murthy, C.S.R., 94
Muthukrishnan, K., 103, 109, 131, 133

N

Nappa, A., 20
Naraine, R., 18
Naslund, M., 15
Nebot, E., 112–113
Newman, P., 111–112, 114
Ng, A.Y., 113
Niemi, A., 42
Niessen, T., 58

Nieto, J., 113
Nobel, C., 268
Norrman, K., 15
Novak, T.P., 299

O

Odylzko, A., 254
Olson, E., 112, 114
Ostrovsky, M., 300–301

P

Papalilo, D., 42
Pappas, T.N., 84
Partridge, C., 273
Pauwels, K., 302
Pavlidis, Y., 114
Pavlidou, F.N., 52–54
Perelman, M., 243
Perkins, C., 55, 82–83, 94–95, 97
Peterson, J., 5, 14
Peterson, R., 109
Pickard, V.W., 237, 240–241, 250, 271
Png, I., 308
Poindexter, J.C., 309
Polit, S., 273
Porta, T.L., 43
Pregibon, D., 305

Q

Quelch, J., 290

R

Racic, R., 43
Rahman, M., 305
Ramanathan, R., 273
Ramsdell, B., 15
Rantakokko, J., 156
Rao, R., 58
Redfern, A., 108–109
Reiley, D., 294, 297, 299

Renaudin, V., 111
Ren, Z., 53
Rescorla, E., 14
Reynolds, B., 42
Ricknäs, M., 281
Ritson, M., 293
Rizzo, L., 84, 91–92
Robertson, P., 115
Romero, R., 108
Rosenberg, J., 5, 14, 46
Rossi, D., 53
Ross, K.W., 55, 82–83, 95, 97
Roth, D., 262
Rubin, P.H., 310
Rus, D., 109
Rutz, O.J., 301
Ryu, W., 53–54

S

Salsano, S., 42
Santivanez, C., 273
Sarkar, S., 296
Sarvary, M., 300, 302
Sasse, A., 55, 82, 95
Sat, B., 53, 82
Schiller, D., 242, 244
Schmittlein, D.C., 301
Schulz, D., 121
Schulzrinne, H.G., 5, 9, 14, 53
Schuster, G.M., 84–85
Schwarz, M., 300–301
Scott, B., 273
Segan, S., 268
Selepak, M.J., 107–108
Sendelbach, T.E., 107
Sengar, H., 42, 53
Shapiro, C., 294
Shin, B.C., 84
Sicari, S., 53–54
Sicker, D.C., 42
Sidhu, I., 84–85
Silk, A.J., 304–306
Simon, D., 126
Singel, R., 253

Singh, N., 42
Singh, S., 109, 111–114, 127
Skog, I., 156
Smith, M.D., 301
Soberman, D., 295
Song, S., 305
Song, T.L., 142
Spangler, W.E., 295
Sparks, R., 5, 14
Steingart, D., 108
Stevens, M., 113
Stolterman, E., 267
Stone, B., 262
Strömbäck, P., 156
Sukkarieh, S., 112, 114
Sun, D., 113
Sze, H.P., 53–54, 94

T

Tamma, B.R., 94
Tardif, J.-P., 114
Teller, S., 112, 114
Thouin, F., 58
Thrun, S., 112–113, 139
Tomé, P., 111
Topolski, R., 254, 278
Torvinen, V., 42
Traynor, P., 43
Trusov, M., 302
Tsai, T., 42
Tucker, C., 289, 293–294, 296, 299–301, 303, 305–307, 310
Turow, J., 308

U

Urban, G.L., 298

V

van Schewick, B., 239, 259
Varian, H.R., 300–301, 308
Veltri, L., 42
Villas-Boas, J.M., 308
Villas-Boas, M., 295

W

Waclawsky, J.C., 269
Waclawsky John, G., 253–254
Wah, B.W., 53, 82
Waldman, F., 53–54
Walker, D., 303
Wang, H., 42, 53
Wang, P.J., 53, 82
Wang, X., 1, 43, 47
Wan, S., 135
Watson, A., 55, 82, 95
Wedel, M., 305
Wegbreit, B., 113
White, A., 301, 307
Widyawan, W., 111
Wiesenfeld, B., 305
Wijesekera, D., 42, 47, 53
Wilbur, K.C., 300, 304–305
Wilson, I.W., 305
Wilson, J., 108–109
Wing, D., 46
Wirkander, S.-L., 156
Woodman, O., 111
Worrell, M., 108
Wright, P., 108–109
Wu, H., 297
Wu, Y.-S., 42

Y

Yalak, O., 111
Yang, D., 300
Yang, S., 301
Yang, X., 43, 47, 94
Yao, S., 301
Yip, D.C.S., 53–54
Yochai, B., 240–241, 246, 271
Yuk, S.W., 84

Z

Zetter, K., 253
Zhai, F., 84
Zhang, J., 303, 305
Zhang, Q., 84
Zhang, R., 1, 43, 47
Zhang, W., 109, 111, 113, 127
Zhang, X., 301
Zhang, Y.Q., 84
Zheng, X., 84
Zhou, P., 58
Zhu, C., 94
Zhu, W.W., 84
Zhu, Y., 300, 305
Zingirian, N., 52
Zubcsek, P.P., 302

Subject Index

A

Auditing tools, BT
  BlueScanner, Network Chemistry
    description, 221–222
    device discovery, 222–223
    polling requests, 222
  Bluesniff, The Shmoo Group
    BD_ADDR, 224–225
    description, 223–224
    time-intensive process, 225
  BlueSweep, AirMagnet, 223, 224
  BT_audit, trifinite.org
    description, 226
    open ports, 227
    PSM_scan tool, 227
    RFCOMM_scan tool, 227
  BTScanner, Pentest Ltd, 225–226
  goals, 218–219
  potential attackers, 218
  Wardriving
    goal, 221
    Wi-Fi networks, 219–221

B

Bluetooth (BT)
  applications, 162–164, 236
  auditing tools
    BlueScanner, Network Chemistry, 221–223
    Bluesniff, The Shmoo Group, 223–225
    BlueSweep, AirMagnet, 223
    BT_audit, trifinite.org, 226–227

    BTScanner, Pentest Ltd, 225–226
    goals, 218–219
    potential attackers, 218
    Wardriving, 219–221
  basics, wireless
    design features, 165
    factors, differentiating, 165
    technologies, 165–168
  Class-2, 235
  core specifications and profile usage models, 164
  exploitation
    categories, malware, 200–208
    limitations, 189
    malware firsts, 207–218
    potential attack impacts, 189–190
    propagation limitations, 197–200
    protocol, 190–191
    wireless device, 191–197
  market
    BT SIG, 170
    chip built-in and adaptor, 170–171
    In-Stat study, 171–172
  namesake origin, 169
  network topology
    active member address (AM_ADDR), 177–178
    channel-hopping pattern, 175
    clock synchronization, piconet, 176
    FHS packet, 177
    frequency-hopping pattern, 175–176
    PAN, 173
    piconet, 174–175

Bluetooth (BT) (Continued)
  power management, 176–177
  sniff, hold and park, 178
  official specification revisions, 171
  packet format
    access code, 183, 184, 186
    EDR, 187–188
    frequency-hopping algorithm, 187
    header, 186–187
    network traffic, 183
    payload, 187
    SCO links, 183
    transfer, devices, 187
  protocol stack and profiles
    adopted/upper-level, 180–182
    components, five core, 179–180
    human interface device, 179
    implementations, 181–182
    interface, 183
    operating system (OS), 178–179
  security recommendations
    antivirus application and update, 234–235
    applications installation, 234
    device's defaults, 229
    discoverable mode, 229–230
    pairing, 230–232
    pass-phrase, 232–234
    precautions, 228
    services, 230
  specifications and compliance
    Baseband, 169
    protocol stack, 169–170
    radio frequency (RF) transceiver, 169
    SIG, 170
  transmission range
    nondirectional communication, 189
    PAN, 188
    power classification, 188–189
    wireless medium, 172–173
Broadband Truth-in-Labeling, 275–276
BT. *See* Bluetooth

C

Call forwarding set up
  Gizmo hijacking
    611 call session, 32
    MITM, 31
    RTP server, 30–31
    softphone system, 30
    SSL/TLS connection, 31
  Vonage, manipulation
    bogus call forwarding number, 29–30
    international phone number, 30
    MITM intercepts, 29
    RTP event packets, 28
Children's Online Privacy Protection Act (COPPA), 307

D

DDoS. *See* Distributed denial of service
Digital feudalism
  Broadband connectivity, 282–283
  critiquing, internet
    corporate encroachment, 243–244
    "digital capitalism", 244
  DRM rules (*see* Digital rights management rules)
  enclosure, OSI dimensions
    application layer problems, 263–266
    data link and network layer problems, 250–259
    physical layer problems, 245–250
    presentation layer problems, 262–263
    session layer problems, 261–262
    transport layer problems, 259–261
  Internet, political economy
    broadband, 241
    challenges, 240–241
    policy debates, 241
  new paradigm
    contemporary policy debates, 280
    "layer 8", 281–282
    OSI hourglass model, 280–281

telecommunication networks, layer 1
 monopolies, 281
open technology
 AT&T telephone network, 267–268
 and closed, 266–267
 limitations, closed network, 268–270
 Web 2.0 applications, 267
OSI model, 245
public space and "enclosure"
 description, 242
 digital commons, debates, 242
 ownership shifting, 243
telecommunications, policy recommendation
 data link and network layer solution, 274–278
 FDR, 270–271
 physical layer solution, 271–274
 session, presentation and application layer solutions, 279–280
 transport layer solutions, 278
Digital rights management (DRM) rules, 239–240
Distributed denial of service (DDoS), 206
DNS. *See* Domain name system
Domain name system (DNS)
 description, 264
 hijacking, 264–265
 NXDomain response, 264
 redirection, 264–265
 SIP phone, 35–36
 spoofing attack
  experimental results and analysis, 41–42
  message flow, 38–41
 VoIP traffic, 16
 Vonage phone
  query/response, 37–38
  spoofing attack, 34–35

E

"Economics 101", 242
Emergency response
 Bayesian filter implementations, 156

constraints and limitations
 high-tech systems, 108
 human error, 108
embedded sensing and computation, 104
evaluation methods, 156–157
firefighter, 105–106
implementation, median relative position accuracy, 150
initialization, sensor positions, 151–152
lifelines, 106–107
location and navigation support, 105
loop closure, 156
multimodal sensing and algorithm
 inertial PDR, zero-velocity updates, 116–119
 Kalman filtering (*see* Kalman filtering)
 pedestrian SLAM, 125–126
 sensor initialization, 126–128
 ultrasonic ranging and angulation, 119–121
multiple sensors
 errors, 155
 ultrasonic measurements, 156
NIOSH, 105, 106
orientation, tracking
 cross-correlation, 154
 foot-mounted sensor, 153
PASS, 107
Pathfinder system, 107
reliability, 108–109
SLAM (*see* Simultaneous localization and mapping)
system evaluation
 experiment description, 128–133
 inertial PDR, 133–135
 Kalman filter, sensor positions, 135–136
 SLAM, 137–149
technologies and systems
 Dräger patent, 110
 flipside RFID, 111
 LifeNet, 109

Emergency response (Continued)
  map matching, RFID, 111
  particle filters, map matching, 111
  PPL, 110
  Relate Trails project, 109–110
  SmokeNet, 109
  Thales indoor positioning, 110
Erasure coding
  description, 84
  multiple voice packets, 86
  packet loss recovery, 85
  redundant packets, 87
  scheme, 92
  time, software implementation, 92
Exploitation, BT
  categories, malware
    BlueBug, 201–202
    BlueJack, 202–203
    BlueSnarf, 203–204
    Car Whisperer, 205
    DoS attack, 206–208
    HeloMoto, 200–201
  malware firsts
    Inqtana, 217–218
    OSs, 207–208
    proof-of-concept, 207
    RedBrowser, 216–217
    Symbian OS, 208–213
    Windows CE, 213–216
  potential attack, 189–190
  propagation limitations
    complexity, device, 197
    device configuration, 197–198
    hardware/software, 198–199
    platform, device, 200
    user interaction, 199–200
  protocol
    attacks, 190–191
    BT SIG, 190
  wireless device
    device spoofing, 195–197
    downloads/BT transfers, 193–194
    network data transfer protocols, 191–192
    SMS and MMS, 192–193

  sniffing, 191
  social engineering, 194–195

F

Firms, online targeting
  behavioral, 296
  cookie, 297
  "local shared object", 298
  "real-time", 297
  retargeting, 296
  types, 295–296
  "warrantless wiretapping", 298
  Web bugs, 297
Frequency-hopping spread spectrum (FHSS)
  algorithm, 172–173

G

Gizmo softphone, 30–32

H

HeadSLAM, 115

I

Industrial, scientific and medical (ISM) radio
  band, 172–173
Industry, online advertising
  accountability, 304
  effects, traditional
    alcohol advertising, 306
    television network, 305
  two-sided platforms and pricing
    advertisers and viewers, 305
    features, 304–305
    Web, 305
Infrared Data Association (IrDA), 168, 194
Inqtana, 217–218
International Mobile Equipment Identity
  (IMEI), 255
IP multimedia subsystem
  3G and 4G cellular networks, 254
  telecommunications, 253–254
IrDA. *See* Infrared Data Association

SUBJECT INDEX

J

Java 2 Micro Edition (J2ME)
    BT Browser, 221
    RedBrowser, 216–217

K

Kalman filtering
    description, 121–122
    Jacobian matrices, 123
    PDR measurements, 122, 123
    process and measurement function, 123
    sensor coordinates, orientations and measurements, 124
    ultrasonic measurement, 125
    ultrasound sensor nodes, 123

L

Lasco, 211
Light weight piggybacking
    coding and decoding time, software implementation, 92
    computation time, 91
    destination gateway, steps, 89, 90
    parameter value selection, guidelines, 93
    simulation model, 93–99
    small delay, 91
    smaller redundancy, 91
    source gateway, steps, 87–89
    uniform packet size, 92

M

MAC. *See* Media access controller
Malware categories, BT
    BlueBug attack
        limitation, 202
        namesake, 202
        pairing process, 201–202
    BlueJack
        electronic business card, 202–203
        limitations, 203
        proximity advertising, 203
    BlueSnarf attacks
        vs. BlueBug attacks, 203
        limitations, 204
        OBEX push profile (OPP), 204
    Car Whisperer, 205
    DoS attack
        DDoS, 206
        limitations, 207
        resources, 206
Man-in-the-middle (MITM)
    call forwarding setup
        hijacking gizmo, 31
        registration hijacking, 21
        SIP phones and Vonage servers, 29
    Gizmo softphone, 30–32
    remote attack (*see* Remote MITM attack)
    unauthorized call redirection, 27
    voice pharming (*see* Voice pharming)
    VoIP calls, 25
    Vonage, manipulation, 29
Media access controller (MAC)
    and BD_ADDR, 224–225
    connectivity, 254–255
    description, 254
    spoofing, 255
MMS. *See* Multimedia Messaging service
Multimedia Messaging service (MMS)
    Commwarrior, 211–212
    and SMS, 192–193

N

NAT. *See* Network address translation
National Institute for Occupational Safety and Health (NIOSH), 105, 106, 108
Network address translation (NAT)
    router, 35
    SIP phone, 25, 41
    traversal, 45–46
    VoIP phones, 22

O

"Official Bluetooth Wireless Info Site", 170
Online advertising
  agencies, 311
  features, 290, 292
  industry, measurability and targetability
    accountability, 304
    effects, traditional, 305–306
    two-sided platforms and pricing, 304–305
  regulation
    calls, 311
    content and placement, contentious products, 306
    copyright and digital rights protection, 306–307
    policy attention, children, 307
    privacy, 308–310
    targeting technology, 307
  revenue, 291
  vs. traditional
    firms targeting, 295–298
    measurability, 290, 292–294
    pre-internet targeting, 294–295
    targetability, 290, 292, 294
  types
    classifieds, lead generation, sponsorship and email, 303
    display, 298–300
    search engine, 300–302
    social media, 302–303
Open technology, digital feudalism
  AT&T telephone network, 267–268
  and closed, 266–267
  limitations, closed network
    Apple and AT&T, 268–269
    Bluetooth functionality, 268
    data communications, 269
    "last-mile" links, 270
    municipal and enterprise 802.11 wireless, 269–270
  Web 2.0 applications, 267

Opportunistic spectrum access (OSA), 272–273
OSI dimensions, layer problems
  application
    cripple and communication, 263
    DNS hijacking, 264–265
    H.264 and online video, 265–266
  data link and network
    blocking video, 257–259
    copyright enforcement vs. fair use, 256
    IMEI (see International Mobile Equipment Identity)
    IMS (see IP multimedia subsystem)
    Internet protocol addresses, 250–253
    MAC (see Media access controller)
    tampering and packet forging, 257
  physical
    network, defined, 245–246
    open access and common carriage, 246–247
    spectrum resources, 247–250
  presentation
    ASCII and MIME, 263
    text display protocols, 262
  session
    limits, 262
    Web pages, 261
  transport
    cable network operators, 259
    description, 259
    port blocking, 259–261
OSI model, 245

P

Packet loss recovery
  definition, 82
  piggybacking (see Piggybacking)
  shared
    decoding, 85
    erasure coding, 84, 85
    multiple voice packets, 86
  voice packets, 81–82

Parked member address (PM_ADDR), 178
PASS. *See* Personal alert safety system
PDR. *See* Pedestrian dead-reckoning
Pedestrian dead-reckoning (PDR)
  inertial, zero-velocity updates
    dead-reckoning, 116
    distance drift, 118
    stance and swing phase, 117–118
    transformation, sensor to world coordinates, 117
    ZUPTs, 119
  predeployed RFID, 111
  sensors, inertial, 115
Personal alert safety system (PASS), 107, 108
Phone-to-phone configuration, Internet telephony
  computer-to-computer configuration, 52–53
  internet usage, statistics, 53
  long-distance, 54
  packet loss recovery
    definition, 82
    light weight, 87–99
    piggybacking, 82–83
    shared, 84–86
    voice packets, 81–82
  service coverage and voice quality, 55
  Skype, 53
  sparse gateway
    complexity, 62–65
    cost-effectiveness, 56, 58
    generalization, 73–75
    model, optimization, 58–59
    numerical results, 75–81
    operations research (OR) problems, 58
    overall two-stage heuristic algorithm, 73
    problem formulation, optimization, 60–62
    saved and additional cost, 56
    service provider, 55
    two-stage heuristic algorithm, 65–72
Piggybacking
  description, 82
  light weight
    coding and decoding time, software implementation, 92
    computation time, 91
    delay, small, 91
    destination gateway, steps, 89, 90
    parameter value selection, guidelines, 93
    simulation model, 93–99
    smaller redundancy, 91
    source gateway, steps, 87–89
    uniform packet size, 92
  packet loss recovery, one voice stream, 83
  recovery delay, 82
  Skype, 82
PM_ADDR. *See* Parked member address
PPL. *See* Precision personnel location
Precision personnel location (PPL), 110
Public switched telephony network (PSTN)
  AT&T SIP phone calls, 23–24
  invite and 200 ok messages, 7, 8
  Vonage SIP phone and, 26

R

Radio frequency identification (RFID), 111, 168
RedBrowser
  Java 2 Micro Edition (J2ME), 216–217
  target device, 217
Redirection, VoIP
  callee side call
    AT&T phone, 27
    SIP, 27
    unauthorized, MITM, 27–28
  caller side call, 28
  third party, 26
Remote MITM attack
  DNS spoofing
    analysis, 41–42
    message flow, 39
    register message, 39

Remote MITM attack (Continued)
    remote attacker, 40
    timeline, round of attack, 40
    Vonage SIP phone, 38
  message flow, normal startup/reboot
    attack and vulnerable window, 40–41
    DNS query, 35–36
    SIP register message, 37
  network setup
    NAT routers, 35
    testbed setup, 36
  VoIP traffic, 34
  Vonage SIP phone, vulnerabilities
    DNS query/response, implementation, 37–38
    invite message, 38
RFID. *See* Radio frequency identification

S

Secure real-time transport protocol (SRTP), 15–16
Security, BT
  applications
    antivirus and update, 234–235
    installation, 234
  device's defaults, 229
  discoverable mode, 229–230
  pairing
    description, 230–231
    devices, 231
    location, 231–232
    review, devices, 232
  pass-phrase
    authentication, 233–234
    manufacturer-issued, 233
    mechanisms, 232
    pairing, secure location, 232–233
    perzmutations, length, 233
  precautions, 228
  services, 230
Session initiation protocol (SIP)
  description, 5

digest authentication, 13–14
functionalities, 5
messages
  call setup and tear down, flow, 7
  fields, 7
  flow, 9, 24
  invite, 8
  200 ok, 8
  request, 6
  response, 6–7
  servers, 6
  Vonage, 25, 37–38
Short message service (SMS)
  and MMS, 192–193
  RedBrowser, 217
Signaling security mechanisms, VoIP
  authentication
    response, 14
    secret password, 13
  IPsec, 15
  S/MIME, 15
  transport, 15–16
  transport layer security (TLS), 14
Simulation model, light weight piggybacking
  computational complexity, comparison, 97
  high packet loss rate, performance, 97, 99
  interdependency, 93
  performance metrics, 94–95
  vs. piggybacking, 95–96
  shared packet loss recovery, 97, 98
Simultaneous localization and mapping (SLAM)
  aerial, 114–115
  EKF-SLAM, 112
  FastSLAM, 113
  graphSLAM, 113
  motivation, 111
  multimodal pedestrian
    inertial sensors, 115
    ultrasound sensors, 116
  pedestrian, 115, 125–126
  robotics, 112

SUBJECT INDEX

sensing modalities
  range-only and bearing-only sensors, 114
  robot positioning, 113
  stereo cameras, 113–114
system evaluation
  cumulative bearing error distributions, 146–147, 150–151
  cumulative range error distributions, 144–145, 148–149
  data sets, 144
  deployment, sensor, 141
  drift, 140
  errors, median bearing, 154–155
  initialization, 141–142
  innovations, 144
  large-scale distortion, 146
  map, sensor positions, 138
  median range errors, 152–153
  near sensors, cumulative range error distributions, 148–149
  offline methods, 139
  output visualization, 137
  positions, sensors, 137, 139–143
  symmetry ambiguity, 147–148
SLAM. *See* Simultaneous localization and mapping
SMS. *See* Short message service
Spam over internet technology (SPIT), 19
Sparse telephone gateway
  facility location and nonlinear integer programming problem, 58
  first numerical experiment
    expected revenue, performance evaluation, 75
    mean execution time, 76, 78
    parameter values, 75–76
    performance, 77, 78
    traffic patterns, 76
  long-distance, Internet telephony, 57
  operations research (OR) problems, 58
  optimization
    complexity, problem, 62–65

    cost components and charging scheme, 58, 59
    cost-effectiveness, 56, 58
    generalization, 73–75
    overall two-stage heuristic algorithm, 73
    problem formulation, 60–62
    traffic load, 59
    two-stage heuristic algorithm, 65–72
  saved and additional cost, 56
  second numerical experiment
    mean execution time, 79, 81
    nonuniformly generated traffic, 78–80
    uniformly generated traffic, 80
  service provider, 55
SPIT. *See* Spam over internet technology
Symbian OS
  Cabir worm
    description, 210
    malware installation and transfer, 210
    MMS, 211
  Cardtrap
    description, 212
    goal, 213
    installation, 212–213
    malware installation, 213
  Commwarrior
    description, 211–212
    impact, 212
    propagation, BT and MMS, 212
  description, 208
  design features, 208
  interfaces, 210
  Lasco
    vs. Cabir worm, 211
    SIS file, 211
Synchronous connection-oriented (SCO) link packet, 178

T

Telecommunications
  data link and network layer solution
    Broadband Truth-in-Labeling, 275–276

Telecommunications (Continued)
    FCC's open Internet principles and rulemaking, 276–278
    ICANN, 276
    IMS, 274–275
    ISPs, 274
    open architecture and source driver, 274
    private network, 274
    FDR, 270–271
    physical layer solution
        common carriages, 271
        federal spectrum regulation, 272
        OSA, 272–273
        TV broadcast bands, 273–274
        Wi-Fi, 272
    session, presentation and application layer solutions
        application neutrality, 279
        codices, 279–280
        interoperability, 279
        open protocol and standards, 279
    transport layer solutions, 278
Traditional *vs.* online advertising
    firms target, 295–298
    measurability, 292–294
    pre-internet targeting, 294–295
    targetability, 294
Two-stage heuristic algorithm
    first stage
        algorithm 1, 2 and 3, 66–70
        gateway j, 66
        third constraint, relaxation, 65–66
    generalization, 73–75
    overall, 73
    second stage
        algorithm 4, 71–72
        service, city pairs, 71
Types, online targeting
    classifieds, lead generation, sponsorship and email, 303
    display
        "click-through rates", 299
        description, 298
        electronic billboard, 299
        measurability and targetability, 300
    search engine
        algorithmic/sponsored, 301–302
        auctions, 300–301
        features, 300
    social media
        Facebook, 302–303
        "word-of-mouth", 302

U

Uniform resource identifiers (URIs), 6, 22
Universal serial bus (USB), 168, 170, 173, 194

V

Vertex cover problem, 63–65
Voice over Internet protocol (VoIP)
    attacks
        manipulating and hijacking, call forwarding set up
        redirection, 26–28
        registration hijacking, 20–22
        remote MITM, 34–42
        transparent detour, 22–26
        voice pharming, 32–34
    built-in security mechanisms
        signaling, 13–15
        transport, 15–16
    callerID, TD AMERITRADE, 3–4
    Citibank checking account, 4
    description, 3
    Google stock options, 3, 4
    reliability, 3
    research, security
        MMS, 43
        SCIDIVE, 42
        SIP digest authentication, 42
    security requirements
        authenticity, 11
        confidentiality, privacy and anonymity, 12
        integrity, 12

SUBJECT INDEX

public, 11
temper resistance and availability, 12–13
security threats
  annoyance, 19
  botnet, 20
  call eavesdropping, 17–18
  call hijacking, 18
  fraud, 18
  interception and modification, 18
  involuntary involvement, 19
  service stealing and disruption, 17
signaling, 5
SIP (*see* Session initiation protocol)
technical challenges, securing
  E-911, 44–45
  firewall traversal, 45
  multiple protocols, 44
  NAT traversal, 45–46
  open architecture, 43
  real-time constraints, 43–44
transport, RTP
  functionalities, 9–10
  packet format, 10–11
  protocols, 9
vulnerabilities, 16–17
Voice pharming
hypothetical attack
  bank customer, 34
  bogus IVR, 33
  Citibank, 33
MITM, 32
VoIP callers, 33
VoIP. *See* Voice over Internet protocol

W

Windows CE (WinCE)
Brador
  launching tasks, 215–216
  transfer methods, 216
description, 213
Duts
  description, 214
  file transfer methods, 214–215
  virus installation, 214
Windows Mobile and Smartphone platforms, 214
Wireless local area network (WLAN), 167, 219
Wireless personal area network (WPAN)
  BT, 173–174
  Ultra-wideband and BT, 167
  Wibree, 167–168
  ZigBee, 167
Wireless wide area network (WAN), 166
Wireless wide region area network (WRAN), 165–166

Y

YouTube, 239, 265–266

Z

Zero velocity updates (ZUPTs)
  advanced algorithms, 119
  PDR algorithm, 117
ZUPTs. *See* Zero velocity updates

Contents of Volumes in This Series

Volume 60

Licensing and Certification of Software Professionals
    DONALD J. BAGERT
Cognitive Hacking
    GEORGE CYBENKO, ANNARITA GIANI, AND PAUL THOMPSON
The Digital Detective: An Introduction to Digital Forensics
    WARREN HARRISON
Survivability: Synergizing Security and Reliability
    CRISPIN COWAN
Smart Cards
    KATHERINE M. SHELFER, CHRIS CORUM, J. DREW PROCACCINO, AND JOSEPH DIDIER
Shotgun Sequence Assembly
    MIHAI POP
Advances in Large Vocabulary Continuous Speech Recognition
    GEOFFREY ZWEIG AND MICHAEL PICHENY

Volume 61

Evaluating Software Architectures
    ROSEANNE TESORIERO TVEDT, PATRICIA COSTA, AND MIKAEL LINDVALL
Efficient Architectural Design of High Performance Microprocessors
    LIEVEN EECKHOUT AND KOEN DE BOSSCHERE
Security Issues and Solutions in Distributed Heterogeneous Mobile Database Systems
    A. R. HURSON, J. PLOSKONKA, Y. JIAO, AND H. HARIDAS
Disruptive Technologies and Their Affect on Global Telecommunications
    STAN MCCLELLAN, STEPHEN LOW, AND WAI-TIAN TAN
Ions, Atoms, and Bits: An Architectural Approach to Quantum Computing
    DEAN COPSEY, MARK OSKIN, AND FREDERIC T. CHONG

Volume 62

An Introduction to Agile Methods
    DAVID COHEN, MIKAEL LINDVALL, AND PATRICIA COSTA
The Timeboxing Process Model for Iterative Software Development
    PANKAJ JALOTE, AVEEJEET PALIT, AND PRIYA KURIEN
A Survey of Empirical Results on Program Slicing
    DAVID BINKLEY AND MARK HARMAN
Challenges in Design and Software Infrastructure for Ubiquitous Computing Applications
    GURUDUTH BANAVAR AND ABRAHAM BERNSTEIN

Introduction to MBASE (Model-Based (System) Architecting and Software Engineering)
    DAVID KLAPPHOLZ AND DANIEL PORT
Software Quality Estimation with Case-Based Reasoning
    TAGHI M. KHOSHGOFTAAR AND NAEEM SELIYA
Data Management Technology for Decision Support Systems
    SURAJIT CHAUDHURI, UMESHWAR DAYAL, AND VENKATESH GANTI

Volume 63

Techniques to Improve Performance Beyond Pipelining: Superpipelining, Superscalar, and VLIW
    JEAN-LUC GAUDIOT, JUNG-YUP KANG, AND WON WOO RO
Networks on Chip (NoC): Interconnects of Next Generation Systems on Chip
    THEOCHARIS THEOCHARIDES, GREGORY M. LINK, NARAYANAN VIJAYKRISHNAN, AND MARY JANE IRWIN
Characterizing Resource Allocation Heuristics for Heterogeneous Computing Systems
    SHOUKAT ALI, TRACY D. BRAUN, HOWARD JAY SIEGEL, ANTHONY A. MACIEJEWSKI, NOAH BECK,
        LADISLAU BÖLÖNI, MUTHUCUMARU MAHESWARAN, ALBERT I. REUTHER, JAMES P. ROBERTSON,
        MITCHELL D. THEYS, AND BIN YAO
Power Analysis and Optimization Techniques for Energy Efficient Computer Systems
    WISSAM CHEDID, CHANSU YU, AND BEN LEE
Flexible and Adaptive Services in Pervasive Computing
    BYUNG Y. SUNG, MOHAN KUMAR, AND BEHROOZ SHIRAZI
Search and Retrieval of Compressed Text
    AMAR MUKHERJEE, NAN ZHANG, TAO TAO, RAVI VIJAYA SATYA, AND WEIFENG SUN

Volume 64

Automatic Evaluation of Web Search Services
    ABDUR CHOWDHURY
Web Services
    SANG SHIN
A Protocol Layer Survey of Network Security
    JOHN V. HARRISON AND HAL BERGHEL
E-Service: The Revenue Expansion Path to E-Commerce Profitability
    ROLAND T. RUST, P. K. KANNAN, AND ANUPAMA D. RAMACHANDRAN
Pervasive Computing: A Vision to Realize
    DEBASHIS SAHA
Open Source Software Development: *Structural Tension in the American Experiment*
    COSKUN BAYRAK AND CHAD DAVIS
Disability and Technology: Building Barriers or Creating Opportunities?
    PETER GREGOR, DAVID SLOAN, AND ALAN F. NEWELL

Volume 65

The State of Artificial Intelligence
    ADRIAN A. HOPGOOD
Software Model Checking with SPIN
    GERARD J. HOLZMANN

Early Cognitive Computer Vision
    Jan-Mark Geusebroek
Verification and Validation and Artificial Intelligence
    Tim Menzies and Charles Pecheur
Indexing, Learning and Content-Based Retrieval for Special Purpose Image Databases
    Mark J. Huiskes and Eric J. Pauwels
Defect Analysis: Basic Techniques for Management and Learning
    David N. Card
Function Points
    Christopher J. Lokan
The Role of Mathematics in Computer Science and Software Engineering Education
    Peter B. Henderson

Volume 66

Calculating Software Process Improvement's Return on Investment
    Rini Van Solingen and David F. Rico
Quality Problem in Software Measurement Data
    Pierre Rebours and Taghi M. Khoshgoftaar
Requirements Management for Dependable Software Systems
    William G. Bail
Mechanics of Managing Software Risk
    William G. Bail
The PERFECT Approach to Experience-Based Process Evolution
    Brian A. Nejmeh and William E. Riddle
The Opportunities, Challenges, and Risks of High Performance Computing in Computational Science and Engineering
    Douglass E. Post, Richard P. Kendall, and Robert F. Lucas

Volume 67

Broadcasting a Means to Disseminate Public Data in a Wireless Environment—Issues and Solutions
    A. R. Hurson, Y. Jiao, and B. A. Shirazi
Programming Models and Synchronization Techniques for Disconnected Business Applications
    Avraham Leff and James T. Rayfield
Academic Electronic Journals: Past, Present, and Future
    Anat Hovav and Paul Gray
Web Testing for Reliability Improvement
    Jeff Tian and Li Ma
Wireless Insecurities
    Michael Sthultz, Jacob Uecker, and Hal Berghel
The State of the Art in Digital Forensics
    Dario Forte

Volume 68

Exposing Phylogenetic Relationships by Genome Rearrangement
    Ying Chih Lin and Chuan Yi Tang

Models and Methods in Comparative Genomics
  GUILLAUME BOURQUE AND LOUXIN ZHANG
Translocation Distance: Algorithms and Complexity
  LUSHENG WANG
Computational Grand Challenges in Assembling the Tree of Life: Problems and Solutions
  DAVID A. BADER, USMAN ROSHAN, AND ALEXANDROS STAMATAKIS
Local Structure Comparison of Proteins
  JUN HUAN, JAN PRINS, AND WEI WANG
Peptide Identification via Tandem Mass Spectrometry
  XUE WU, NATHAN EDWARDS, AND CHAU-WEN TSENG

Volume 69

The Architecture of Efficient Multi-Core Processors: A Holistic Approach
  RAKESH KUMAR AND DEAN M. TULLSEN
Designing Computational Clusters for Performance and Power
  KIRK W. CAMERON, RONG GE, AND XIZHOU FENG
Compiler-Assisted Leakage Energy Reduction for Cache Memories
  WEI ZHANG
Mobile Games: Challenges and Opportunities
  PAUL COULTON, WILL BAMFORD, FADI CHEHIMI, REUBEN EDWARDS, PAUL GILBERTSON, AND
  OMER RASHID
Free/Open Source Software Development: Recent Research Results and Methods
  WALT SCACCHI

Volume 70

Designing Networked Handheld Devices to Enhance School Learning
  JEREMY ROSCHELLE, CHARLES PATTON, AND DEBORAH TATAR
Interactive Explanatory and Descriptive Natural-Language Based Dialogue for Intelligent Information Filtering
  JOHN ATKINSON AND ANITA FERREIRA
A Tour of Language Customization Concepts
  COLIN ATKINSON AND THOMAS KÜHNE
Advances in Business Transformation Technologies
  JUHNYOUNG LEE
Phish Phactors: Offensive and Defensive Strategies
  HAL BERGHEL, JAMES CARPINTER, AND JU-YEON JO
Reflections on System Trustworthiness
  PETER G. NEUMANN

Volume 71

Programming Nanotechnology: Learning from Nature
  BOONSERM KAEWKAMNERDPONG, PETER J. BENTLEY, AND NAVNEET BHALLA
Nanobiotechnology: An Engineer's Foray into Biology
  YI ZHAO AND XIN ZHANG

Toward Nanometer-Scale Sensing Systems: Natural and Artificial Noses as Models for Ultra-Small, Ultra-Dense Sensing Systems
    BRIGITTE M. ROLFE
Simulation of Nanoscale Electronic Systems
    UMBERTO RAVAIOLI
Identifying Nanotechnology in Society
    CHARLES TAHAN
The Convergence of Nanotechnology, Policy, and Ethics
    ERIK FISHER

Volume 72

DARPA's HPCS Program: History, Models, Tools, Languages
    JACK DONGARRA, ROBERT GRAYBILL, WILLIAM HARROD, ROBERT LUCAS, EWING LUSK, PIOTR LUSZCZEK, JANICE MCMAHON, ALLAN SNAVELY, JEFFERY VETTER, KATHERINE YELICK, SADAF ALAM, ROY CAMPBELL, LAURA CARRINGTON, TZU-YI CHEN, OMID KHALILI, JEREMY MEREDITH, AND MUSTAFA TIKIR
Productivity in High-Performance Computing
    THOMAS STERLING AND CHIRAG DEKATE
Performance Prediction and Ranking of Supercomputers
    TZU-YI CHEN, OMID KHALILI, ROY L. CAMPBELL, JR., LAURA CARRINGTON, MUSTAFA M. TIKIR, AND ALLAN SNAVELY
Sampled Processor Simulation: A Survey
    LIEVEN EECKHOUT
Distributed Sparse Matrices for Very High Level Languages
    JOHN R. GILBERT, STEVE REINHARDT, AND VIRAL B. SHAH
Bibliographic Snapshots of High-Performance/High-Productivity Computing
    MYRON GINSBERG

Volume 73

History of Computers, Electronic Commerce, and Agile Methods
    DAVID F. RICO, HASAN H. SAYANI, AND RALPH F. FIELD
Testing with Software Designs
    ALIREZA MAHDIAN AND ANNELIESE A. ANDREWS
Balancing Transparency, Efficiency, AND Security in Pervasive Systems
    MARK WENSTROM, ELOISA BENTIVEGNA, AND ALI R. HURSON
Computing with RFID: Drivers, Technology and Implications
    GEORGE ROUSSOS
Medical Robotics and Computer-Integrated Interventional Medicine
    RUSSELL H. TAYLOR AND PETER KAZANZIDES

Volume 74

Data Hiding Tactics for Windows and Unix File Systems
    HAL BERGHEL, DAVID HOELZER, AND MICHAEL STHULTZ
Multimedia and Sensor Security
    ANNA HAĆ

Email Spam Filtering
    ENRIQUE PUERTAS SANZ, JOSÉ MARÍA GÓMEZ HIDALGO, AND JOSÉ CARLOS CORTIZO PÉREZ
The Use of Simulation Techniques for Hybrid Software Cost Estimation and Risk Analysis
    MICHAEL KLÄS, ADAM TRENDOWICZ, AXEL WICKENKAMP, JÜRGEN MÜNCH,
    NAHOMI KIKUCHI, AND YASUSHI ISHIGAI
An Environment for Conducting Families of Software Engineering Experiments
    LORIN HOCHSTEIN, TAIGA NAKAMURA, FORREST SHULL, NICO ZAZWORKA,
    VICTOR R. BASILI, AND MARVIN V. ZELKOWITZ
Global Software Development: Origins, Practices, and Directions
    JAMES J. CUSICK, ALPANA PRASAD, AND WILLIAM M. TEPFENHART

Volume 75

The UK HPC Integration Market: Commodity-Based Clusters
    CHRISTINE A. KITCHEN AND MARTYN F. GUEST
Elements of High-Performance Reconfigurable Computing
    TOM VANCOURT AND MARTIN C. HERBORDT
Models and Metrics for Energy-Efficient Computing
    PARTHASARATHY RANGANATHAN, SUZANNE RIVOIRE, AND JUSTIN MOORE
The Emerging Landscape of Computer Performance Evaluation
    JOANN M. PAUL, MWAFFAQ OTOOM, MARC SOMERS, SEAN PIEPER, AND MICHAEL J. SCHULTE
Advances in Web Testing
    CYNTRICA EATON AND ATIF M. MEMON

Volume 76

Information Sharing and Social Computing: Why, What, and Where?
    ODED NOV
Social Network Sites: Users and Uses
    MIKE THELWALL
Highly Interactive Scalable Online Worlds
    GRAHAM MORGAN
The Future of Social Web Sites: Sharing Data and Trusted Applications with Semantics
    SHEILA KINSELLA, ALEXANDRE PASSANT, JOHN G. BRESLIN, STEFAN DECKER,
    AND AJIT JAOKAR
Semantic Web Services Architecture with Lightweight Descriptions of Services
    TOMAS VITVAR, JACEK KOPECKY, JANA VISKOVA, ADRIANMOCAN, MICK KERRIGAN, AND DIETER FENSEL
Issues and Approaches for Web 2.0 Client Access to Enterprise Data
    AVRAHAM LEFF AND JAMES T. RAYFIELD
Web Content Filtering
    JOSÉMARÍA GÓMEZ HIDALGO, ENRIQUE PUERTAS SANZ, FRANCISCO CARRERO GARCÍA, AND MANUEL DE
    BUENAGA RODRÍGUEZ

Volume 77

Photo Fakery and Forensics
    HANY FARID

Advances in Computer Displays
    JASON LEIGH, ANDREW JOHNSON, AND LUC RENAMBOT
Playing with All Senses: Human–Computer Interface Devices for Games
    JÖRN LOVISCACH
A Status Report on the P Versus NP Question
    ERIC ALLENDER
Dynamically Typed Languages
    LAURENCE TRATT
Factors Influencing Software Development Productivity—State-of-the-Art and Industrial Experiences
    ADAM TRENDOWICZ AND JÜRGEN MÜNCH
Evaluating the Modifiability of Software Architectural Designs
    M. OMOLADE SALIU, GÜNTHER RUHE, MIKAEL LINDVALL, AND CHRISTOPHER ACKERMANN
The Common Law and Its Impact on the Internet
    ROBERT AALBERTS, DAVID HAMES, PERCY POON, AND PAUL D. THISTLE

Volume 78

Search Engine Optimization—Black and White Hat Approaches
    ROSS A. MALAGA
Web Searching and Browsing: A Multilingual Perspective
    WINGYAN CHUNG
Features for Content-Based Audio Retrieval
    DALIBOR MITROVIĆ, MATTHIAS ZEPPELZAUER, AND CHRISTIAN BREITENEDER
Multimedia Services over Wireless Metropolitan Area Networks
    KOSTAS PENTIKOUSIS, JARNO PINOLA, ESA PIRI, PEDRO NEVES, AND SUSANA SARGENTO
An Overview of Web Effort Estimation
    EMILIA MENDES
Communication Media Selection for Remote Interaction of *Ad Hoc* Groups
    FABIO CALEFATO AND FILIPPO LANUBILE

Volume 79

Applications in Data-Intensive Computing
    ANUJ R. SHAH, JOSHUA N. ADKINS, DOUGLAS J. BAXTER, WILLIAM R. CANNON, DANIEL G. CHAVARRIA-MIRANDA, SUTANAY CHOUDHURY, IAN GORTON, DEBORAH K. GRACIO, TODD D. HALTER, NAVDEEP D. JAITLY, JOHN R. JOHNSON, RICHARD T. KOUZES, MATTHEW C. MACDUFF, ANDRES MARQUEZ, MATTHEW E. MONROE, CHRISTOPHER S. OEHMEN, WILLIAM A. PIKE, CHAD SCHERRER, ORESTE VILLA, BOBBIE-JO WEBB-ROBERTSON, PAUL D. WHITNEY, AND NINO ZULJEVIC
Pitfalls and Issues of Manycore Programming
    AMI MAROWKA
Illusion of Wireless Security
    ALFRED W. LOO
Brain–Computer Interfaces for the Operation of Robotic and Prosthetic Devices
    DENNIS J. MCFARLAND AND JONATHAN R. WOLPAW
The Tools Perspective on Software Reverse Engineering: Requirements, Construction, and Evaluation
    HOLGER M. KIENLE AND HAUSI A. MÜLLER

Volume 80

Agile Software Development Methodologies and Practices
    LAURIE WILLIAMS
A Picture from the Model-Based Testing Area: Concepts, Techniques, and Challenges
    ARILO C. DIAS-NETO AND GUILHERME H. TRAVASSOS
Advances in Automated Model-Based System Testing of Software
Applications with a GUI Front-End
    ATIF M. MEMON AND BAO N. NGUYEN
Empirical Knowledge Discovery by Triangulation in Computer Science
    RAVI I. SINGH AND JAMES MILLER
StarLight: Next-Generation Communication Services, Exchanges, and Global Facilities
    JOE MAMBRETTI, TOM DEFANTI, AND MAXINE D. BROWN
Parameters Effecting 2D Barcode Scanning Reliability
    AMIT GROVER, PAUL BRAECKEL, KEVIN LINDGREN, HAL BERGHEL, AND DENNIS COBB
Advances in Video-Based Human Activity Analysis: Challenges and Approaches
    PAVAN TURAGA, RAMA CHELLAPPA, AND ASHOK VEERARAGHAVAN